Proceedings in Technology Transfer

The book series "Proceedings in Technology Transfer" provides a valuable platform for researchers, practitioners, and policymakers to share their experiences and insights related to technology transfer. Our aim is to facilitate knowledge exchange between academic researchers and industry professionals, accelerating the introduction of new technologies and driving economic growth.

Our series publishes books and conference proceedings that focus on original research findings, best practice recommendations, emerging trends, and technology evaluations. The series covers all topics which explore the transfer of technology in various industry sectors that require innovations and economic growth.

Join us in advancing the technology transfer and its impact on innovation. Publish your proceedings with us and offer insights and expertise to professionals in your field.

Hari Prasad Dasari · Chinta Sankar Rao ·
Vaishakh Nair
Editors

Proceedings of the 1st National Conference on Climate Resilience and Environmentally Sustainable Technologies

NITK-CREST 2025, 27 Febuary – 01 March,
Surathkal, India

Springer

Editors
Hari Prasad Dasari
Chemical Engineering Department
National Institute of Technology Karnataka
Surathkal
Mangaluru, Karnataka, India

Chinta Sankar Rao
Chemical Engineering Department
National Institute of Technology Karnataka
Surathkal
Mangaluru, Karnataka, India

Vaishakh Nair
Chemical Engineering Department
National Institute of Technology Karnataka
Surathkal
Mangaluru, Karnataka, India

ISSN 2948-2321 ISSN 2948-233X (electronic)
Proceedings in Technology Transfer
ISBN 978-981-95-3217-9 ISBN 978-981-95-3218-6 (eBook)
https://doi.org/10.1007/978-981-95-3218-6

This Springer imprint is published by the registered company Springer Nature Singapore Pte Ltd.
The registered company address is: 152 Beach Road, #21-01/04 Gateway East, Singapore 189721, Singapore

Dedication

This conference proceeding is gratefully dedicated to Malur Bharathaiyengar Saidutta, an alumnus of the 1981 batch and a faculty member in the Department of Chemical Engineering at NITK, Surathkal. He earned his Ph.D. in Chemical Engineering from IIT Bombay in 1990, his M.Tech in Chemical Engineering from Anna University in 1983, and his B.Tech in Chemical Engineering from Mysore University in 1981. He joined the erstwhile KREC as a faculty member in 1993 and has served the Institute for over 31 years. His research interests include process optimization, reaction engineering, and environmental engineering. He has numerous publications in reputed journals and has contributed significantly to the field through his work on process modeling and simulation, waste treatment, and energy-efficient processes. M.B. Saidutta has also been involved in various academic and industrial projects, showcasing his expertise in chemical engineering applications. He has also been instrumental in several alumni initiatives that have immensely contributed to the betterment of the Institute.

Malur Bharathaiyengar Saidutta

Foreword

The growing population and the use of natural resources are exacerbating the environmental strain. Consequently, humanity is inclined to observe nature and endeavor to safeguard it from degradation. The 1st National Conference on Climate Resilience and Environmentally Sustainable Technologies (NITK-CREST 2025) represents a burgeoning initiative and a significant event that convenes a collaborative assembly of experts, researchers, and professionals. NITK-CREST 2025 was a platform for fostering transformative insights for idea dissemination, exchanging pioneering research, and promoting ecological sustainability.

This document encapsulates the core of NITK-CREST 2025, examining innovative ideas for fostering resilient communities and sustainable ecosystems. The participants engaged in meaningful dialogue, unveiled a transformative idea, and played a pivotal role in shaping a sustainable future for generations to come.

This proceedings collection encapsulates the essence of NITK CREST, including peer-reviewed papers that include a diverse array of research in climate resilience and ecologically sustainable technology, serving as a valuable resource for experts and aspiring researchers. I extend sincere appreciation to the organizers for their exemplary efforts and to the participants for their significant contributions to advancing meaningful research. They have generated articles on both present advancements and prospective research trajectories through their collaboration. The book will undoubtedly be illuminating and motivating, and I anticipate it will promote the adoption of Sustainable and Climate-Resilient Technologies.

August 2025 M. B. Saidutta

Editorial Preface

The National Institute of Technology Karnataka, Surathkal, hosted the Environmentally Sustainable Technologies conference from February 27th to March 1st, 2025. The conference addressed a variety of sustainable and resilient technologies. Organized by the Department of Chemical Engineering at NITK, Surathkal. The event is sponsored by the Anusandhan National Research Foundation, Cezen Tech, Mangalore Refinery and Petrochemicals Limited – Oil and Natural Gas Corporation, 3D Engineering Automation LLP, Cambridge University Press, APAR Industries Ltd., De Gruyter Brill, Sri Durga Laboratory Equipment Supplies, BHAANVI enviro and agro India Pvt. Ltd., and World Scientific, and is organized in collaboration with the Karnataka State Pollution Control Board. The three-day event was attended by delegates from various regions of India, as well as keynote sessions featuring professionals from IIT Madras, IIT BHU Varanasi, IIT Kharagpur, IIT Bombay, IIT Roorkee, CSIR-NIIST Thiruvananthapuram, DRDO Bangalore, CWRDM Kozhikode, Chitkara University, KSPCB, Global Foundaries, H.E.A.R.T, and APAR Industries Ltd. Our objective in publishing the book is to enable readers to actively participate in the conference's ethos beyond the event. The event facilitated interaction among students, professionals, and researchers who are engaged in the field of climate challenges and are committed to sustainable modernization. During the event, they have also shared their ideas, impending research directions, and discussions. An accessible account of the event's essence will be provided by this book.

NITK CREST-2025 was attended by a large number of participants from various regions of India. The event featured keynote sessions, sponsor talks, technical seminars on fire, space, air, water, earth (materials), and artificial intelligence, as well as poster presentations.

August 2025

Hari Prasad Dasari
Chinta Sankar Rao
Vaishakh Nair

Acknowledgments

We extend our sincere gratitude to all writers and reviewers for their substantial efforts. We like to express our profound appreciation to our Patron, B. Ravi (Director, NITK), and Chairman, I. Regupathi (Head of the Department of Chemical Engineering, NITK), for their guidance and encouragement.

Special thanks to the Convenors:

Chinta Sankar Rao	Chemical Engineering, NITK, Surathkal
Vaishakh Nair	Chemical Engineering, NITK, Surathkal
Ranjeet Kumar Sahu	Mechanical Engineering, NITK, Surathkal
Anupama Surenjan	Civil Engineering, NITK, Surathkal
Nikhil K S	Electronics and Communication Engineering, NITK, Surathkal
Sourav Kanti Addya	Computer Science & Engineering, NITK, Surathkal
Lipak Kumar Sahoo	Metallurgical and Materials Engineering, NITK, Surathkal

Also, acknowledge the significant support of the Advisory Committee:

Subhash C. Yaragal	Deputy Director, NITK, Surathkal
Udaya Bhat	Dean (R&C), NITK, Surathkal
Dwarakish G S	Dean (Academics), NITK, Surathkal
Gangadharan K V	Dean (P&D), NITK, Surathkal
T. Laxminidhi	Dean (FW), NITK, Surathkal
A Chitharanjan Hegde	Dean (SW), NITK, Surathkal
Shreekantha S. Rao	Dean (ACR), NITK, Surathkal

We are thankful to the Steering Committee members for their active involvement and insights:

Basavaraju Manu	Civil Engineering, NITK, Surathkal
S. M. Murigendrappa	Mechanical Engineering, NITK, Surathkal
Kumkum Banerjee	Metallurgical and Materials Engineering, NITK, Surathkal
Harsha Vardhan	Mining Engineering, NITK, Surathkal
Manu Basavaraju	Computer Science & Engineering, NITK, Surathkal
Ashvini Chaturvedi	PIC – CE, NITK, Surathkal
Keyur Raval	PIC – CRF, NITK, Surathkal
Vidya Shetty K	Chemical Engineering, NITK, Surathkal

Hari Prasad Dasari	Chemical Engineering, NITK, Surathkal
Ram Mohana Reddy Guddeti	Information technology, NITK, Surathkal
B Venkatesa Perumal	Electrical and Electronics Engineering, NITK, Surathkal
Shyam Lal	Electronics and Communication Engineering, NITK, Surathkal
Mohit P Tahiliani	Computer Science & Engineering, NITK, Surathkal
Pruthviraj U	Water Resources and Ocean Engineering, NITK, Surathkal
Babloo Chaudhary	Civil Engineering NITK, Surathkal

Deeply appreciate the respected members of the National Advisory Committee for their invaluable academic support and motivation:

Tanmay Basak	Indian Institute of Technology, Madras
R. Vinu	Indian Institute of Technology, Madras
B. C. Meikap	Indian Institute of Technology, Kharagpur
V. Sivasubramanian	National Institute of Technology, Calicut
Uday Bhaskar Babu	National Institute of Technology, Warangal
Soumya K Ghosh	Indian Institute of Technology, Kharagpur
Viswanathan N. Nurni	Indian Institute of Technology, Bombay
Abhilash	Council of Scientific and Industrial Research-National Metallurgical Laboratory, Jamshedpur
Dipin S. Pillai	Indian Institute of Technology, Kanpur
Deepak Kumar Ojha	Indian Institute of Technology, Roorkee
Hemanth Kumar Tanneru	Indian Institute of Petroleum and Energy, Vishakhapatnam

Thanks to the keynote speakers

Tanmay Basak	Indian Institute of Technology, Madras
P C Abhilash	Indian Institute of Technology Bhubaneswar
TCSM Gupta	Apar Industrial Limited
Manoj P Samuel	Centre for Water Resources Development and Management, Kozhikode
Shri. Padmanand Warrior	Warrier H.E.A.R.T, Founder
Ravikrishnan Vinu	Indian Institute of Technology, Madras
Venkata Vanuku	Global Foundaries
Chenna Borra	Indian Institute of Technology, Kharagpur
Pradeep R Nair	Indian Institute of Technology, Bombay
Somashekhar S Hiremath	Indian Institute of Technology, Madras
Shri. Dilip Kumar Dalei	Defence Research and Development Organization

Savithri Sivaram — Council of Scientific and Industrial Research - National Institute for Interdisciplinary Science and Technology, Thiruvananthapuram
H Lakshmi Kantha — Karnataka State Pollution Control Board

Chairs for the Technical sessions

Tanmay Basak
C Abhilash
Rajashekhar Busigari
Gangamma S
Darshak Trivedi
G N Kumar
Hemanth Kumar Tanneru
Vidya Shetty K
Ravikrishna Vinu
Venkata Vankuru
Hari Prasad Dasari
Pradeep R Nair
Somashekhar S Hiremath
Ananthanarayana V S

Co-chairs for the Technical sessions

P S Suvin
Nabanita Naskar
Jitendra Kumar
Poornima G Hiramath
Anand Prabhu
Ishwar Chandra
Chandavarkar
Meera Bai S
Jagannathan T K
Suresha B L
Rajesh Gratian D
Saumen Mandal
Shyam Lal
Ganesh R Chate
We sincerely thank the faculty of the Department of Chemical Engineering, NITK Surathkal, for their invaluable support and guidance.
Vidya Shetty K
Raj Mohan B
P E Jagadeesh Babu
Prasanna B D

I Regupathi
Keyur Raval
Hari Prasad Dasari
Hari Mahalingam
Gangamma S
Jagannathan T K
Ashraf Ali
Chinta Sankar Rao
Vaishakh Nair
S Jitendra Pal
Maneesh Kumar Poddar
Mohan Lal Meena
M Rajasekaran
Gratefully acknowledge the assistance of the following student volunteers, who were provided their support in different organization areas, including:

Registration

Sedevino Sophia, S Srilakshmi, Haritha Hari, Nandini I V, Ritisha Digvijay Kale, Malavika P V, Jahanvi G Bhat, Sreeresmi M

Stage Arrangements, Displays, Lights and Sounds

Suyash Harsh, Dharshini Jenifer, Anusha B Kawatal, Hemanth Kumar Bajaj, Tejas

General Arrangements

Sahana Vijay Kumar, Chaitra Chandrakant Shanbhag, Sheeja Mary Patrick, Tejanshu Sekhar Sahu, Lekha M, Hemant Kumar Bajaj, Mithun Rathod, Thanushree, Prateek

Poster and Exhibition Stalls

Harmeet Singh, Nandini I V, Hemant Kumar Bajaj, Shreya G Prasad, Sarvesha Babu M, Prajwal B R, Vanitha, Shree Ranjini H K, Bhavyashree C, Anshika Gupta

Stage Management

Preethi, Suphalitha Manpreeth, Nikhil Siringi, Krishnaveni S Makanura, Parvathy Purushothaman
Photography and Videography:
Janumpally Sushanth Reddy, Gadepu Praneeth, Bandaru S E Rangaiah Naidu, Dhilipan L, Yash Pareshbhai Suvagiya

Media Publicity

Ananya Mandal, Nishant Patil

Transport and Reception Committee

Prapthi S Mally, Gokul M, Santosh Amarapure, Wangjam Deepak, Hemant Kumar Bajaj, Danish, Divyamshu, Prithwindra Sarkar, Shanmugasundaram Vignesh, Ashish Manash

Accommodation (Participants)

Tejanshu Sekhar Sahu, Athul Akhileshan, Sriram V, Dhakate Mrudal Namdev, Meghna Sarma

Accommodation (Keynote Speakers)

Nandini I V, Vishnu Soman, Shubhanshu Bharat, Pankaj Deori, Prajwal B R, Ayush Kumar

Food and Refreshments

Sahana Vijay Kumar, Chaitra Chandrakant Shanbhag, Sheeja Mary Patrick, Suyash Harsh, Prapthi S Mally, Tubin Manuel, Vishal Kannaujiya, Harsha S, Jonathan Paul James

Preparation of Certificates

Janumpally Sushanth Reddy, Gadepu Praneeth, Ritisha Digvijay Kale, Malavika P V

Technical Sessions

Rahulkumar Shirasangi, Nandini I V, Prakhyath, Nishant Patil, Prapthi S Mally, Sattineedi Siri Chandana, Samiksha Suman Hemant, Anusha B Kawatal, Shrilega M, Adarsh, Lekshmi S B, Sriram V, Navya B Shetty

Gala Dinner and Music

Chaitra Chandrakant Shanbhag, Sheeja Mary Patrick, Nikhil Siringi, Rahulkumar Shirasangi, Dharshini Jenifer, Chandankhede Pranali Vinod, Kesava Moorthy N

Emcee (MC)

S Srilakshmi, Nishant Patil, Ritisha Digvijay Kale
Also, deeply thankful to our sponsors for their kind support, which significantly contributed to the successful organization of this conference.

Anusandhan National Research Foundation

Cezen Tech
MRPL-ONGC
3D Engineering Automation LLP
APAR Industries Ltd.
Sri Durga Laboratory Equipment Supplies
BHAANVI enviro and agro India Pvt. Ltd.
Cambridge University Press
De Gruyter Brill
World Scientific.

Organization

Conference chair

I. Regupathi	Department of Chemical Engineering, NITK, Surathkal

Conference Conveners

Chinta Sankar Rao	Department of Chemical Engineering, NITK, Surathkal
Vaishakh Nair	Department of Chemical Engineering, NITK, Surathkal
Ranjeet Kumar Sahu	Department of Mechanical Engineering, NITK, Surathkal
Anupama Surenjan	Department of Civil Engineering, NITK, Surathkal
Nikhil K S	Electronics and Communication Engineering, NITK, Surathkal
Sourav Kanti Addya	Computer Science & Engineering, NITK, Surathkal
Lipak Kumar Sahoo	Metallurgical and Materials Engineering, NITK, Surathkal

Publicity Committee Chairs

Vaishakh Nair	Department of Chemical Engineering, NITK, Surathkal
Ranjeet Kumar Sahu	Department of Mechanical Engineering, NITK, Surathkal

National Advisory Committee

Tanmay Basak	Indian Institute of Technology, Madras
R. Vinu	Indian Institute of Technology, Madras
B. C. Meikap	Indian Institute of Technology, Kharagpur
V. Sivasubramanian	National Institute of Technology, Calicut

Uday Bhaskar Babu	National Institute of Technology, Warangal
Soumya K Ghosh	Indian Institute of Technology, Kharagpur
Viswanathan N. Nurni	Indian Institute of Technology, Bombay
Abhilash	Council of Scientific and Industrial Research-National Metallurgical Laboratory, Jamshedpur
Dipin S. Pillai	Indian Institute of Technology, Kanpur
Deepak Kumar Ojha	Indian Institute of Technology, Roorkee
Hemanth Kumar Tanneru	Indian Institute of Petroleum and Energy, Vishakhapatnam

Contents

Earth (materials)

Fire (energy)

Contributors

M. S. Abhish National Institute of Technology Karnataka Surathkal, Mangalore, Karnataka, India

Ch. Akash QIS College of Engineering and Technology, Ongole, AP, India

Sanjith S. Anchan The National Institute of Engineering, Mysuru, Karnataka, India

M. Anil QIS College of Engineering and Technology, Ongole, AP, India

Ishwar Chandra Sir M. Visvesvaraya Institute of Technology, Affiliated to VTU, Belagavi, Bangalore, Karnataka, India

Jagadeesha J. Chinagudi Indian Space Research Organization, NRSC-RRSC-Bangalore, Bangalore, India

Sharada Mani Dalu QIS College of Engineering and Technology, Ongole, AP, India

Dhanush Devadiga Department of Water Resource and Ocean Engineering, National Institute of Technology Karnataka Surathkal, Mangaluru, Karnataka, India

Rawal Diganjit Government Polytechnic Kolhapur, Kolhapur, Maharashtra, India

G. Dinesh Department of Mining Engineering, National Institute of Technology Karnataka Surathkal, Mangalore, India

Saikat Dutta Department of Chemistry, National Institute of Technology Karnataka, Surathkal, Mangalore, Karnataka, India

R. Gayathri QIS College of Engineering and Technology, Ongole, AP, India

N. Gnanasekaran Indian Institute of Technology, Tirupati, Andhra Pradesh, India

T. Gopikrishnan Kailas Department of Mechanical Engineering, National Institute of Technology Karnataka, Surathkal, Mangalore, Karnataka, India

Nitin Gotiya Department of Mechanical Engineering, Indian Institute of Technology Madras, Chennai, Tamil Nadu, India

M. Govinda Raj National Institute of Technology Karnataka, Surathkal, Karnataka, India

H. Guruprasad Bhat Department of Mining Engineering, National Institute of Technology Karnataka, Mangaluru, Karnataka, India

Punithraj Gururaj Department of Civil Engineering, National Institute of Technology Calicut, Kozhikode, Kerala, India

Somashekhar S. Hiremath Department of Mechanical Engineering, Indian Institute of Technology Madras, Chennai, Tamil Nadu, India

M. Jayanth The National Institute of Engineering, Mysuru, Karnataka, India

Ritisha Digvijay Kale Department of Chemical Engineering, National Institute of Technology Karnataka Surathkal, Mangalore, Karnataka, India

T. G. Kanakadas Department of Chemical Engineering, National Institute of Technology Karnataka Surathkal, Mangalore, Karnataka, India

S. Kumar Reddy Department of Mining Engineering, National Institute of Technology Karnataka, Mangaluru, Karnataka, India

S. B. Lekshmi Department of Chemical Engineering, National Institute of Technology Karnataka Surathkal, Mangalore, Karnataka, India

Vasudeva Madav Department of Mechanical Engineering, National Institute of Technology Karnataka, Surathkal, Mangalore, Karnataka, India

P. V. Malavika Department of Chemical Engineering, National Institute of Technology Karnataka Surathkal, Mangalore, Karnataka, India

Surjendu Manna RCC Institute of Information Technology, Kolkata, West Bengal, India

Mohan Lal Meena Department of Chemical Engineering, National Institute of Technology Karnataka Surathkal, Mangalore, Karnataka, India

Anisha Mohapatra Research Center for Applied Sciences, Academia Sinica, Taipei, Taiwan

V. Mounika QIS College of Engineering and Technology, Ongole, AP, India

Ch. S. N. Murthy National Institute of Technology Karnataka Surathkal, Mangalore, Karnataka, India

C. P. Nesara The National Institute of Engineering, Mysuru, Karnataka, India

R. Perumalla National Institute of Technology Karnataka, Surathkal, Karnataka, India

Maneesh Kumar Poddar Department of Chemical Engineering, National Institute of Technology Karnataka Surathkal, Mangalore, Karnataka, India

M. H. Prashanth National Institute of Technology Karnataka Surathkal, Mangalore, Karnataka, India

A. Preethi Department of Civil Engineering, National Institute of Technology Calicut, Kozhikode, Kerala, India

Samanyu P. Rai Department of Water Resource and Ocean Engineering, National Institute of Technology Karnataka Surathkal, Mangaluru, Karnataka, India

N. Ramesh School of Applied Sciences, REVA University, Bangalore, Karnataka, India

Chinta Sankar Rao Department of Chemical Engineering, National Institute of Technology Karnataka Surathkal, Mangalore, Karnataka, India

Sandi Kumar Reddy National Institute of Technology Karnataka Surathkal, Mangalore, Karnataka, India

Binoy Krishna Roy National Institute of Technology, Silchar, Assam, India

C. Guru Santosh Department of Chemical Engineering, National Institute of Technology Karnataka Surathkal, Mangalore, Karnataka, India

Indumathi Shivamallappa The National Institute of Engineering, Mysuru, Karnataka, India

Ch. Silpa QIS College of Engineering and Technology, Ongole, AP, India

Shreyang Shekhar Singh Department of Chemical Engineering, National Institute of Technology Karnataka Surathkal, Mangalore, Karnataka, India

Vishnu Soman Department of Chemical Engineering, National Institute of Technology Karnataka Surathkal, Mangalore, Karnataka, India

Pruthviraj Umesh Department of Water Resource and Ocean Engineering, National Institute of Technology Karnataka Surathkal, Mangaluru, Karnataka, India

Anima Upadhyay Sir M. Visvesvaraya Institute of Technology, Affiliated to VTU, Belagavi, Bangalore, Karnataka, India

P. Varalakshmi National Institute of Technology Karnataka Surathkal, Mangalore, Karnataka, India

K. Venkateshwarlu Department of Mining Engineering, National Institute of Technology Karnataka Surathkal, Mangalore, India

Hemant Verma Research Center for Applied Sciences, Academia Sinica, Taipei, Taiwan

B. Kiranpreethi Engineering Scholars, The National Institute of Engineering, Mysuru, 570008 India

Abbreviations

μ	Octahedral Factor
μA	Micro Amps
μs	Microsecs
1D	One-dimensional
1H-NMR	Proton nuclear magnetic resonance
2D	Two-dimensional
3D	3 Times the Diameter
A	Coded variable for neem gum in the quadratic model
A	Catchment area (roof) (m^2).
a_1	Total Specific Surface Area
a_2	External Surface Area
AB	Administrative Building
ACH	Air Change per Hour
ACZ	Agro-Climatic Zone
ADAM	Adaptive Moment Estimation (Optimizer)
ADC	Analog to Digital Converter.
Adj-R^2	Adjusted R-squared
Ag_2O	Silver Oxide
AHP	Analytic Hierarchy Process
AI	Artificial Intelligence
Al	Aluminium
ANN	Artificial Neural Network
AO	Atomic Oxygen
a_p	Pore-specific Surface Area
AP	Andhra Pradesh
API	Application Programming Interface
$a_{s,BET}$	BET-specific Surface Area
ASC-HRDC	Academic Staff College (Human Resource Development Centre)
ATR	Attenuated Total Reflectance
B	Coded variable for cross-linker in the quadratic model
BCS	Black Cotton Soil
BDSW	Biomethanated Distillery Spent Wash
BET	Brunauer–Emmett–Teller (Surface Area Analysis Method)
BIF	Banded Iron Formations
BIS	Bureau of Indian Standards
BJH	Barrett–Joyner–Halenda (Pore Size Distribution Method)
BM-A	B M Sri Auditorium
BOD	Biological Oxygen Demand
BS	Building Sample
BS-VI	Bharath stage VI
C	Calcium

C	Coded variable for ultrasonication duration in the quadratic model
c	Cohesiveness
C	Runoff coefficient (dimensionless)
CAD	Computer-Aided Diagnosis
CaO	Calcium Oxide
CART	Classification and regression tree
CCMC	Computer centre maharaja college
CCN	Cloud condensation nuclei
CH	Crowfard Hall
CH_4	Methane
CIST	Centre for Information Science and Technology
Cl	Chloride
CMIP5	Coupled Model Inter comparison Project Phase 5
CMSP	Coastal Marine Spatial Planning
CNN	Convolutional Neural Network
CNT	carbon nanotubes
CO	Carbon Monoxide
CO2	Carbon Dioxide
COD	Chemical Oxygen Demand
CPHEEO	Central Public Health and Environmental Engineering Organisation
CPU	Central Processing Unit
Cr	Chromium
CUF	Capacity Utilization Factor
CVD	Chemical Vapour Deposition
CWC	Central Water Commission.
D	Tunnel Diameter
DC	Direct Current
DEM	Digital Elevation Model
DF	Decrement Factor
DGPS	Differential Global Positioning System
DI	Deionised
DIY projects	Do It Yourself projects
DL	Deep Learning
DO	Dissolved Oxygen
DOC	Department of criminology
DOS-B&Z	DOS of Studies in Botany and Zoology
DOS-BC	DOS of bio chemistry
DOS-BT	DOS In Biotechnology
DOS-C	DOS of Commerce
DOS-Ch	DOS of Chemistry
DOS-CS	DOS In Computer Science
DOS-ES	DOS of studies in environmental science
DOS-GE	DOS of Geology
DOS-GG	DOS of studies in genetics and geonomics
DOS-JM	DOS of journalism and mass communication
DOS-M	DoS of Mathematics

DOS-MB	DOS of molecular biology
DOS-OC	DOS of organic chemistry
DOS-P	DOS of Physics
DOS-SA	DOS of Studies in Sanskrit
DOS-SMI	DOS of Studies in Microbiology
DOS-SO	DOS of Sociology
DOS-SS	DOS of Social Studies
DOS-ST	DOS of Statistics
DoS-SW	DOS of Student Welfare
d_p	Mean Pore Diameter
$d_{p,peak}$	Pore Size Distribution Peak Diameter
DPR	Detailed Project Report
Dr B R A H	Dr B R Ambedkar Hostel
DS Impv	Depression storage for Impervious surfaces
DS Perv	Depression storage for pervious surfaces
DSC	Differential Scanning Calorimeter
DSM	Desertification Status Maps
DT	Decision Tree
DTG	Differential Thermogravimetric
EA	Axial Stiffness
EC	eddy covariance
EC	Electrical Conductivity
ECM	Electrochemical Machining
E_d	Soil Modulus of Deformation
EDM	Electric Discharge Machining
EDTA	Ethylene Diamine Tetra acetic acid
EDX	Energy Dispersive X-ray
EEPROM	Electrically Erasable Programmable Read-Only Memory
E_g	Bandgap
EI	Flexural Rigidity
ELT	extract load transform
EMPRI	Environmental Monitoring and Policy Research Institute
ES-μ-ECDM	Electrolyte Sonicated Micro Electrochemical Discharge Machining
ESA	European Space Agency
ETL	Electron Transport Layer
EWAM	Equal Weighted Arithmetic Mean
F	F-value in ANOVA analysis
F1-score	F1 Score (harmonic mean of precision and recall)
FA^+	Formamidinium
FCC	Fluid Catalytic Cracking
FDM	Fused Deposition Modeling
Fe	Iron
FEEP	Field Emission Electric Propulsion
FEM	Finite Element Method
FF	Fill Factor
FM	Folklore museum

FN	Food science and nutrition
FOS	Factor of Safety
FRM	Flood Risk Management
FSI	Forest Survey of India
FTIR	Fourier Transmission Infrared Radiation
FTO	Fluorine-doped Tin Oxide
FY	Financial Year
G	Shear Modulus
GANs	Generative Adversarial Networks
GB	Gradient Boosting
GBDTs	gradient-boosted decision trees
GBM	Gradient Boosting Machines
GBR	Gradient Boosting Regressor
GC-HRMS	Gas chromatography-high resolution mass spectrometry
GCM	Global Climate Model
GC-MS	Gas chromatography-mass spectrometry
GDP	Gross Domestic Product
GEV	Generalised Extreme Value
GFDL-ESM2G	Geophysical Fluid Dynamics Laboratory - Earth System Model version 2G
GHG	Greenhouse Gas
GHI	Global Horizontal Irradiance
GHz	Giga Hertz
GIS	Geographic Information System
GKI	Geospatial Knowledge Infrastructure
GND	Ground
GNN	graph neural network
GoK	Government of Karnataka
GP	Gaussian Process
GPIO	General Purpose Input/Output
GPR	Gaussian Process Regression
GTR	Gradient Tree Boosting
GWh	Gigawatt-hour
H/C	Hydrogen-to-carbon ratio
H2	Hydrogen gas
ha	Hectares
HCl	Hydrochloric acid
HD	Hitachi Display
HDPE	High-density polyethylene
HEC-RAS	Hydrologic Engineering Center – River Analysis System
HH	Horizontal transmit and Horizontal receive
HPV	Human Papillomavirus
HTL	Hole Transport Layer
HUSY	Hydrogen Ultra-Stable Y Zeolite
HVAC	Heating, Ventilation, and Air Conditioning
HZSM	Hydrogen-exchanged Zeolite Socony Mobil

I/O	Input Output
IC	Internal Combustion
ICAR	Indian Council of Agricultural Research
ICP-AES	Inductively coupled plasma atomic emission spectroscopy
ICRISA	International Crops Research Institute for the Semi-Arid Tropics
ICT	Information and Communication Technology
IDF	Intensity-Duration-Frequency
IEA	International Energy Agency
IMD	Indian Meteorological Department
IoT	Internet of Things
IPCC	Intergovernmental Panel on Climate Change
IRR	internal rate of return
IS	Indian Standard
ISIC	International Skin Imaging Collaboration
ISRO	Indian Space Research Organisation
ITO	Tin-doped Indium Oxide
JAEO	Joint Attention-based Enhanced Optimization
J_{sc}	Short Circuit Current Density
K	Potassium
K	Width Parameter
K_{sat}	Hydraulic Conductivity
K_0	Earth Pressure at Rest
KB	Kilobytes
kHz	Kilo Hertz
KOH	Potassium Hydroxide
KSNDMA	Karnataka State Natural Disasters Monitoring Agency
KSRSAC	Karnataka State Remote Sensing Applications Centre
KUSUM	Kisan Urja Suraksha evam Utthaan Mahabhiyan
kW	Kilowatts
kWh	Kilowatt-hour
kWp	Kilowatt-peak (peak power output of a solar system)
L2	L2 Regularization
LCC	lambert conforml conical
LCD	Liquid Crystal Display
LDPE	Low-density polyethylene
LEDs	Light Emitting Diodes
LEM	Limit Equilibrium Method
LEO	low Earth orbit
LM-BPNN	Levenberg-Marquardt Backpropagation Neural Network
LoRA	Long Range
LPG	Liquefied Petroleum Gas
LR	Linear Regression
LSA	Land Suitability Analysis
LULC	Land Use Land Cover
mA	Milli amps
MA^+	Methylammonium

MAD	Mean Absolute Deviation
MAE	Mean Absolute Error
MAPE	Mean Absolute Percentage Error
MASW	Multichannel Analysis of Surface Wave
MB	Megabytes
MCM	Mohr-Coulomb Model
MCM	Mobil Composition of Matter
Mg	Magnesium
MgO	Magnesium Oxide
MHP	Metal Halide Perovskite
MHz	Mega Hertz
ML	Million liters
ML	Machine Learning
MLC	Maximum Likelihood Classification
MLP	Multi-Layer Perceptron
MLPNN	Multilayer Perceptron Neural Network
mm	Milli metre
MMU	Minimum Mapping Unit
MNRE	Ministry of New and Renewable Energy
MP	Micropore Plot (MP-Plot)
MPW	Municipal Plastic Waste
MRFO	Manta Rays Foraging Optimizer
MRR	Material Removal Rate
MSE	Mean Squared Error
MT	Metric tonnes
MW	Megawatts
MWCNT	Multi-Walled Carbon Nanotube
MWh	Megawatt-hour
N Impv	Manning's roughness coefficient for impervious surfaces
N Perv	Manning's roughness coefficient for the previous surface
Na	Sodium
NaCl	Sodium chloride
NAD	No Apparent Degradation
NaOH	Sodium hydroxide
NBSS&LUP	National Bureau of Soil Survey & Land Use Planning
NCP	National Carbon Project
NGEs	Non-Government Entities
NGOs	Non-Governmental Organizations
Ni	Nickel
n-i-p	n-type / intrinsic / p-type
NIR	Near-Infrared Region
NIST	National institute of standards and technology
NITK	National Institute of Technology Karnataka
NN	Neural Network
NO_2	Nitrogen dioxide
NOAA	National Oceanic and Atmospheric Administration

NPV	net present value
NRDB	National Resources Database
NRSC	National Remote Sensing Centre
O	Oxygen
O2	Oxygen gas
OAT	One- factor-at-a-time
OB	Overburden
OLI	Operational Land Imager
OPC	Ordinary Portland Cement
p	p-value in statistical significance testing
PAH	Polycyclic Aromatic Hydrocarbons
Pb	Lead
PCA	Principal Component Analysis
PCBM	[6, 6]-Phenyl-C61-butyric acid methyl ester
PCE	Power Conversion Efficiency
PCM	Phase Change Material
PCSWMM	Personal Computer Storm Water Management Model
PDFM	Population Dynamics Foundation Model
PE	Polyethylene
PEIL	Passive Energy Innovation Laboratory
PET	Polyethylene terephthalate
PH2	PH2 Dermoscopic Image Dataset
p-i-n	p-type / intrinsic / n-type
PLA	Polylactic Acid
PM	Particulate matter
PP	Polypropylene
PPISR	Poornaprajna Institute of Scientific Research
ppm	parts per million
PPO	Polypropylene pyrolysis oil
PS	Polystyrene
PSC	Perovskite Solar Cell
PTAA	Poly[bis(4-phenyl)(2,4,6-trimethylphenyl)amine]
PV	Photovoltaic
Q	Quality Rating
QC	Quality Checking
R^2	coefficient of determination (Regression)
RAM	Random Access memory
RBFNN	Radial Basis Function Neural Network
R-CNN	Region-based Convolutional Neural Network
RCP	Representative Concentration Pathway
ReLU	Rectified Linear Unit
RF	Random Forest
RFR	Random Forest Regressor
RISC	Reduced Instruction Set Computer
RK4	Fourth order Runge Kutta
RMSE	Root Mean Square Error

RNNs	Recurrent Neural Networks
RSM	Response Surface Methodology
SAAN	Sequential Artificial Neural Network
SAC	Space Applications Centre.
SAH	solar air Heater
SB	Sugarcane Bagasse
SCAPS	Solar Cell Capacitance Simulator
Scw	Subcatchement width
SE	Squeeze-and-Excitation
SEM	Scanning Electron Microscopy
SERS	surface enhanced Raman spectroscopy
SGD	Stochastic Gradient Descent
SHAP	SHapley Additive exPlanations
Si	Silica
SKY	Suryashakti Kisan Yojana
SO_2	Sulfur dioxide
SP	Sand Particles
Spiro-OMeTAD	2,2',7,7'-Tetrakis[N,N-di(4-methoxyphenyl)amino]-9,9'-spirobifluorene
SPT	Standard Penetration Test
Sq.km	Square kilometres
SRAM	Static Random Access Memory
SRTM	Shuttle Radar Topography Mission
STP	Standard Temperature and Pressure
SVAF	soil vegetation atmosphere flux
SVM	Support Vector Machine
SWMM	Storm Water Management Model
t	Tolerance Factor
t	Thickness
TCO	Transparent Conducting Oxide
TDS	Total Dissolved Solids
TEM	Transmission Electron Microscope
TG	Thermogravimetric
TIRS	Thermal Infrared Sensor
TTL	Transistor-Transistor Logic
TWh	Terawatt-hour
U	Total Displacement
U	Uranium
U	Surface Settlement
UAS	University of agricultural sciences
UG1	Underground 1
UG2	Underground 2
US	United States.
USA	United States of America
USB	Universal Serial Bus
USGS	United States Geological Survey

UTM	Universal Testing Machine
UV	ultraviolet
Ux	Horizontal Displacement
Uy	Vertical Displacement
V	Volts
v	Poisson's Ratio
V	Lining Displacement
VCC	Voltage at the Common Collector
VG	Van Genuchten
V_m	Monolayer Volume
V_{oc}	Open-Circuit Voltage
V_p	Pore Volume
Vs	Volume of the Settlement
VTU	Visvesvaraya Technological University
VUV	vacuum ultraviolet
VWC	Volumetric Water Content
W	Unit Weight
WBERC	West Bengal Electricity Regulatory Commission
WBGEDCL	West Bengal Green Energy Development Corporation Limited
WBPDCL	West Bengal Power Development Corporation Limited
WBSEDCL	West Bengal State Electricity Distribution Company Limited
WHO	World Health Organization
WQI	Water Quality Index
WRIS	Water Resources Information System
WT	Water temperature
WVP	Water Vapor Permeability
WVTR	Water Vapor Transmission Rate
X1	Concentration of Neem gum (g) (independent process variable)
X2	Concentration of Cross-linker (g) (independent process variable)
X3	Duration of ultrasonication (min) (independent process variable)
XGB	Extreme Gradient Boosting
XGBR	Extreme Gradient Boosting Regressor
XRD	X-ray diffraction
y	Lateral Distance
Z_0	Depth of the Neutral Axis
Zn	Zinc
ZnO	zinc oxide
β0	Intercept coefficient in the regression equations
β1	Coefficient for main effect of A
β11	Quadratic coefficient for A^2
β12	Interaction coefficient between variables A and B
β13	Interaction coefficient between variables A and C
β2	Coefficient for main effect of B
β22	Quadratic coefficient for B^2
β23	Interaction coefficient between variables B and C
β3	Coefficient for main effect of C

β33	Quadratic coefficient for C^2
γ	Soil Unit Weight
φ	Friction Angle
ψ	Dilatancy Angle

Earth (materials)

Optimization of Chitosan-Neem Gum Based Biodegradable Edible Film Production Parameters Using Response Surface Methodology

Vishnu Soman and Maneesh Kumar Poddar(✉)

Department of Chemical Engineering, National Institute of Technology Karnataka Surathkal, Mangalore, Karnataka, India
maneesh.poddar@nitk.edu.in

Abstract. The study utilizes the Response Surface Methodology (RSM) which aims to optimize the production parameters of chitosan-neem gum based edible packaging films. The factors taken into account were neem gum concentration, cross-linker concentration and duration of ultrasonication with the responses being tensile strength, thickness, swelling and Water Vapor Permeability (WVP). Significant effects of these factors on the responses are evident through the analysis. Neem gum, cross-linker and duration along with their interactions had an evident impact on the tensile strength. Thickness was seen to be decreased with increased cross-linker concentration while swelling was affected by both cross-linker and duration of ultrasonication. Water Vapor Permeability (WVP) was also impacted by neem gum, duration and their interactions. The final regression equations derived from the RSM provided a quantitative outlook into these relationships. These models hence obtained facilitated the identification of optimum process conditions for improved mechanical and barrier properties Hence, the application of RSM proved effective for process optimization in the edible film preparation.

Keywords: Chitosan · Neem gum · Biodegradable films · Optimization · Tensile strength

1 Introduction

The alarming effects of petroleum based plastic pollution have driven considerable interest in developing biodegradable alternatives especially in the field of food packaging due to their biodegradability and lower environmental impact [1–3]. Further studies are being conducted on enhancing the longevity of packaged food items as well as edibility to reduce the amount of waste generated. It is noteworthy that, chitosan which is a polysaccharide derived from chitin is widely explored for the synthesis of biodegradable polymer films but has certain shortcoming such as poor mechanical, thermal, and moisture barrier properties [4, 5].

The present work focusses on a blended polymer of chitosan and neem gum (a natural resin obtained from the bark of *Azadirachta indica*) where the latter has good

H. P. Dasari et al. (Eds.): NITK-CREST 2025, PTT, pp. 3–11, 2026.
https://doi.org/10.1007/978-981-95-3218-6_1

antimicrobial properties as well as film forming capability [6–8]. Neem gum is readily soluble in water, whereas chitosan dissolves only in organic acids such as acetic acid by means of protonation of amine groups present in chitosan. The ionic strength of chitosan can be influenced by the presence of natural gums, such as gum Arabic, alginate and kappa-carrageenan etc. The binding constant of chitosan with these gums is reduced with an increase in ionic strength and the pH and polymer ratio can affect the formation and solubility of these complexes [9]. Likewise, neem gum also seems to have a similar effect of the chitosan solution in acetic acid and affects the rheology of the film forming solution hence results in difficulty in the process of solvent casting. Chitosan can be modified by cross-linking process where its solubility, biological activity and mechanical properties may be improved [10–12]. Therefore, it is crucial to cross-link chitosan and neem gum for achieving improved structural and functional properties which is a crucial factor in designing food packaging films.

It is very evident from the literature that the properties of chitosan films depend significantly on its composition. Concentration of chitosan, plasticizers, essential oils, nanofillers, clay etc. play a vital role in shaping the film characteristics [13–15]. There are numerous studies where response surface methodology (RSM) is utilized to optimize the formulation of chitosan-based food packaging films. Thakur et al. [16] found that a combination of 1% concentration, pea starch at 1.5% and glycerine at 0.5% yielded films with desirable properties. Similarly, Singh et al. [17] identified 2.0% chitosan, 0.75% glycerol and a drying temperature of 40 °C as the optimal conditions for the development of films. Ferreira et al. [18] incorporated cassava starch, gelatin and grape pomace extracts into the films, respectively and evaluating their physicochemical properties. Therefore, the key objective of the present work is to establish an optimized methodology for the formulation of chitosan-based packaging films which incorporates neem gum and NaOH (as a cross-linker) and varying the duration of ultrasonication using the Response Surface Methodology (RSM).

2 Materials and Methods

2.1 Materials

Chitosan (90% deacetylated), sodium hydroxide and acetic acid were obtained from Sisco Research Lab, India. Neem gum was sourced locally from a vendor in Tamil Nadu.

2.2 Methodology

2.2.1 Film Casting

A 1% chitosan solution (w/v) was prepared by dissolving chitosan powder (medium molecular weight, 90% deacetylation) 1% acetic acid. The chitosan solution was sonicated at 60 °C. Neem gum slurry was prepared by dissolving it in graded concentrations in distilled water. Both the slurries of chitosan and neem gum were mixed. Sodium hydroxide (NaOH) cross-linker was then added to the solution with concentrations ranging from 0.01 g to 0.05 g per 100 ml of the solution. The concentration of neem gum was also varied from 0.1 g to 0.5 g per 100 ml of the solution. This solution was then subjected to ultrasound-assisted mixing for durations ranging from 15 to 60 min and then poured into petriplates and dried in a vacuum oven at 60 °C overnight and peeled out subsequently after the stipulated time period for further analysis.

2.2.2 Film Characterization

Thickness
Digital micrometer was used to obtain the film thickness [16]. Each film sample was positioned under the measuring probe, and the thickness was noted. These thickness measurements were then used to calculate the water vapor permeability (WVP) of the film samples.

Swelling Index
Pre-weighed films, were cut into square shape with a dimension of 2.5 cm and were individually placed in petri dishes with distilled water at room temperature 8 h. After every hour, the films were carefully removed and excess water on their surfaces was removed using blotting paper. The swollen films were weighed and then re-immersed into distilled water. The percentage swelling was calculated using the formula:

$$Swelling(\%) = \frac{W_f - W_i}{W_i}(X1)00$$

where W_f and Wi are weight of the film before and after soaking, respectively.

Water Vapour Transmission Rate
The water vapor transmission rate (WVTR) was calculated using a modified ASTM E96-95 standard method. Sqaure shaped films of 2 cm dimensions were tied tightly on the tops of test tubes filled with 10 ml of distilled water and were then placed in a desiccator. Weight of the tubes were noted on a daily basis. The weight loss (Δm) was plotted as a function of time (t) and the slope was determined. The WVTR was calculated using the formula:

$$WVTR = \frac{\Delta m.X}{A.\Delta t.\Delta P}$$

where, ($\Delta m/\Delta t$) is the rate of moisture uptake rate, A is the surface area of the film, T is the thickness of the films and ΔP is the water vapor pressure difference across the film.

Tensile Strength of Films
Universal Testing Machine (UTM) was employed for the tensile testing. Film samples were prepared as per ASTM D882 guidelines.

2.2.3 Response Surface Methodology

Design-Expert 8.0.7.1 software was employed to implement the Box-Behnken design with 17 experimental runs for this study. The factors considered for the study were:

1. Factor (X_1): Concentration of Neem gum (g) from 0.10 to 0.50 with coded values of −1.000 to 1.000 corresponding to the minimum and maximum levels.
2. Factor X_2: Cross-linker (g) ranging from 0.01 to 0.05 with coded values of −1.000 to 1.000.
3. Factor X_3: Duration of ultrasonication (min) between 15.00 and 60.00 with coded values of −1.000 to 1.000.

Three level studies were conduected for assessing the linear, quadratic and interaction effects.The mean and standard deviation of each of the mentioned factors were calculated based on their respective ranges. Each run was conducted following the designated coded values for the factors with the responses recorded and analyzed to determine the optimum values (Table 1).

Table 1. Box-Behnken design for optimization

Run	Independent variables			Responses			
	X_1	X_2	X_3	Tensile strength (Mpa)	Thickness (mm)	Swelling (%)	WVP
1	0.3	0.03	37.5	6.4	0.0697	12.61	2.15
2	0.3	0.03	37.5	6.4	0.0751	12.79	2.56
3	0.5	0.05	37.5	6.11	0.1013	14.33	3.5
4	0.3	0.05	15	2.2	0.0859	87.24	3.7
5	0.3	0.01	60	5.6	0.0822	32.87	1.47
6	0.1	0.01	37.5	9.7	0.0538	84.97	2.54
7	0.3	0.01	15	1.2	0.0863	79.06	2.97
8	0.1	0.03	15	2.6	0.0225	63.7	3.11
9	0.1	0.05	37.5	42	0.0311	12.13	1.3
10	0.5	0.01	37.5	11.2	0.1362	83.11	1.72
11	0.5	0.03	15	2.1	0.1247	89.01	2.35
12	0.3	0.03	37.5	9.4	0.0974	72.3	1.59
13	0.5	0.03	60	5.3	0.1341	12.62	3.1
14	0.1	0.03	60	39.2	0.0654	10.84	1.35
15	0.3	0.03	37.5	6.4	0.0753	11.59	1.31
16	0.3	0.05	60	4.3	0.071	12.76	1.47
17	0.3	0.03	37.5	7.1	0.0678	12.33	1.96

The following second-order (quadratic) equation represents the final model for tensile strength in terms of coded (A, B, C) and actual factors (Neem gum, Cross-linker, Duration) obtained through RSM.

Response,

$$Y = \beta_0 + \beta_1 A + \beta_2 B + \beta_3 C + \beta_{12} AB + \beta_{13} AC + \beta_{23} BC + \beta_{11} A^2 + \beta_{22} B^2$$

Here, $\beta_0, \beta_1, \ldots, \beta_{22}$ represent the coefficients of the model.

Statistical Analysis
Analysis of variance (ANOVA), R^2 and the adjusted R^2 (Adj-R^2) were employed to evaluate the model validity using Design-Expert 8.0.7.1 software.

3 Results and Discussions

3.1 Model Fitting

1. Tensile strength

   The Stepwise Regression and ANOVA analyses indicate that the model obtained for tensile strength is highly significant and well fits the data. Coefficients for each factor and their interactions are significant ($p < 0.05$) with the R^2 value at 0.9965 which explains 99.65% of the variation. Predicted R^2 (0.9857) is almost close to the Adjusted R^2 (0.9930). Lack of Fit is not significant ($p = 0.7508$) indicates a good model fit. The ANOVA table confirms that the model is significant (F-value of 285.08, $p < 0.0001$) with all main effects (A, B, C) and interactions (AB, AC, BC) significant ($p < 0.0001$). Adequate Precision (49.881) is greater than the desired value of 4 showing a reliable signal-to-noise ratio. Thickness of films.
2. Thickness of the films

   In case of the thickness of the films, the ANOVA analysis proves that the model is significant (F-value of 29.12, $p < 0.0001$) with Neem gum (X1) being the only significant factor ($p < 0.0001$). Both Cross-linker (X2) and Duration of ultrasonication (X3) are not significant ($p > 0.05$). A p value of 0.4683 means that the Lack if Fit is not significant and therefore the model fits very well to the experimental data. The Predicted R^2 value is 0.7680 is in close agreement with the Adjusted R^2 of 0.8406. The Adequate Precision value is 16.092 indicates an adequate signal-to-noise ratio. The coefficients for Neem gum, Cross-linker and Duration of sonication are provided with Neem gum showing a positive effect on the film thickness.
3. Swelling index

   TANOVA analysis for the swelling of the films indicates a significant model (F-value of 6.14, $p = 0.0078$), with Cross-linker (X2) and Duration of ultrasonication (X3) being the significant factors ($p < 0.05$). Neem gum (X1) is not a significant factor ($p = 0.6959$). Here also, the model fits adequately to the data because the Lack of Fit is not significant ($p = 0.6776$). The Predicted R-squared value is 0.3603, which is almost equal to the Adjusted R-squared value of 0.4909. The Adequate Precision value is 8.570 indicating an adequate signal-to-noise ratio. The coefficients for Neem gum, Cross-linker, and Duration are provided with Cross-linker and Duration showing negative effects on swelling while Neem gum has a positive effect.

4. The ANOVA analysis for WVP proves the model to be significant with an F-value =4.65 and a p-value = 0.0163. Among the individual factors, Duration of ultrasonication (X3) is significant with a p-value of 0.0096. However, Neem gum (X1), Cross-linker (X2) and the interaction terms X1, X2 and X3 are not significant since their $p > 0.05$. The Lack of Fit is not significant either, with a p-value of 0.4239 which suggests a good model to data fit. The Adjusted R-squared is 0.5781, implying that the model might be improved. The Predicted R-squared value is 0.2161 which is lower than expected and may indicate some issues with the model or data. The signal-to-noise ratio is significant since the adequate precision value is 8.090. The coefficients for Neem gum, Cross-linker, and Duration are given with Duration showing a negative effect on WVP while Neem gum and the interaction terms have positive effects.

3.2 Prediction by Empirical Model

The empirical model derived from Box–Behnken design resulted in a second-order polynomial equation which relates the process parameters and the outcomes, including tensile strength, thickness, swelling, and water vapor permeability (WVP). 3D contour plots visualizes the interaction between independent variables. The model obtained is as given below:

$$Y_{\text{Tensile}} = -8.25714 + 13.30965(X_1) - 221.48684X_2 + 0.50000X_3 - 2336.87500(X_1)X_2 - 1.85556(X_1)X_3 + 16.33333X_2X_3 + 138.97697(X_1)^2 + 11272.69737{X_2}^2$$

$$Y_{\text{Thickness}} = 0.081 + 0.040(X_1) - 8.650\text{E-}003X_2 + 4.162\text{E-}003X_3$$

$$Y_{\text{Swelling}} = 41.43 + 3.43(X_1) - 19.19X_2 - 31.24X_3$$

$$Y_{\text{WVP}} = 2.24 + 0.30(X_1) + 0.16X_2 - 0.59X_3 + 0.75(X_1)X_2 + 0.63(X_1)X_3 - 0.18X_2X_3$$

Thickness of the films increased with neem gum (+0.040) and duration (+4.162E-3) but decreased with the concentration of the cross-linker (−8.650E-3) due to restricted polymer mobility. Neem gum reduced tensile strength (−8.60) while cross-linker (+7.33) and duration (+9.75) improved it. Negative interaction effects were seen for X1X2 (−9.35) and X1X3 (−8.35) while X2X3 increased strength (+7.35). Squared terms of X1 and X2 also increased strength. There was an increase in swelling with neem gum (+3.43) but decreased with cross-linker (−19.19) and duration (−31.24). Duration reduced the value of WVP (−0.59) (Fig. 1).

The swelling response increased with neem gum concentration (+3.43) but decreased with higher cross-linker (−19.19) and duration (−31.24). For WVP, duration of ultrasonication gave the strongest negative effect (−0.59 followed by very small contributions from neem gum (+0.30) and cross-linker (+0.16). The combined effects were notable as interaction terms AB, AC, and BC showed coefficients of +0.75, +0.63, and –0.18, respectively.

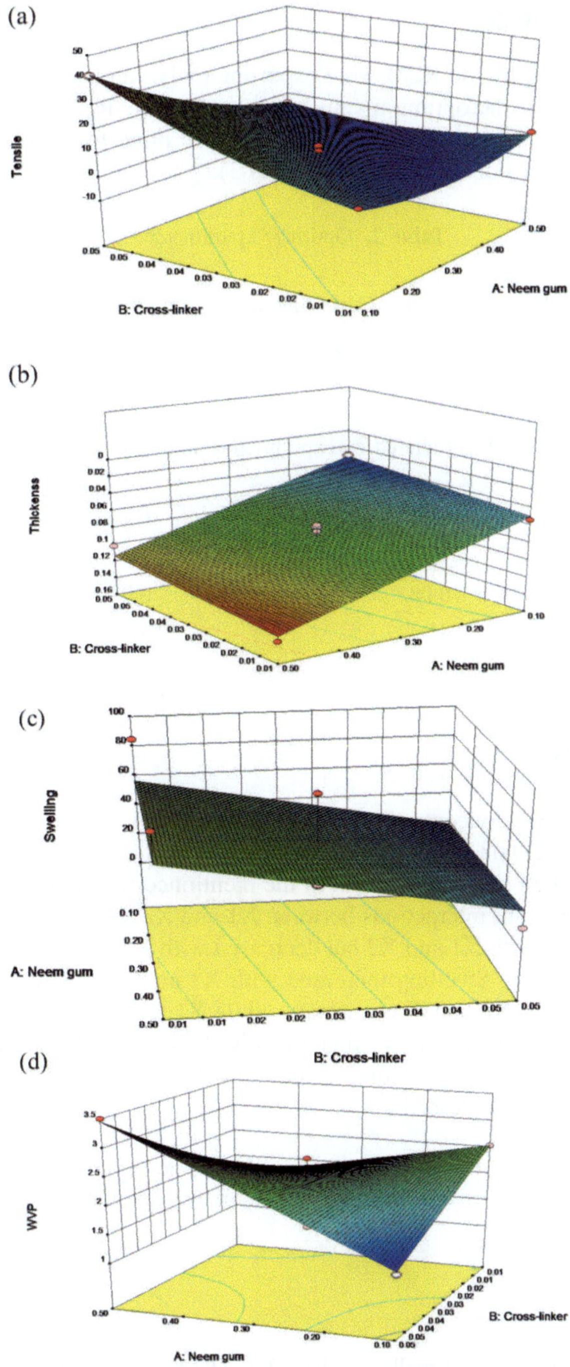

Fig. 1. RSM model graphs for the responses (X1) Tensile strength (X2) Thickness (X3) Swelling (d) WVP

3.3 Validation of Model

The models predicted by the RSM were experimentally validated. Actual experiments were conducted as per the optimum values obtained by the RSM model. Finally, comparison of the experimental and predicted data were done for model validation. The optimum values were found to be (Tables 2 and 3).

Table 2. Optimized parameters

Neem Gum (g)	Cross-linker (g)	Duration (min)
0.30	0.03	37.50

Table 3. Predicted vs. experimental values of reponses

Response	Predicted Value	Experimental value
Tensile strength	7.140	11
Thickness	0.081	0.086
Swelling	41.427	37.243

4 Conclusions

The effect of neem gum concentration (X1), cross-linker concentration (X2) and duration of ultrasonication (X3) on the physio-mechanical properties of the chitosan-neem gum films was studied well using the RSM. All the mentioned factors influenced the tensile strength, with significant interactions between X1 and X2, X1 and X3, and X2 and X3. Thickness increased with X1 and X3 but decreased with X2 likely because of restrcited polymer chain mobility. Swelling increased with X1 and decreased with both X2 and X3. For WVP, X1 and X3 had direct effects, while X1 and X2, along with X1 and X3, showed measurable interactions. The final models obtained by RSM outlined how much each factor contributed to the responses which justifies that the RSM model is reliable for the film production process.

Declaration of Conflicts of Interest. The authors declare no conflicts of interest and received no external funding. Institutional facilities were used for the research.

References

1. Chiellini, E., Barghini, A., Cinelli, P., Ilieva, V.I.: Overview of environmentally compatible polymeric materials for food packaging. Environ. Compat. Food Packag. **1**(1), 371–395 (2008)
2. Cruz, R.M.S., et al.: Bioplastics for food packaging: environmental impact, trends and regulatory aspects. Foods. **11**(19), 3087 (2022)

3. Sid, S., Mor, R.S., Kishore, A., Sharanagat, V.S.: Bio-sourced polymers as alternatives to conventional food packaging materials: a review. Trends Food Sci. Technol. **115**, 87–104 (2021)
4. Elsabee, M.Z., Abdou, E.S.: Chitosan-based edible films and coatings: a review. Mater. Sci. Eng. C. **33**(4), 1819–1841 (2013)
5. Kausar, A.: Scientific potential of chitosan blending with different polymeric materials: a review. J. Plast. Film Sheeting. **33**(4), 384–412 (2017)
6. Uma, A.P.: Antimicrobial activity in pulp extract of Neem (Azadirachta indica Linn.). Int. J. Curr. Res. Biol. Med. **2**(5), 43–45 (2017)
7. Coventry, E., Allan, E.J.: Microbiological and chemical analysis of Neem (Azadirachta indica) extracts: new data on antimicrobial activity. Phytoparasitica. **29**(5), 441–450 (2001)
8. Senthilkumar, P., Rashmitha, S., Veera, P., Ignatious, C., SaiPriya, C., Samrot, A.: Antibacterial activity of Neem extract and its green synthesized silver nanoparticles against *Pseudomonas aeruginosa.* J. Pure Appl. Microbiol. **12**(2), 969–974 (2018)
9. Rabelo, R.S., Tavares, G.M., Prata, A.S., Hubinger, M.D.: Complexation of chitosan with gum Arabic, sodium alginate and κ-carrageenan: effects of pH, polymer ratio and salt concentration. Carbohydr. Polym. **223**, 115120 (2019)
10. Xu, C., Lei, C., Meng, L., Wang, C., Song, Y.: Chitosan as a barrier membrane material in periodontal tissue regeneration. J Biomed Mater Res B Appl Biomater. **100B**(5), 1435–1443 (2012)
11. Shahid-ul-Islam, Butola, B.S.: Recent advances in chitosan polysaccharide and its derivatives in antimicrobial modification of textile materials. Int. J. Biol. Macromol. **121**, 905–912 (2019)
12. Sanchez-Salvador, J.L., Balea, A., Monte, M.C., Negro, C., Blanco, A.: Chitosan grafted/cross-linked with biodegradable polymers: a review. Int. J. Biol. Macromol. **178**, 325–343 (2021)
13. Homez-Jara, A., Daza, L.D., Aguirre, D.M., Muñoz, J.A., Solanilla, J.F., Váquiro, H.A.: Characterization of chitosan edible films obtained with various polymer concentrations and drying temperatures. Int. J. Biol. Macromol. **113**, 1233–1240 (2018)
14. Yoshida, C.M.P., Oliveira Junior, E.N., Franco, T.T.: Chitosan tailor-made films: the effects of additives on barrier and mechanical properties. Packag. Technol. Sci. **22**(3), 161–170 (2009)
15. Casariego, A., et al.: Chitosan/clay films' properties as affected by biopolymer and clay micro/nanoparticles' concentrations. Food Hydrocoll. **23**(7), 1895–1902 (2009)
16. Thakur, R., et al.: Use of response surface methodology (RSM) to optimize pea starch–chitosan novel edible film formulation. J. Food Sci. Technol. **54**(8), 2270–2280 (2017)
17. Singh, T.P., Chatli, M.K., Sahoo, J.: Development of chitosan-based edible films: process optimization using response surface methodology. J. Food Sci. Technol. **52**(5), 2530–2543 (2015)
18. Ferreira, A.S., Nunes, C., Castro, A., Ferreira, P., Coimbra, M.A.: Influence of grape pomace extract incorporation on chitosan films properties. Carbohydr. Polym. **113**, 490–499 (2014)

Characterization of Polypropylene Pyrolysis Oil (PPO) Obtained by High-Temperature Pyrolysis of End-of-Life Polypropylene

T. Gopikrishnan Kailas[1], Saikat Dutta[2], and Vasudeva Madav[1](✉)

[1] Department of Mechanical Engineering, National Institute of Technology Karnataka, Surathkal, Mangalore 575025, Karnataka, India
vasu@nitk.edu.in

[2] Department of Chemistry, National Institute of Technology Karnataka, Surathkal, Mangalore 575025, Karnataka, India

Abstract. Synthetic organic polymers, specifically high-volume thermoplastics in commodity products, are ubiquitous in industrialized societies. The rapidly growing market of these thermoplastics (e.g., polypropylene) has created huge amounts of plastic waste globally. The heavy reliance of the global economy on fossilized feedstocks (e.g., petroleum) has necessitated the search for alternative energy sources. Polypropylene (PP) has a hydrocarbon backbone that can be depolymerized by thermal cracking under pyrolytic conditions into diesel-range fuels. The process forms liquid fuels, carbonized solids (char), and combustible gases. The end-of-life PP is one of the most common plastic waste (volume-wise) found in municipal solid waste (MSW). The PP wastes can be pyrolyzed to produce liquid fuels suitable for furnaces and diesel engines. However, the composition of polypropylene pyrolysis oil (PPO) must be studied to optimize the pyrolytic process conditions and also to design suitable upgrading technology. The present study reports an extensive characterization of the pyrolysis oil obtained by the industrial-scale pyrolysis of end-of-life PP. PPO produced from shredded PP is pyrolyzed in a semi-continuous screw kiln reactor at temperatures in the range of 600–700 °C and analyzed using Fourier transform infrared (FT-IR) spectroscopy, nuclear magnetic resonance spectroscopy (NMR) and gas chromatography-mass spectrometry (GC-MS) analysis. The ^{1}H-NMR analysis revealed that the PPO sample contained 58.88, 27.41, and 13.74 vol% of paraffins, olefins, and aromatics respectively. GC-MS analysis showed the presence of high molecular weight compounds in PPO that may result in more CO_2 and CO emissions than conventional diesel fuel. Many of the physicochemical and thermal properties of PPO were similar to that of diesel. However, the poor storage and thermal stability of PPO warrants suitable upgrading for better combustion and performance when used in diesel engines.

Keywords: Plastic waste · Polypropylene · Pyrolysis · Transportation fuel · Waste to energy

H. P. Dasari et al. (Eds.): NITK-CREST 2025, PTT, pp. 12–25, 2026.
https://doi.org/10.1007/978-981-95-3218-6_2

1 Introduction

It would not be an overstatement to claim that we now live in the age of plastics, where numerous items made of plastics are used in our daily lives because of their low production cost, lightweight, high longevity, synthetic versatility, and control over their properties by synthetic design when compared with other classes of materials [1]. With the growing use of plastics coupled with their non-biodegradability and insufficient infrastructure for their recycling, a major share of the end-of-life plastic items end up in landfilling or scattered, leading to serious pollution in all parts of the biosphere [2]. Global plastic production has been growing at 9% on average every year since 1950 and reached 359 million tonnes in 2018. The demand for plastics is expected to double by 2025 and may exceed three times its current levels by 2050 [3]. The world produces a whopping 2.01 billion tonnes of waste annually, of which 12% comes from plastic waste. The non-biodegradable nature of plastic makes the process of degradation of the accumulated waste plastic decades to centuries to complete making it a huge environmental concern [3]. This plastic pollution is managed by using several strategies like recycling (mechanical, chemical), disposal at the landfill, incineration, microbial degradation, and converting into fuels by pyrolysis [4, 5]. All the high-volume thermoplastics, such as low-density polyethylene (LDPE), high-density polyethylene (HDPE), and polyethylene terephthalate (PET), are produced from petroleum-derived monomers. All the polymers have a hydrocarbon backbone that can, in principle, be thermally cracked (breaking of $C - C$ bonds in the main chain) into transportation fuel-range, hydrocarbon-based molecules under pyrolytic conditions. At present, the major share of energy and fuel demand in the world is met through fossil fuels (e.g., petroleum) [6, 7]. The conversion of waste thermoplastics into liquid fuels by pyrolysis can alleviate the waste management issues with waste plastics while generating energy and fuels. Therefore, pyrolysis is a viable alternative to incineration and materials recycling of end-of-life plastics [8]. Pyrolysis of end-of-life thermoplastics produces three primary product streams: (1) hydrocarbon-rich liquid, (2) carbonized solid char, and (3) combustible gases [5, 9]. Factors that influence the yield and composition of the products include the type of waste plastic(s) in the feed, reactor design, residence time of the vapors inside the reactor, temperature, pressure, and the catalyst used (if any) [10].

LDPE, HDPE, PP, PS, PVC, and PET are the most common plastics in plastic waste [1]. LDPE, PP, and PS are not recycled in India as they are economically unviable to be recycled [11]. PP is a versatile thermoplastic used in the food packaging industry, medical industry, computer parts, various containers, and automotive applications due to its good chemical resistance, low density, and excellent mechanical properties [12]. Williams et al. conducted pyrolysis of HDPE, LDPE, PP, PS, PET, and PVC in a fixed bed batch reactor at a temperature of 700 °C and heating rate of 25 °C/min and found that PP gave the maximum oil yield (84.44%) and PVC (31.69%) the minimum oil yield. Analysis of the pyrolysis oils showed that the composition of oil from HDPE, LDPE, PP, and PVC is mainly aliphatic. PS has a more complex composition, mostly consisting of aromatic components, and PET gives mainly oxygenated and aromatic compounds [9]. Several studies have shown the suitability of using plastic pyrolysis oil in diesel engines [5, 13–16]. Thermal and catalytic pyrolysis of waste PP in a stainless-steel batch reactor at temperatures of 400–500 °C produced oil that can be utilized as an engine fuel [10].

Pyrolysis of waste LDPE, HDPE, PP, and PS in a bench-scale fixed-bed reactor at 500 °C with varying heating rates resulted in the production of liquid hydrocarbons. PS predominantly yielded aromatic hydrocarbons within the gasoline range, while PP, LDPE, and HDPE primarily produced aliphatic hydrocarbons, rendering them more suitable for diesel engine applications [17]. Mangesh et al. pyrolyzed LDPE, HDPE, PP, and PS to produce oil and compared them for their application in the diesel engine. They found that PPO has physicochemical properties closer to diesel. Gas chromatography-mass spectrometry (GC-MS) and Fourier transform-infrared spectroscopy (FT-IR) analysis reveals lower carbon number compounds in PPO making it better than the pyrolysis oil obtained from other feed plastics for application in diesel engine [18]. PP's high energy density (46.4 MJ/kg) makes it a suitable candidate to replace conventional fossil fuels [19]. PP is one of the best polymers that affords higher amounts of liquid product (i.e., pyrolysis oil) during pyrolysis. PP is easily degraded due to the presence of tertiary carbons in the hydrocarbon chain [20]. After production, plastic materials enter the supply chain and are extensively utilized over several years. During this period, they are exposed to a variety of environmental conditions, which results in significant degradation before they eventually return for recycling. This prolonged exposure can lead to the deterioration of their structural and chemical properties, complicating the recycling process and reducing the material's potential for reuse. The combined effects of UV radiation exposure [21] and the presence of tertiary carbon [20] in the polymer backbone can significantly degrade the quality of plastic waste, yet these impacts remain underexplored in the current literature [22].

The novelty of this study lies in its focus on extensively characterizing PPO produced from high-temperature pyrolysis of end-of-life PP, which has undergone significant degradation due to environmental exposure and the susceptibility of PP to undergo oxidation due to the presence of tertiary carbons. The degradation impacts of PP remain underexplored in the current literature. While pyrolysis presents a viable solution for addressing plastic waste and generating alternative fuels, the complexities arising from the degradation of PP during its lifecycle necessitate further investigation. While PPO shares some physicochemical properties with diesel, this research highlights the need for further optimization and upgradation to make PPO suitable for use in internal combustion engines.

2 Material and Methods

2.1 Production of Polypropylene Pyrolysis Oil (PPO)

Polypropylene pyrolysis oil (PPO) was produced by pyrolyzing end-of-life PP sourced from local industries handling PP-based products. Shredded end-of-life PP was thermally pyrolyzed at 600–700 °C in a semi-continuous screw kiln reactor (Fig. 1) at Pyrotask Energy, Mumbai. The uncondensed combustible gases formed as byproducts are combusted to heat the reactor, and the carbon-rich char remaining in the reactor is collected.

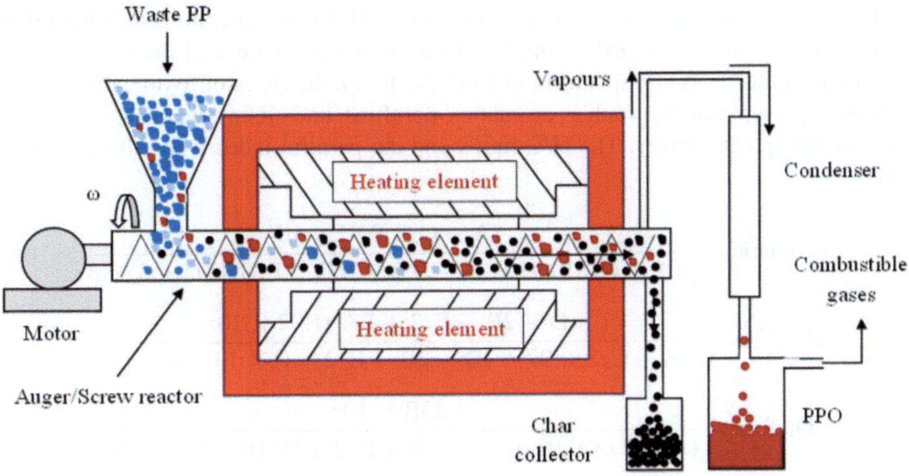

Fig. 1. Schematic diagram of an auger/screw reactor pyrolysis [23]

2.2 Characterization of PPO

The physicochemical properties of PPO, including density, viscosity, calorific value, cetane index, flash point, fire point, sulfur content, and ash content, are studied. An elemental analyzer (FLASH EA 1112 series) revealed the elemental composition of PPO. Inductively Coupled Plasma-Atomic Emission Spectrometry (ICP-AES) was employed using the ARCOS simultaneous ICP spectrometer to qualitatively and quantitatively estimate the various metallic elements present in PPO.

Fourier transform infrared (FT-IR) spectroscopy is a powerful technique that provides information about the functional groups in a molecule and their relative proportions, offering a unique fingerprint in the spectrum that distinguishes the compound from others based on its absorption patterns. The FT-IR spectrum of the PPO sample was obtained using a 3000 Hyperion Microscope with Vertex 80 FT-IR System with data processing and library search facilities available. The ^{1}H-NMR spectrum of PPO was collected in a Jeol ECZR Series 600 MHz NMR spectrometer by dissolving in deuterated chloroform ($CDCl_3$). The ^{1}H-NMR data helps identify the functional groups, classify the types of molecules (e.g., aromatics, paraffins, oxygenates), and estimate their relative proportions. The gas chromatography-high resolution mass spectrometry (GC-HRMS) analysis was performed to find the chemical composition of PPO using Accu-TOF GCV7890, Joel GC-HRMS analyzer. The EB-5 column was used, and GC oven temperature was programmed to reach 60 °C and held for 1 min, then the temperature was increased from 60 °C to 220 °C at a ramping rate of 3 °C/min, then from 220–280 °C at a ramping rate of 15 °C/min. The National Institute of Standards and Technology (NIST) library was used to identify the compounds from the mass spectra.

In this study, we use the correlations given by [24] to estimate the vol% of paraffin, olefin, and aromatic in the PPO sample. From the integral values of the various types of protons in the ^{1}H-NMR spectrum of PPO (Table 1), the types of hydrocarbon-based molecules present in PPO, such as aromatics, paraffins, and olefins, were analyzed qualitatively and quantitatively. The H/C ratios and iso-paraffin index were also estimated [25].

$$\text{Aromatics} = \frac{(A + C/3)0.97 \times 10^2}{(A + C/3)0.97 + (D - 2B + E/2 + F/3)1.02 + 3.33B} \quad (1)$$

$$\text{Paraffins} = \frac{(D - 2B + E/2 + F/3)1.02 \times 10^2}{(A + C/3)0.97 + (D - 2B + E/2 + F/3)1.02 + 3.33B} \quad (2)$$

$$\text{Olefins} = \frac{3.33B \times 10^2}{(A + C/3)0.97 + (D - 2B + E/2 + F/3)1.02 + 3.33B} \quad (3)$$

$$H/C = \frac{A + B + C + D + E + F}{(A + C/3)1.28 + (D - 2B + E/2 + F/3)1.02 + 3.42B} \quad (4)$$

$$\text{Iso-paraffin Index} = (CH_3)/(CH_2) = (F/3)/(E/2) \quad (5)$$

Table 1. Spectral regions of NMR [26]

	Proton type	Chemical shift region (ppm)
A	Ring aromatics	6.6–8.0
B	Olefin	4.5–6.0
C	α-Methyl	2.0–3.0
D	Methine (paraffins)	1.5–2.0
E	Methylene (paraffins)	1.0–1.5
F	Methyl (paraffins)	0.6–1.0

3 Results and Discussion

3.1 Physicochemical Characteristics of Polypropylene Pyrolysis Oil (PPO)

Various physicochemical properties of PPO were studied and compared with diesel (Table 2). The density of PPO at 15 °C is slightly lower than that of diesel, signifying slightly lower energy density in PPO than diesel. The viscosity of PPO is slightly lower than diesel but within the desired range and hence should provide acceptable combustion characteristics, but may influence atomization and combustion efficiency [27].The gross calorific value of PPO was found to be marginally lower than that of diesel [14]. Additionally, the significantly lower cetane index of PPO suggests potential challenges with ignition delay, higher peak pressure and combustion quality, requiring potential additives for optimization [28]. These results can reduce the performance of the engine compared to that of diesel. Flash point, fire point, and pour point of PPO are lower than the ideal value of diesel which can lead to storage and transportation problems. Sulphur in fuels used in internal combustion engines presents significant environmental and mechanical challenges. PPO contains high sulphur content, higher than the allowed BS-VI norm of 10 ppm. Higher concentration sulphur compounds can generate sulphur oxides (SO_x) during combustion, which contribute to acid rain and air pollution. Additionally, sulphur oxides can corrode engine components and deactivate catalysts used in emission control systems, reducing their effectiveness. Reducing sulphur content in fuels is essential to mitigate these impacts and meet environmental regulations [29–31].

Table 2. Physicochemical properties of polypropylene pyrolysis oil (PPO) and diesel.

Property	PPO	Diesel [18]	unit
Density @ 15 °C	0.819	0.837	g/ml
Kinematic viscosity @ 40 °C	2.34	2.61	cst
Gross calorific value	10251	10897	kcal/kg
Calculated cetane index	41.09	52	–
Flash point	48	72	°C
Fire point	54	82	°C
Pour point	<−6	<−15	°C
Total sulfur content	259	10	mg/kg
Ash content	0.13	0.01	%
Conradson carbon residue	0.43	0.18	%

The increased ash content in PPO can lead to accelerated wear and tear on engine components as it accumulates in critical areas such as piston rings, liners, and valve seats, leading to abrasive wear. This accumulation reduces the effectiveness of lubrication, resulting in increased friction and heat, which further accelerates component degradation. Over time, this can significantly shorten engine life and increase maintenance costs [32, 33]. The study demonstrates that although PPO shares similar properties with diesel, further refining and upgrading are necessary to reduce environmental impact, prevent engine wear, and achieve comparable combustion efficiency and performance in internal combustion engines.

3.2 ICP-AES and elemental analysis of PPO

Qualitative analysis of PPO was done using ARCOS, Simultaneous ICP Spectrometer. The result shows that PPO contains barium (Ba), bromine (Br), calcium (Ca), iron (Fe), magnesium (Mg), manganese (Mn), phosphorus (P), silica (Si), tin (Sn) and zinc (Zn) in trace amounts. Combustion of plastic pyrolysis oil, particularly from waste streams, not only releases gaseous pollutants like CO, CO_2, and NO_x but also serves as a significant source of hazardous trace metals such as arsenic, chromium, copper, manganese, and zinc, which contribute to environmental and health issues [34]. Trace metals present in fuels, even in small quantities, can significantly impact combustion processes by contributing to ash formation [35] and catalyst poisoning [36]. During combustion, these metals can accumulate in engine components, leading to increased wear and tear, clogging and hence increased maintenance needs, further emphasizing the need for post-treatment of PPO to enhance its performance and compatibility with engine systems.

The elemental analysis of PPO reveals that the carbon content in PPO is 84.77%, the hydrogen content is 12.84%, the nitrogen content is 1.5%, and the oxygen content is 0.9%. The values are similar to those reported in the literature [37]. Oxygenates may come from partially-oxidized waste PP and partial oxidation of relatively reactive components (e.g., terminal olefins) in PPO during pyrolysis and storage. Oxygenates in fuel can improve combustion, lubricity and reduce emissions but can lead to increased corrosivity and reduced thermal stability, necessitating further refinement or treatment to improve the fuel quality [38, 39].

3.3 FT-IR analysis of PPO

The large peaks at 2956 cm^{-1}, 2923 cm^{-1}, 2870 cm^{-1}, and 2852 cm^{-1} come from the C–H stretching (symmetric and antisymmetric) vibrations from methyl ($–CH_3$) and methylene ($–CH_2–$) groups present in paraffins, long-chain olefins and alkylated aromatic compounds in PPO. The C–H bending (in-plane and out-of-plane) vibrations appeared at 1460 cm^{-1} and 1377 cm^{-1}. The small peak at 3074 cm^{-1} comes from the Csp^2–H stretching from olefins and aromatic hydrocarbons. The smaller peaks at 1720 cm^{-1} and 1649 cm^{-1} correspond to oxygenated compounds containing ester/carboxylic acid and aldehyde moieties. A relatively narrow, moderate absorption at 1650 cm^{-1} indicates olefinic unsaturation (C=C). The methylene (CH_2) C–H symmetric stretching vibrations appeared at 2851 cm^{-1}, and asymmetric stretching vibrations appeared at 2923 cm^{-1}. The methylene (CH_2) C-H bending vibrations appeared at 1459 cm^{-1}, and rocking vibrations appeared at 740 cm^{-1}. A series of absorptions between 670 cm^{-1}–850 cm^{-1} indicates aromatic C–H out-of-plane bending on the aromatic ring. Absorptions between 950 cm^{-1} – 1150 cm^{-1} show in-plane C–H bending of aromatics. All these absorption peaks indicate the presence of aromatic compounds in the PPO sample. A broad absorption peak at 3425 cm^{-1} shows hydrogen-bonded –OH stretch, indicating the presence of alcohols. The presence of moisture in the PPO sample may also be responsible for the peak (Fig. 2).

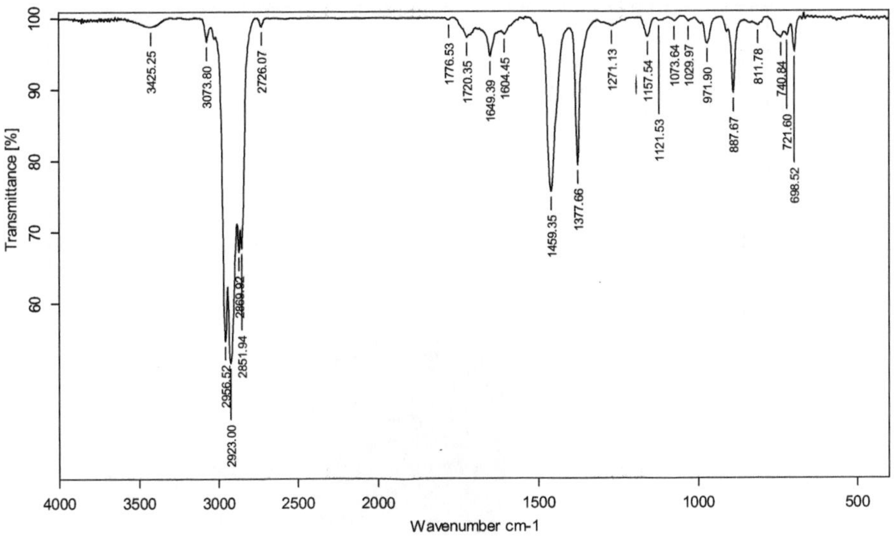

Fig. 2. The FT-IR spectrum of PPO.

The FT-IR analysis of PPO indicates a mix of hydrocarbons and oxygenated compounds, which impacts its fuel properties. PPO contains largely paraffinic hydrocarbons, olefins, and aromatics. A small amount of oxygenates in the form of alcohols and carbonyl compounds is also confirmed. While the hydrocarbon content suggests energy potential, the presence of unsaturated hydrocarbons, oxygenates, and moisture can reduce stability (gum formation), increase corrosivity, and lead to higher emissions [38–41]. These factors highlight the need for further upgradation to improve PPO's suitability as a fuel.

3.4 ^{1}H-NMR analysis of PPO

Quantitative analysis of hydrocarbon compositions of PPO was carried out by ^{1}H-NMR spectroscopy using the correlations provided by [24, 25]. From the integrated areas of the NMR spectral regions given by them, the hydrocarbon types, including aromatics, paraffins, and olefins, as well as the H/C ratios and iso-paraffin index, can be estimated. It was found that the PPO sample contained 58.88, 27.41, and 13.74 vol% of paraffins, olefins, and aromatics, respectively. Analysis also found H/C and Iso-paraffin index as 1.74 and 1.44 respectively. The presence of olefins and aromatics can increase the combustion delay, lead to higher engine temperature, and lower brake thermal efficiency in diesel engines [14, 18, 19, 42]. As shown in Fig. 3, the results indicate that PPO mostly contains around 60% paraffins and 30% olefins. The higher amount of aromatics compared to literature [43] may be due to the higher temperature at which PP is pyrolyzed in the reactor leading to further cracking and aromatization reactions. Moreover, aromatic-rich polymers (e.g., PS) as an impurity in the waste PP feed cannot be ruled out. The unsaturated compounds found in the oil prevents PPO from getting diesel-like combustion results; these indicate the need for further upgradation for better utilization of PPO in the diesel engine.

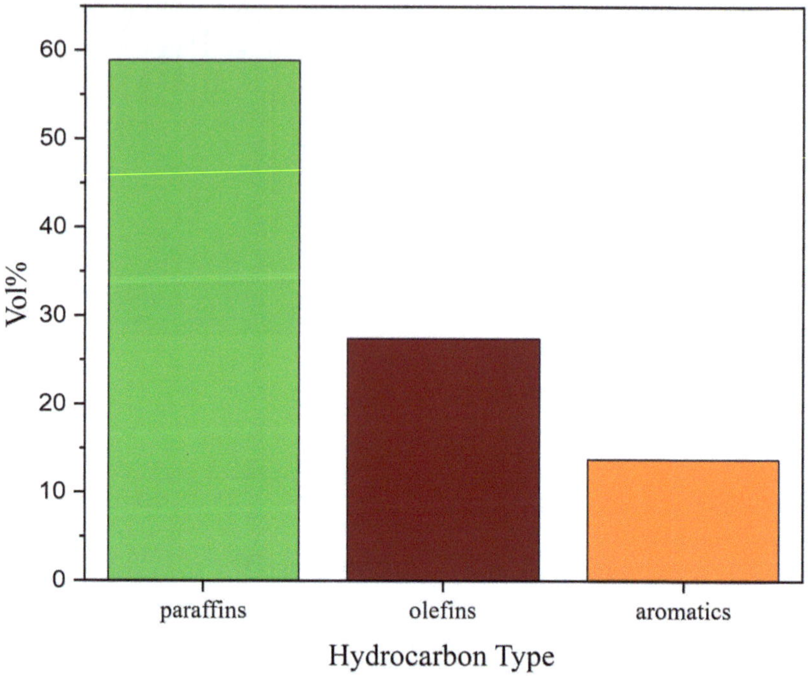

Fig. 3. Molecular composition (vol%) of paraffin, olefins, and aromatics in PPO

3.5 GC-HRMS analysis of PPO

Identification of the major compounds in the PPO mixture was carried out using the GC-HRMS analysis. The GC chromatogram of PPO is shown in Fig. 4. The major compounds in PPO were cross-checked against the NIST library, and the verified ones are listed in Table 3. The identified compounds can be grouped as aromatics, olefins, naphthenes, alcohols, esters, and paraffins. The peak list from the GC-MS analysis of PPO reveals a variety of compounds that influence its properties as a fuel. The presence of unsaturated hydrocarbons like 2,4-Dimethyl-1-heptene and 1-Octene, 3,7-dimethyl- indicates potential reactivity, leading to stability issues [38, 39]. Aromatics such as Ethylbenzene and α-Methylstyrene can increase soot formation and emissions [44, 45]. Alcohols like 1-Decanol, 2-hexyl- contribute to oxygen content, affecting combustion efficiency [40, 41]. Heavier compounds, including long-chain alcohols and esters, suggest higher boiling points, which could lead to incomplete combustion and higher emissions [46], reinforcing the need for further refining of PPO for use as a fuel. There are some similarities in PPO and diesel composition, but heavier paraffins (higher carbon number) in PPO may result in higher CO and CO_2 emissions than diesel fuel. Therefore, PPO needs to be further upgraded before being used as an alternative fuel for diesel.

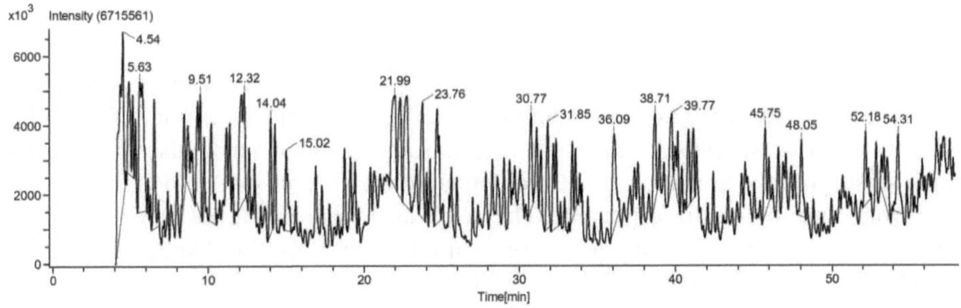

Fig. 4. Gas chromatogram of PPO

Table 3. Compounds in PPO (GC-HRMS)

Sl.No.	Compound	Formula
1	2,4-Dimethyl-1-heptene	C_9H_{18}
2	Ethylbenzene	C_8H_{10}
3	Cyclopentane, 1,2,3,4,5-pentamethyl-	$C_{10}H_{20}$
4	Benzene, (1-methylethyl)-	C_9H_{12}
5	α-Methylstyrene	C_9H_{10}
6	1-Octene, 3,7-dimethyl-	$C_{10}H_{20}$
7	Octane, 3,3-dimethyl-	$C_{10}H_{22}$
8	Cyclohexane, 1,1-dimethyl-2-propyl-	$C_{11}H_{22}$
9	1-Decene, 2,4-dimethyl-	$C_{12}H_{24}$
10	1-Nonene, 4,6,8-trimethyl-	$C_{12}H_{24}$
11	(2,4,6-Trimethylcyclohexyl) methanol	$C_{10}H_{20}O$
12	2-Isopropyl-5-methyl-1-heptanol	$C_{11}H_{24}O$
13	2-Hexyl-1-octanol	$C_{14}H_{30}O$
14	1-Hexadecanol, 3,7,11,15-tetramethyl-	$C_{20}H_{42}O$
15	11-Dodecen-1-ol, 2,4,6-trimethyl-, (R, R, R)-	$C_{15}H_{30}O$
16	Benzene, 1,1′-(1,3-propanediyl) bis-	$C_{15}H_{16}$
17	1-Decanol, 2-hexyl-	$C_{16}H_{34}O$
18	Pentafluoropropionic acid, heptadecyl ester	$C_{20}H_{35}F_5O_2$
19	Trichloroacetic acid, hexadecyl ester	$C_{18}H_{33}Cl_3O_2$
20	1-Decanol, 2-hexyl-	$C_{16}H_{34}O$
21	1-Octacosanol	$C_{28}H_{58}O$
22	1-Heptacosanol	$C_{27}H_{56}O$
23	Tricosyl heptafluorobutyrate	$C_{27}H_{47}F_7O_2$
24	Cyclotetradecane,1,7,11-trimethyl-4-(1-methylethyl)-	$C_{20}H_{40}$

4 Conclusion

The PPO obtained from pyrolysis of end-of-life PP contains largely paraffins with noticeable amounts of alkylated aromatics and olefins. The aromatics may come from other polymers in the waste plastic feed, such as PS. Some aromatics may also form during the pyrolysis process. Some oxygenated compounds are also present in PPO, which may come from partially oxidized plastics. They may also form by partial oxidation of the olefins during pyrolysis or storage. As seen in FT-IR and ^{1}H-NMR spectra, the presence of carbonyl compounds also hints at the partial oxidation of olefins in PPO. The poor thermal and oxidative stability of PPO could compromise its use as a fuel in the engine. Therefore, a suitable upgrading strategy should target removing or converting the reactive compounds in PPO into stable compounds. Some of the physicochemical properties of PPO are within the range set for diesel. However, other properties like cetane number, flash point, sulfur content, and ash content must be improved. PPO may be used as fuel in furnaces directly or after mild upgrading. However, using PPO in diesel engines warrants significant upgrading, leading to molecular alterations in its chemical compositions.

Acknowledgement. The authors thank Pyrotask Energy, Mumbai, India, their support. Sophisticated analytical instrument facility (SAIF), IIT Mumbai, and Central Research Facility (CRF), NITK, Surathkal, are thanked for assisting in the instrumental analyses. T Gopikrishnan Kailas thanks NITK, Surathkal for the scholarship support.

Funding. The authors thank the Maire Tecnimont Centre for Research in Waste Recycling and Circular Economy (CRWRCR) at NITK, Surathkal, for the funding support.

References

1. Al-Salem, S.M., Antelava, A., Constantinou, A., Manos, G., Dutta, A.: A review on thermal and catalytic pyrolysis of plastic solid waste (PSW). J. Environ. Manag. **197**, 177–198 (2017). https://doi.org/10.1016/j.jenvman.2017.03.084
2. Das, A.K., Sahu, S.K., Panda, A.K.: Current status and prospects of alternate liquid transportation fuels in compression ignition engines: A critical review. Renew. Sust. Energ. Rev. **161**, 112358 (2022). https://doi.org/10.1016/j.rser.2022.112358
3. Kartik, S., Balsora, H.K., Sharma, M., Saptoro, A., Jain, R.K., Joshi, J.B., Sharma, A.: Valorization of plastic wastes for production of fuels and value-added chemicals through pyrolysis – A review. Therm. Sci. Eng. Prog. **32**, 101316 (2022). https://doi.org/10.1016/j.tsep.2022.101316
4. Ilyas, M., Ahmad, W., Khan, H., Nazir, S., Khan, K., Yousaf, S.: Plastic waste as a significant threat to environment – a systematic literature review. Rev. Environ. Health. **33**, 383–406 (2018). https://doi.org/10.1515/reveh-2017-0035
5. Budsaereechai, S., Hunt, A.J., Ngernyen, Y.: Catalytic pyrolysis of plastic waste for the production of liquid fuels for engines. RSC Adv. **9**, 5844–5857 (2019). https://doi.org/10.1039/C8RA10058F
6. Manu, J., Kailas, T.G., Madav, V.: Exploring the synergetic effects of rice husk, cashew shell, and cashew husk biomass blends on fluidized bed gasification for enhanced hydrogen production. J. Clean. Prod. **419**, 137991 (2023). https://doi.org/10.1016/J.JCLEPRO.2023.137991

7. Manu, J., Madav, V.: Numerical modeling of rice husk gasification in fluidized bed gasifier for sustainable biofuel production. Case Stud. Therm. Eng. **39**, 102429 (2022). https://doi.org/10.1016/j.csite.2022.102429
8. Williams, P.T., Slaney, E.: Analysis of products from the pyrolysis and liquefaction of single plastics and waste plastic mixtures. Resour. Conserv. Recycl. **51**, 754–769 (2007). https://doi.org/10.1016/j.resconrec.2006.12.002
9. Williams, E.A., Williams, P.T.: The pyrolysis of individual plastics and a plastic mixture in a fixed bed reactor. J. Chem. Technol. Biotechnol. **70**, 9–20 (1997). https://doi.org/10.1002/(SICI)1097-4660(199709)70:1<9::AID-JCTB700>3.0.CO;2-E
10. Onwudili, J.A., Insura, N., Williams, P.T.: Composition of products from the pyrolysis of polyethylene and polystyrene in a closed batch reactor: effects of temperature and residence time. J. Anal. Appl. Pyrolysis. **86**, 293–303 (2009). https://doi.org/10.1016/j.jaap.2009.07.008
11. Centre for Science and Environment: Managing plastic waste in India. (2020).
12. Abbas-Abadi, M.S., Haghighi, M.N., Yeganeh, H., McDonald, A.G.: Evaluation of pyrolysis process parameters on polypropylene degradation products. J. Anal. Appl. Pyrolysis. **109**, 272–277 (2014). https://doi.org/10.1016/j.jaap.2014.05.023
13. Mani, M., Subash, C., Nagarajan, G.: Performance, emission and combustion characteristics of a DI diesel engine using waste plastic oil. Appl. Therm. Eng. **29**, 2738–2744 (2009). https://doi.org/10.1016/j.applthermaleng.2009.01.007
14. Kaimal, V.K., Vijayabalan, P.: A detailed study of combustion characteristics of a DI diesel engine using waste plastic oil and its blends. Energy Convers. Manag. **105**, 951–956 (2015). https://doi.org/10.1016/j.enconman.2015.08.043
15. Lamani, V.T., Yadav, A.K., Kumar, G.N.: Combustion, performance, and tail pipe emissions of common rail diesel engine fueled with waste plastic oil-diesel blends. J. Therm. Sci. Eng. Appl. **10** (2018). https://doi.org/10.1115/1.4039965
16. Kalargaris, I., Tian, G., Gu, S.: The utilisation of oils produced from plastic waste at different pyrolysis temperatures in a DI diesel engine. Energy. **131**, 179–185 (2017). https://doi.org/10.1016/j.energy.2017.05.024
17. Panda, A.K., Singh, R.K.: Experimental optimization of process for the Thermo-catalytic degradation of waste polypropylene to liquid fuel. Adv. Ener. Eng. **1** (2013)
18. Mangesh, V.L., Padmanabhan, S., Tamizhdurai, P., Ramesh, A., Zhang, X., Lei, H.: Experimental investigation to identify the type of waste plastic pyrolysis oil suitable for conversion to diesel engine fuel. J. Clean. Prod. **246**, 119066 (2020). https://doi.org/10.1016/j.jclepro.2019.119066
19. Kalargaris, I., Tian, G., Gu, S.: Experimental characterisation of a diesel engine running on polypropylene oils produced at different pyrolysis temperatures. Fuel. **211**, 797–803 (2018). https://doi.org/10.1016/j.fuel.2017.09.101
20. Hayashi, J.I., Nakahara, T., Kusakabe, K., Morooka, S.: Pyrolysis of polypropylene in the presence of oxygen. Fuel Process. Technol. **55**, 265–275 (1998). https://doi.org/10.1016/S0378-3820(98)00047-2
21. Ranjan, V.P., Goel, S.: Recyclability of polypropylene after exposure to four different environmental conditions. Resour. Conserv. Recycl. **169**, 105494 (2021). https://doi.org/10.1016/J.RESCONREC.2021.105494
22. Kailas, T.G., A R, A., Dutta, S., Madav, V.: Novel adsorption-based upgradation of end-of-life polypropylene pyrolysis oil using carbonised rice husk. Energy Convers. Manag. **25**, 100824 (2025). https://doi.org/10.1016/J.ECMX.2024.100824
23. Qureshi, K.M., Kay Lup, A.N., Khan, S., Abnisa, F., Wan Daud, W.M.A.: A technical review on semi-continuous and continuous pyrolysis process of biomass to bio-oil. J. Anal. Appl. Pyrolysis. **131**, 52–75 (2018). https://doi.org/10.1016/j.jaap.2018.02.010

24. Myers, M.E., Stollsteimer, J., Wims, A.M.: Determination of hydrocarbon-type distribution and hydrogen/carbon ratio of gasolines by nuclear magnetic resonance spectrometry. Anal. Chem. **47**, 2010–2015 (1975). https://doi.org/10.1021/ac60362a020
25. Myers, M.E., Stollsteimer, J., Wims, A.M.: Determination of gasoline octane numbers from chemical composition. Anal. Chem. **47**, 2301–2304 (1975). https://doi.org/10.1021/ac60363a015
26. Karagöz, S., Yanik, J., Uçar, S., Song, C.: Catalytic coprocessing of low-density polyethylene with VGO using metal supported on activated carbon. Energy Fuels. **16**, 1301–1304 (2002). https://doi.org/10.1021/ef020064q
27. Kalargaris, I., Tian, G., Gu, S.: Combustion, performance and emission analysis of a DI diesel engine using plastic pyrolysis oil. Fuel Process. Technol. **157**, 108–115 (2017). https://doi.org/10.1016/j.fuproc.2016.11.016
28. Wathakit, K., Sukjit, E., Kaewbuddee, C., Maithomklang, S., Klinkaew, N., Liplap, P., Arjharn, W., Srisertpol, J.: Characterization and impact of waste plastic oil in a variable compression ratio diesel engine. Energies. **14**, 2230 (2021). https://doi.org/10.3390/EN14082230
29. Oke, E.A.: Examining the effectiveness of deep eutectic solvents in removal of sulfur from fuel oil: A mini review. Chem. Afr. **1–14** (2024). https://doi.org/10.1007/S42250-024-00989-0/FIGURES/9
30. Chowdhury, R., Pedernera, E., Reimert, R.: Trickle-bed reactor model for desulfurization and dearomatization of diesel. AIChE J. **48**, 126–135 (2002). https://doi.org/10.1002/aic.690480113
31. Mohan, A., Dutta, S., Madav, V.: Characterization and upgradation of crude tire pyrolysis oil (CTPO) obtained from a rotating autoclave reactor. Fuel. **250**, 339–351 (2019). https://doi.org/10.1016/j.fuel.2019.03.139
32. Aravelli, K., Heibel, A.: Improved lifetime pressure drop management for Robust Cordierite (RC) filters with asymmetric cell technology (ACT). SAE Tech. Paper. **2007**, 776–790 (2007). https://doi.org/10.4271/2007-01-0920
33. Watson, S.A.G.: Lubricant-derived ash – in-engine sources and opportunities for reduction. Ph.D. dissertation. (2010).
34. Morajkar, P.P., Abdrabou, M.K., Raj, A., Elkadi, M., Stephen, S., Ibrahim Ali, M.: Transmission of trace metals from fuels to soot particles: an ICP-MS and soot nanostructural disorder study using diesel and diesel/Karanja biodiesel blend. Fuel. **280**, 118631 (2020). https://doi.org/10.1016/J.FUEL.2020.118631
35. Sappok, A., Santiago, M., Vianna, T., Wong, V.W.: Characteristics and effects of ash accumulation on diesel particulate filter performance: rapidly aged and field aged results. (2009).
36. Shelef, M., Otto, K., Otto, N.C.: Poisoning of automotive catalysts. Adv. Catal. (1979). https://doi.org/10.1016/S0360-0564(08)60059-0
37. Kusenberg, M., Zayoud, A., Roosen, M., Thi, H.D., Abbas-Abadi, M.S., Eschenbacher, A., Kresovic, U., De Meester, S., Van Geem, K.M.: A comprehensive experimental investigation of plastic waste pyrolysis oil quality and its dependence on the plastic waste composition. Fuel Process. Technol. **227**, 107090 (2022). https://doi.org/10.1016/j.fuproc.2021.107090
38. Singh, B., Korstad, J., Sharma, Y.C.: A critical review on corrosion of compression ignition (CI) engine parts by biodiesel and biodiesel blends and its inhibition. Renew. Sust. Energ. Rev. **16**, 3401–3408 (2012). https://doi.org/10.1016/j.rser.2012.02.042
39. Fazal, M.A., Haseeb, A.S.M.A., Masjuki, H.H.: Biodiesel feasibility study: an evaluation of material compatibility; performance; emission and engine durability. Renew. Sust. Energ. Rev. **15**, 1314–1324 (2011). https://doi.org/10.1016/j.rser.2010.10.004
40. Czarnocka, J., Matuszewska, A., Odziemkowska, M.: Autoxidation of fuels during storage. In: Storage Stability of Fuels. IntechOpen (2015). https://doi.org/10.5772/59807

41. Masudi, A., Muraza, O., Jusoh, N.W.C., Ubaidillah, U.: Improvements in the stability of biodiesel fuels: recent progress and challenges. Environ. Sci. Pollut. Res. **30**(6), 14104–14125 (2022). https://doi.org/10.1007/s11356-022-25048-4
42. Qian, Y., Qiu, Y., Zhang, Y., Lu, X.: Effects of different aromatics blended with diesel on combustion and emission characteristics with a common rail diesel engine. Appl. Therm. Eng. **125**, 1530–1538 (2017). https://doi.org/10.1016/J.APPLTHERMALENG.2017.07.145
43. Das, P., Tiwari, P.: Valorization of packaging plastic waste by slow pyrolysis. Resour. Conserv. Recycl. **128**, 69–77 (2018). https://doi.org/10.1016/j.resconrec.2017.09.025
44. Braun-Unkhoff, M., Riedel, U.: Alternative fuels in aviation. CEAS Aeronaut. J. **6**, 83–93 (2015). https://doi.org/10.1007/s13272-014-0131-2
45. Richter, S., Kathrotia, T., Naumann, C., Scheuermann, S., Riedel, U.: Investigation of the sooting propensity of aviation fuel mixtures. CEAS Aeronaut. J. **12**, 115–123 (2021). https://doi.org/10.1007/s13272-020-00482-7
46. Scheepers, P.T.J., Bos, R.P.: Combustion of diesel fuel from a toxicological perspective - I. Origin of incomplete combustion products. Int. Arch. Occup. Environ. Health. **64**, 149–161 (1992). https://doi.org/10.1007/BF00380904

Modeling Soil Moisture and Hydraulic Conductivity in Goan Iron Ore Mines with GeoStudio

R. Perumalla, S. Kumar Reddy(✉), and M. Govinda Raj

National Institute of Technology Karnataka, Surathkal, Karnataka 575025, India
skreddy@nitk.edu.in

Abstract. Slope stability in the iron ore mining areas of Goa is a major issue especially considering the state average rainfall of 3000 mm to 2500 mm per annum. The ratio between extreme rainfall and geotechnical properties of the local soils may cause failure on slopes and pose a risk to both mining and environmental activities. In this study, the module of GeoStudio, SEEP/W, is used in simulating an unsaturated flow of the soils in a mine using the Van Genuchten (VG) model predict Volumetric Water Content (VWC) as well as hydraulic conductivity functions. The Parameters of VG models are soil specific, which allows characterization of the Soil-Water Retention Curve (SWRC) and the respective hydraulic conductivity function. By integrating the functions into SEEP/W the analysis provides the insights to into the spatial and temporal distribution of soil moisture as well as its dynamics, the movement of water in the unsaturated zones. This method facilitates the prediction of soil moisture by providing insights into its spatial and temporal distribution. Seepage pattern and evaluation of possible mechanisms of failure of mine slopes, playing a role in the formulation of effective dewatering and slope stabilizations strategies. The results highlight the need of including unsaturated soil mechanics and high numerical modelling techniques on the geotechnical test of the mining activities.

Keywords: Slope Stability · Open Pit mining · Soil-Water Retention Characteristics · GeoStudio

1 Introduction

Mine slope stability is a very critical criterion in the safety and sustainability of open-pit mining operations. The consequences of slope failures and rock falls may be critically severe as those can lead to loss of life, destruction of the mining equipment and large financial losses. Rainfall-induced instability of slopes is a common notion among the other factors that influence the slope stability. Water infiltration through slope materials increases the pore water pressure and decreases the shear strength that largely increases the possibility of slope failure.

The foremost geographic location of residual soils that are the most prone to failure when their inclination is steep is in tropical areas where heavy-intense rainfall may as

H. P. Dasari et al. (Eds.): NITK-CREST 2025, PTT, pp. 26–42, 2026.
https://doi.org/10.1007/978-981-95-3218-6_3

well cause an increase in pore water pressure and low shear strength. It is believed that the loss of soil suction ensures many failures of the tropical cut slopes by undermining the cohesion of soils leading to slope instability [1, 2]. Thus, the distribution of pore pressure in tropical residual soil, cut slopes determine the functionality of the process of evaluating and reducing the risk of slope failures at mines [3].

Theoretical models have been developed to predict landslide and slope failure susceptibility based on topographic and hydrologic variables. Recent research has predominantly focused on mass strength and stability analysis procedures for slope stability assessment. However, a critical gap remains in understanding the role of groundwater and pore pressure conditions, which significantly influence slope failure mechanisms [4–6].

Despite the independent and substantial impact of groundwater on slope instability, research efforts have disproportionately concentrated on the strength properties of geologic materials. Rainwater infiltration into unsaturated soils increases the degree of saturation, directly influencing shear strength properties and enhancing the likelihood of slope failure. Extensive research has demonstrated that shear strength varies with soil water suction, particularly negative porewater pressure or matric suction, in unsaturated soils. Consequently, precise prediction of the relationship between soil water content and shear strength is critical for assessing slope stability. Improved modeling of these interactions is essential for developing effective geotechnical solutions to mitigate slope instability risks in rainfall-affected regions.

2 Impact of Goa's Climatic Conditions

In Goa's iron ore mining regions, slope stability is a critical concern, particularly given the state's significant annual rainfall ranging from 2,500 mm to 3,400 mm. The interaction between intense precipitation and the geotechnical properties of local soils can lead to slope failures, posing risks to both mining operations and environmental safety. Previous studies have shown that prolonged rainfall events with high cumulative volumes are among the most critical factors influencing significant variations in pore-water pressure within slopes. This is primarily due to the extended duration allowing water to infiltrate deeper into the soil profile, gradually saturating previously unsaturated zones. Among the various factors, extended rainfall events with substantial cumulative precipitation most significantly alter pore-water pressure in slopes. As infiltration progresses, matric suction decreases, reducing effective stress and potentially destabilizing slopes [7, 8]. Conversely, high-intensity rainfall doesn't always cause dramatic pore-water pressure increases. Low surface permeability (due to compaction, crusting, or fine-grained materials) prevents rapid infiltration, leading to surface runoff instead [9]. This limits the depth and extent of pore-water pressure changes during intense, short-duration rainfall. Additionally, antecedent moisture is crucial. Saturated soils from prior rainfall experience rapid pore pressure build-up even with moderate rain, as infiltration capacity is minimal [10, 11]. Dry soils with high initial matric suction have high initial infiltration rates, but the effect on pore pressure is delayed until deeper layers saturate.

The hydrogeological analysis of Goa reveals distinct runoff and soil moisture patterns influenced by its tropical monsoon climate. Post-monsoon data (eighth July 2022) shows significant runoff (186.819 mm), corresponding to peak monsoon rainfall, leading to rapid water accumulation, potential soil erosion, and slope instability, especially in regions like Sanguem, Quepem, and Salcete with varied topography. Runoff declines sharply after August 2022, indicating efficient drainage or high infiltration rates based on soil type detail mapping is shown in Fig. 1.

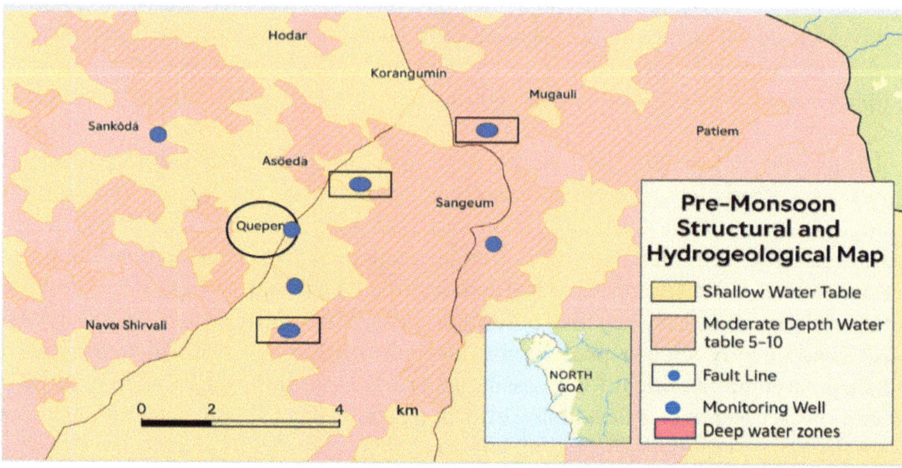

Fig. 1. Post monsoon ground water mapping of Goa in the month of July 2022 (Courtesy: Bhuvan NRSC 2022).

In contrast, pre-monsoon conditions (27th March 2022) exhibit 0 mm runoff, reflecting dry soil with minimal surface water movement and low groundwater recharge, making soils prone to desiccation cracks, particularly in clayey formations as shown in Fig. 2.

Mining activities in Salcete, Quepem, and Sanguem significantly disrupt the natural hydrological balance by altering drainage, compacting soils, and reducing vegetation cover, intensifying runoff and erosion during peak monsoon periods. Recognizing the spatial variability in runoff and subsurface water retention is essential for effective slope stabilization, groundwater management, and sustainable land use planning across Goa's varied landscape. Goa's terrain characterized by undulating topography and shallow groundwater conditions further complicates slope behavior. In regions where the water table is close to the surface, even moderate rainfall can rapidly lead to soil saturation, increasing pore-water pressures and triggering slope movements. The heterogeneous nature of the soil and rock formations in mining belts ranging from lateritic cover to fractured ferruginous quartzites and phyletic clays adds another layer of complexity. These materials vary widely in permeability and strength, influencing both infiltration rates and the rate of pore-pressure dissipation. For instance, silty or clayey materials with low permeability tend to retain water longer, sustaining elevated pore pressures post-rainfall events and increasing the risk of delayed failure.

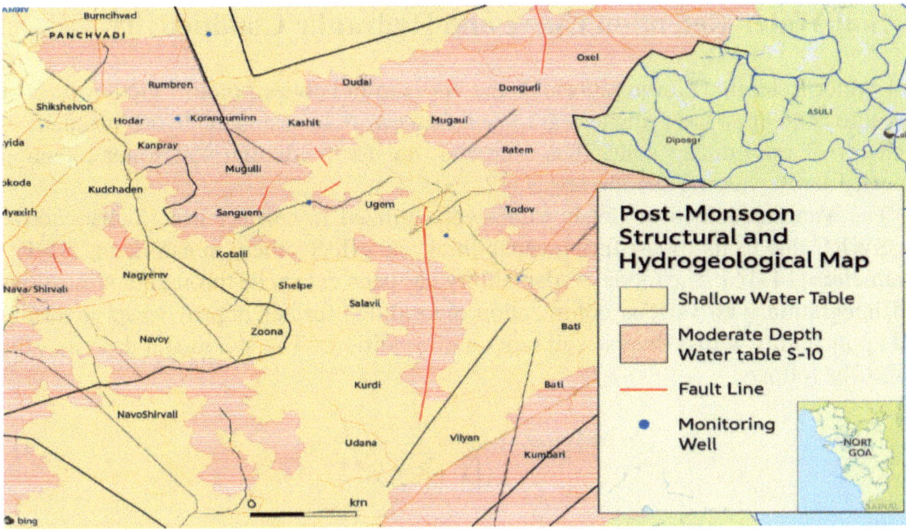

Fig. 2. Pre monsoon ground water mapping of Goa in the month of March 2022 (Courtesy: Bhuvan NRSC 2022).

Moisture content analysis highlights high post-monsoon soil saturation, reducing matric suction and increasing slope instability risks, notably in open-pit mines. Conversely, the pre-monsoon phase shows reduced moisture and elevated matric suction, temporarily enhancing slope stability and providing critical conditions for studying Soil Water Retention Curves (SWRC) in slope stability assessments.

The post-monsoon period consistently exhibits elevated soil moisture levels, indicating higher degrees of saturation. This condition significantly reduces matric suction, a key component of unsaturated soil strength thereby diminishing effective stress and increasing the potential for slope failure. These effects are more pronounced in the weathered and loose overburden strata commonly found in mining slopes.

Figure 3 shows that soil moisture levels in Goa exhibit cyclical variations, with alternating periods of increase and decrease. South Goa consistently retains more moisture than North Goa, likely due to hydrological and geographical differences. The recovery in 2024 may indicate a positive shift in rainfall or soil conditions.

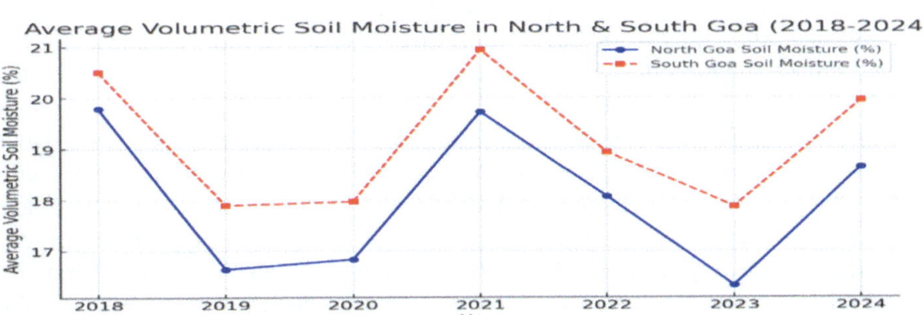

Fig. 3. Volumetric Soil moisture of North and South Goa regions (IWRIS 2022)

3 Soil-Water Retention Curve and Hydraulic Conductivity

Soil Water Retention Curve represents the correlation between matric suction (ψ) and moisture content (w), serving as a graphical framework to assess the hydro-mechanical behavior of unsaturated soils. It is a crucial tool for predicting soil water retention capacity across varying suction levels [12, 13].

The Van Genuchten model is extensively utilized in soil science to characterize the SWRC due to its robust fitting capabilities and effectiveness in estimating SWRC parameters [14–16]. The model's adaptability has been enhanced through modifications and integration with various optimization algorithms, further improving its precision and applicability across diverse soil types and conditions. The VG model describes the SWRC as follows:

$$\sigma(\psi) = \theta_r + \frac{(\theta_s - \theta_r)}{\left[1 + (\propto h)^n\right]^{1-\frac{1}{n}}} \tag{1}$$

where

- θ (ψ) volumetric water content at matric suction h (cm)
- Θ_r residual volumetric water content (m^3/m^3)
- θ_s saturated volumetric water content (m^3/m^3)
- α a parameter related to the inverse of the air entry suction (cm^{-1})
- n a measure of the pore-size distribution
- h matric suction (cm)

3.1 Hydraulic Conductivity of Soil

When rainfall occurs on slopes, the infiltration and seepage behavior are primarily governed by the hydraulic properties of the soil and the rainfall pattern. The spatial variability of soil properties plays a crucial role in influencing infiltration rates and overall slope stability [17, 18] One of the key hydraulic properties, the saturated hydraulic conductivity (Ks), dictates the rate at which water infiltrates into the soil, while rainfall characteristics such as intensity and duration determine the total amount of water available for infiltration [19–22].

Most research assumes that soil hydraulic conductivity is isotropic, meaning it is uniform in all directions. However, in natural environments, soil hydraulic conductivity is typically anisotropic, meaning its permeability differs in different directions. Despite its significance, the effect of hydraulic conductivity anisotropy on slope stability has received limited attention in slope stability studies. The anisotropic nature of soil permeability is generally quantified using the anisotropic ratio (Kx/Ky), where Kx and Ky represent the hydraulic conductivity in the horizontal and vertical directions, respectively. A higher anisotropic ratio indicates greater horizontal permeability, which can significantly influence water movement within slopes, pore-water pressure distribution, and slope failure mechanisms. Considering soil anisotropy in slope stability models is crucial for accurately predicting rainfall-induced slope failures and designing effective drainage or slope reinforcement measures. A numerical study was carried out to assess the influence of hydraulic conductivity anisotropy on the stability of stratified, weakly cemented rock slopes [23]. Their findings highlighted that disregarding the anisotropic nature of hydraulic conductivity can lead to a significant overestimation of the slope safety factor by approximately 40%. This overestimation occurs because assuming isotropic conditions fails to accurately capture the complex seepage patterns and pore-water pressure distributions within the slope.

4 Soil Moisture Modeling

In modern geotechnical and mining engineering, advanced numerical modeling tools such as GeoStudio have revolutionized slope stability analysis by providing a comprehensive framework for simulating complex hydrogeological conditions. GeoStudio integrates various methods of modelling such as limit equilibrium, and finite element procedures, to study slope stability, seepage, and stress strains behavior under varying, site conditions.

GeoStudio plays a crucial role in evaluating moisture movement, pore water pressure variations, and hydraulic conductivity in the context of iron ore mines of Goa, where hydrogeological features would be important determinants of the slope stability. The software allows the simulation of steady-state and transient seepage conditions, such that the engineers can evaluate the water infiltration effects, rainfall, and groundwater fluctuations on mines slopes. This is especially critical where there are areas as far as high monsoonal rainfall, whereby water is accumulated in slope materials thus drastically changing the shear strength causing instability.

Another major issue in designing the slopes of the iron ore mines is that geological variations are different, such as friable ore siliceous clay, and phyllite clay, which have diverse hydraulic conductivity and moister retention. The geo-studio facilitates incorporation data of Soil Water Retention Curve (SWCR) to accurately model the behavior of unsaturated soil and variations of matric suction. These parameters are essential in understanding negative pore pressure effects, which contribute to temporary slope stability but can diminish under drawdown or prolonged wetting conditions.

Furthermore, using GeoStudio for seepage analysis allows for the identification of potential failure zones due to excessive water pressure build-up, which can lead to piping, heave, or liquefaction in mine pit walls. The ability to simulate different loading scenarios, such as sudden drawdown conditions following dewatering activities, provides valuable insights into the response of slopes under changing hydrogeological regimes. By calibrating model predictions with field instrumentation data, such as tensiometer readings and piezometric measurements, a more reliable slope stability assessment can be achieved.

The implementation of numerical modeling through GeoStudio enhances the accuracy of slope stability predictions, aiding in the development of more effective slope reinforcement and drainage strategies. As mining operations in Goa continue to evolve, integrating advanced hydrogeological modeling techniques ensures improved mine safety, reduced environmental impact, and sustainable resource extraction practices.

SEEP/W modeling is a crucial tool in geotechnical engineering for simulating soil moisture movement and seepage behavior under various hydrogeological conditions. Unlike physical modeling, which relies on scaled-down laboratory experiments or full-scale field studies, SEEP/W modeling employs mathematical equations to approximate real-world processes. GeoStudio, particularly SEEP/W, is widely used for modeling water flow through porous media, allowing engineers to analyze seepage, pore-water pressures, and moisture distribution in soil structures [24]. The ability to define boundary conditions, material properties, and transient flow behavior makes SEEP/W an effective tool for predicting groundwater movement and assessing slope stability in mine slopes and embankments.

Proficiency in soil moisture modeling using GeoStudio is not achieved simply by acquiring the software; it requires time, experience, and a deep understanding of numerical techniques. Engineers must develop skills to properly define input parameters, validate results, and interpret model outputs for practical applications. In geotechnical practice, numerical modeling is relatively new, and there is often a lack of awareness regarding its methodologies and limitations [25]. Understanding these fundamental aspects is essential for conducting reliable seepage analyses, particularly in mining environments where hydrogeological conditions significantly impact slope stability. Through continuous practice and refinement, numerical modeling with SEEP/W can become an indispensable tool for optimizing dewatering strategies and improving overall mine slope safety.

In seepage modeling, both **saturated and unsaturated flow conditions** must be considered, requiring a **saturated-unsaturated analysis** approach. Unlike in fully saturated conditions where hydraulic conductivity (K) remains constant, in unsaturated zones, hydraulic conductivity varies as a function of moisture content. As soil desaturates, its hydraulic conductivity decreases, making it necessary to estimate K using a function rather than a fixed value. SEEP/W incorporates sample **volumetric water content (VWC) functions** for various soil textures, enabling users to define the hydraulic properties of different soil types. In this study, the VWC function adopted represents a **silt material** with a saturated water content of 0.5. The corresponding **K function** is estimated using the **Van Genuchten method**, which requires specifying the **saturated hydraulic conductivity (Ksat)** [26]. By integrating these functions, SEEP/W allows for a more accurate representation of the **unsaturated flow regime**, which is essential for assessing pore-water pressures and seepage mechanisms in geotechnical structures.

The application of GeoStudio software in seepage and slope stability analysis has been demonstrated in the study on the Haditha Dam, where SEEP/W and SLOPE/W were utilized to evaluate seepage patterns and factor of safety under different hydraulic conditions [27]. The present study highlights the effectiveness of GeoStudio in modeling groundwater flow through embankment dams, considering both saturated and unsaturated zones.

The numerical modeling of hydraulic conductivity for Phyllite clay was performed using SEEP/W, a module within the GeoStudio software suite. This model incorporates both saturated and unsaturated flow conditions, where hydraulic conductivity (K) is not a constant but a function of matric suction and volumetric water content (VWC). The relationship between hydraulic conductivity and water pressure head was estimated using the Van Genuchten method, which effectively captures the nonlinear behavior of unsaturated soil permeability.

Goa's geological formations have been influenced by tectonic movements, erosion, and long-term weathering, leading to the development of widespread laterite deposits, which are significant for field studies and sample collection. These laterites, derived from iron-rich parent rocks such as basalts, schists, and granites, are predominant across the region. The Bicholim Formation, hosting significant iron ore deposits, comprises Banded Iron Formations (BIFs) with alternating layers of hematite, magnetite, and siliceous material.

Field investigations in Goa's mining belt focus on understanding the structural and hydrogeological controls influencing ore distribution and stability. Geological surveys, core drilling, and geophysical mapping aid in delineating ore bodies and assessing their economic viability. Hydrogeological studies are crucial in evaluating groundwater flow through fractures and faults, which can impact mine stability and water management. Effective dewatering strategies and monitoring of pore water pressure variations are essential to mitigate geotechnical risks associated with open-pit mining.

The processes involved in extraction have significantly been altered by the advancements taking place in the process of mechanized opencast mining or rather precise removal of overburdens and precision in the process of ore recovery. The combination of the remote sensing and geospatial analysis increases mine planning and environmental sustainability. The combined effect of the geological structures together with the hydrogeology and modern mining processes and activities further asserts that constant field studies are needed to optimize utilization of the resource as well as reduce the effects of mining on the environment.

Field investigations were conducted to characterize the physical and geotechnical properties of Friable Ore, Siliceous Ore, and Phyllite in iron ore mining areas. Samples were collected from different mine slopes and subjected to visual inspection for color, texture, and mineral composition. In-situ density measurements were performed using a core cutter and sand replacement method, while specific gravity was determined through field-based pycnometer tests. Swelling behavior was assessed qualitatively based on the presence of clay minerals. Permeability was estimated using field permeameters, and shear strength parameters, including cohesion and internal friction angle, were evaluated through direct shear tests on undisturbed samples as given in Table 1. These investigations provided critical data for assessing slope stability and hydrogeological behavior under varying mining conditions.

Table 1. Physical properties of samples collected from Iron ore mines.

Sl.No	Property	Friable ore	Siliceous	Phyllite
1.	Color	Typically, reddish-brown to dark brown	Usually, gray to dark gray	Generally greenish gray to dark gray
2.	Texture	Loose and crumbly	Loose and crumbly	Loose and crumbly
3.	Mineral Composition	Hematite Goethite, and other iron oxides	Quartz, silicates, and clay minerals	Quartz, mica, and clay minerals
4.	Density (g/cm3)	2.6–3.5	2.2–2.7	2.7–2.9
5.	Specific Gravity	2.7–3.2	2.4–2.8	2.6–2.9
6.	Swelling	Low	Moderate to high	Low to moderate

(*continued*)

Table 1. *(continued)*

Sl.No	Property	Friable ore	Siliceous	Phyllite
7.	Permeability	Moderate	Low	Poor
8.	Cohesion (kPa)	15–20	15–20	12–15
9.	Internal friction angle (degrees)	30°–38°	42°–50°	35°–50°
10.	Porosity	20–35	30–50	5–45

The collected data were further integrated into slope stability models to simulate potential failure scenarios under different rainfall and excavation conditions. The variability in material properties across different lithological units highlighted the need for zone-specific design interventions. For instance, friable ore zones, with lower cohesion and higher permeability, required reinforced slope angles and effective drainage systems to prevent slope failure. In contrast, phyllite-rich sections, characterized by anisotropic strength due to foliation, demanded careful orientation analysis and slope reinforcement using geosynthetics or shotcrete lining. The integration of these site-specific parameters into mine planning not only enhanced operational safety but also informed groundwater management strategies, helping to prevent aquifer contamination and uncontrolled seepage.

Temporal monitoring of pore-water pressure using piezometers and tensiometers was initiated to capture real-time fluctuations in groundwater levels and matric suction, particularly during monsoon and post-monsoon periods. These readings were essential in validating the physical slope models and calibrating numerical simulations for dynamic conditions as shown in Table 2. Seasonal trends in groundwater recharge and discharge were also analyzed to understand their interaction with different lithological zones, particularly where perched water tables were observed in siliceous layers.

Table 2. Volumetric Water content at matric suction from SWRC characteristics

Soil type	Saturated Volumetric Water content (θs)	Residual Water content (θr)	Inverse of air entry suction (α, cm $^{-1}$)	Pore size distribution parameter (n)	Volumetric Water Content at given matric suction (θ(ψ))
Friable ore	0.45	0.05	0.03	1.50	0.347–0.362
Siliceous clay	0.45	0.07	0.01	1.20	0.347–0.362
Phyllite	0.45	0.08	0.45	1.20	0.320–0.400

5 Results of Soil Moisture Modeling

In GeoStudio, soil moisture behavior is represented through saturated and unsaturated hydraulic properties, which govern water movement under varying pore water pressures. The transition from saturated to unsaturated conditions significantly impacts hydraulic conductivity, water retention, and seepage characteristics in geotechnical applications such as mine slope stability.

5.1 Phyllite Soil

The hydraulic conductivity function for Phyllite clay, as depicted in Fig. 4, shows a sharp increase in permeability near zero suction, indicative of the transition from unsaturated to saturated conditions. At high suction values (<−500 kPa), the conductivity decreases significantly, reaching values as low as 10–10 m/s. confirming the extremely low permeability, of Phyllite under dry conditions. This behavior is consistent with fine-grained soils, where the reduction in pore water results in a drastic decline in connectivity between water-filled pores.

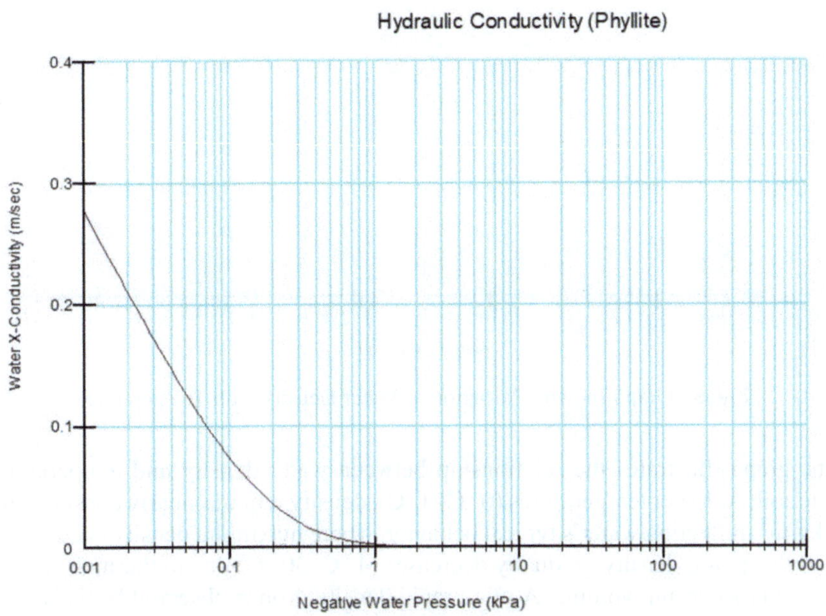

Fig. 4. Hydraulic conductivity of Phyllite clay

The graph illustrates the relationship between water density and temperature for Phyllite soil. At lower temperatures (0 °C–4 °C), density remains nearly constant around 1000 kg/m^3, reflecting water's typical behavior, where maximum density occurs at 4 °C. Beyond this point, density gradually decreases (4 °C–30 °C) due to thermal expansion, reducing mass per unit volume. A more rapid density drop is observed between 30 °C–40 °C, stabilizing around 992 kg/m^3, likely due to increased molecular activity and volume expansion at higher temperatures.

Temperature-dependent variations in water density Fig. 5 were also considered, as changes in temperature can influence fluid properties and, consequently, hydraulic conductivity. The modeling results indicate that the water density remains nearly constant at lower temperatures but decreases significantly at higher temperatures (>30 °C). This aspect is particularly relevant for field applications in regions experiencing seasonal temperature variations.

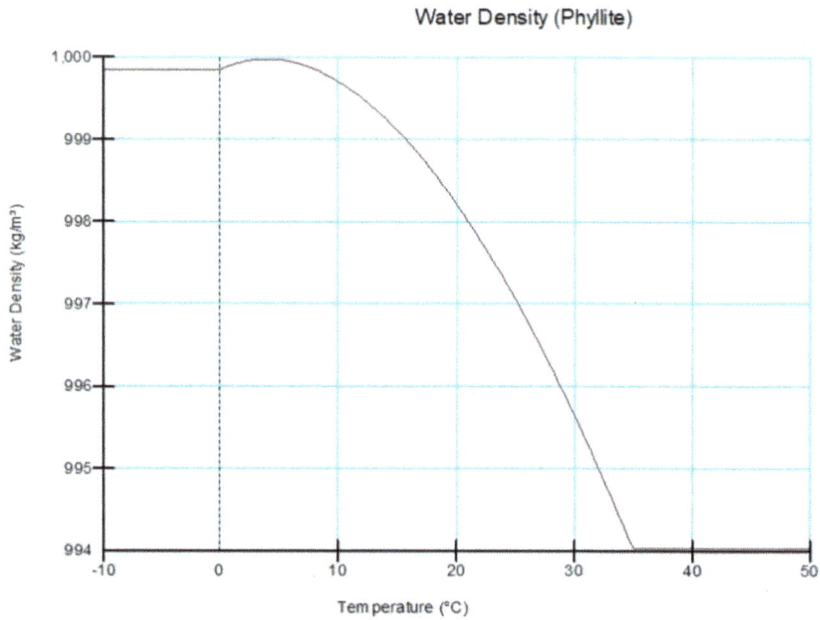

Fig. 5. Graph showing Variation of Water Density with Temperature

The graph illustrates the relationship between water density and temperature for Phyllite soil. At lower temperatures (0 °C–4 °C), density remains nearly constant around 1000 kg/m^3, reflecting water's typical behavior, where maximum density occurs at 4 °C. Beyond this point, density gradually decreases (4 °C–30 °C) due to thermal expansion, reducing mass per unit volume. A more rapid density drop is observed between 30 °C–40 °C, stabilizing around 992 kg/m^3, likely due to increased molecular activity and volume expansion at higher temperatures. Temperature-dependent variations in water density Fig. 5 were also considered, as changes in temperature can influence fluid properties and, consequently, hydraulic conductivity. The modeling results indicate that the water density remains nearly constant at lower temperatures but decreases significantly at higher temperatures (>30 °C). This aspect is particularly relevant for field applications in regions experiencing seasonal temperature variations. The graph illustrates the relationship between water density and temperature for Phyllite soil. At lower temperatures (0 °C–4 °C), density remains nearly constant around 1000 kg/m^3, reflecting water's typical behavior, where maximum density occurs at 4 °C. Beyond this point, density gradually decreases (4 °C–30 °C) due to thermal expansion, reducing mass per unit volume. A more rapid density drop is observed between 30 °C–40 °C, stabilizing around 992 kg/m^3, likely due to increased molecular activity and volume expansion at higher temperatures. Temperature-dependent variations in water density Fig. 5 were also considered, as changes in temperature can influence fluid properties and, consequently, hydraulic conductivity. The modeling results indicate that the water density remains nearly constant at lower temperatures.

5.2 Siliceous Soil Moisture Modeling

Siliceous soils, commonly found in weathered geological formations and mining exhibit unique hydro-mechanical properties due to their mineralogical composition and fine particle structure [28, 29]. These soils often possess low plasticity and high permeability variations, which affect their water retention capacity and pore water pressure behavior under changing moisture regimes. In slope environments, matric suction plays a pivotal role in providing apparent cohesion, thereby enhancing the shear strength of unsaturated siliceous soils [30]. However, during rainfall infiltration or groundwater fluctuations, changes in matric suction can lead to a reduction in effective stress, compromising slope stability [31–33]. The moisture dynamics of siliceous soils under slope conditions can be effectively understood by integrating laboratory-derived hydraulic conductivity and matric suction data into numerical models, thereby enhancing the prediction and management of moisture-driven instability mechanisms in siliceous soil-dominated terrains.

To improve slope design and risk assessment in these soils, it is critical to incorporate the SWCC and unsaturated shear strength parameters into advanced numerical simulations. These models must capture transient seepage, hysteresis effects, and suction-stress relationships, especially under conditions of prolonged rainfall or rapid drawdown. Additionally, field monitoring using tensiometers, piezometers, and moisture sensors can validate modeling outputs and help identify early signs of slope instability.

The analysis of siliceous clay hydraulic properties is based on three key relationships: hydraulic conductivity, volumetric water content (VWC), and the slope of the water retention function with respect to negative water pressure (suction). The hydraulic conductivity vs. suction graph shows that as negative water pressure increases from 0.1 kPa to 100 kPa, the hydraulic conductivity decreases exponentially from 10^{-5} cm/s to nearly 10^{-10} cm/s, indicating a sharp reduction in water movement as the soil dries. Similarly, the VWC vs. suction graph reveals a significant decline in water retention, starting at around 35% VWC at 1 kPa suction, which decreases steadily to approximately 5% at 100 kPa suction, following a typical water retention curve pattern. The slope of the water retention function exhibits a steep decline in the early stages, with the highest rate of change occurring between 1 kPa and 10 kPa, after which the slope stabilizes, suggesting minimal further water release at higher suctions. These trends highlight the crucial influence of soil moisture behavior on slope stability, particularly in mining environments where rapid water loss can affect pore pressure dynamics and overall structural integrity (Figs. 6 and 7).

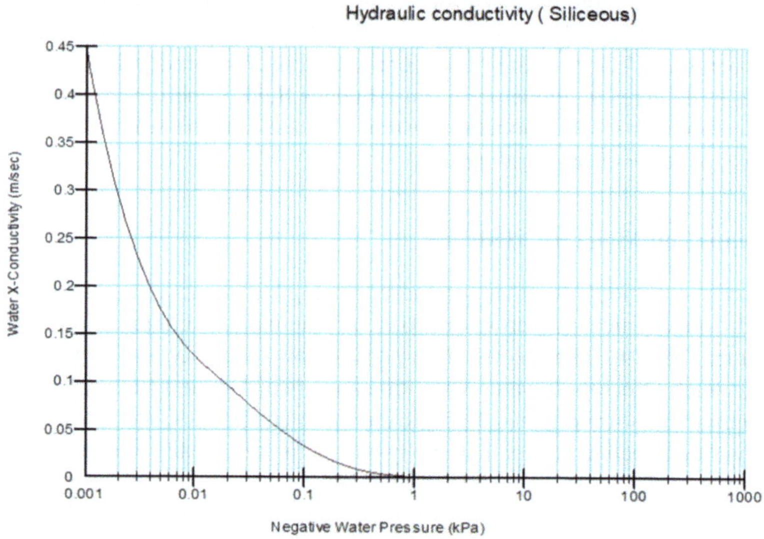

Fig. 6. Graph showing Hydraulic conductivity vs. Negative Water Pressure (kPa) Matric Suction

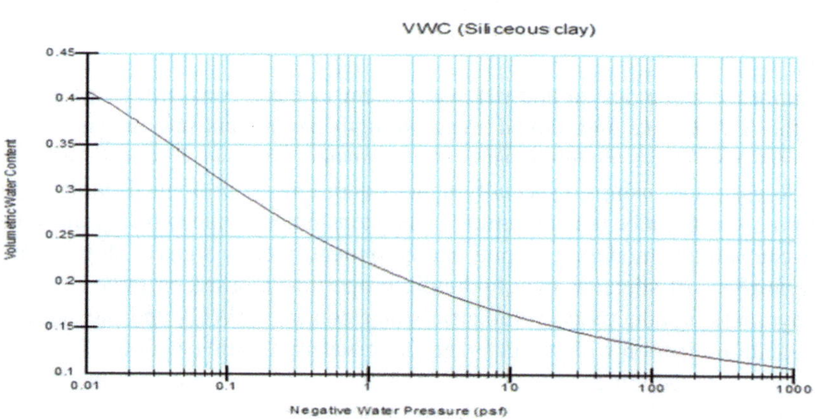

Fig. 7. Graph showing volumetric water content vs. Negative Water Pressure (kPa) Matric Suction

5.3 Friable Ore Moisture Modeling

Friable ore is closely governed by the unsaturated hydro-mechanical behavior of the material, particularly under transient seepage conditions. Friable ore, often encountered in iron ore mining regions, is characterized by its loosely bonded, porous, and weakly cemented structure, which renders it highly sensitive to moisture fluctuations and degradation under wetting. These materials exhibit significant variability in both matric suction and hydraulic conductivity, making their behavior difficult to predict using conventional saturated soil mechanics.

The hydraulic conductivity curve for friable ore under varying negative water pressures (matric suctions) demonstrates a sharp decline in conductivity with increasing suction. Initially, at low suction levels (close to saturation), the hydraulic conductivity remains relatively high, indicating that water can move freely through the interconnected pore network. However, as suction increases typically representing drier conditions the conductivity drops exponentially by several orders of magnitude. This transition highlights the material's shift from a saturated to an unsaturated state, during which water flow becomes increasingly restricted. Such behavior has critical implications for slope stability in open-pit mines and mine waste dumps, especially during drying-wetting cycles induced by seasonal rainfall or dewatering operations. The steep reduction in conductivity with increasing suction implies that friable ore can retain water under unsaturated conditions but offers very limited permeability for further drainage. This can lead to a perched water table and localized pore water pressure buildup when rainfall or infiltration occurs faster than drainage, potentially triggering slope instability (Fig. 8).

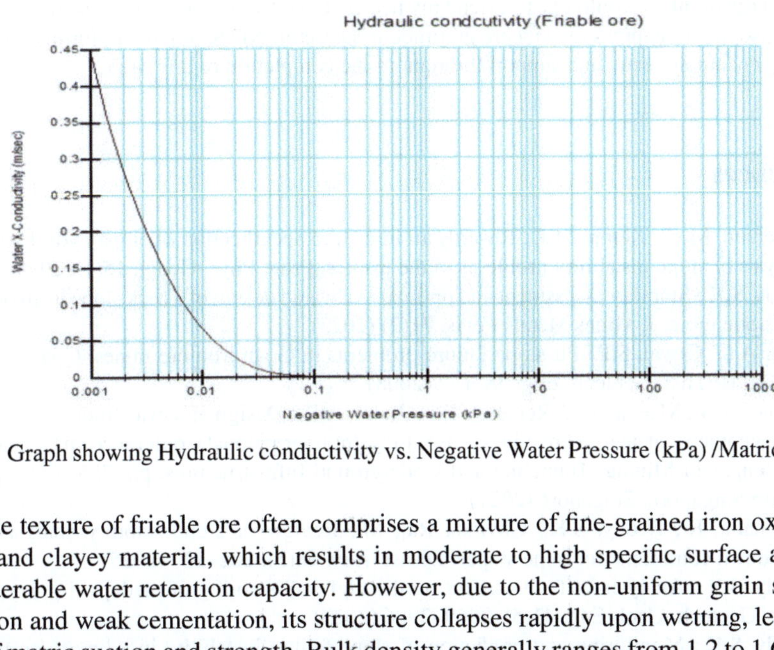

Fig. 8. Graph showing Hydraulic conductivity vs. Negative Water Pressure (kPa) /Matric Suction

The texture of friable ore often comprises a mixture of fine-grained iron oxide particles and clayey material, which results in moderate to high specific surface area and considerable water retention capacity. However, due to the non-uniform grain size distribution and weak cementation, its structure collapses rapidly upon wetting, leading to loss of matric suction and strength. Bulk density generally ranges from 1.2 to 1.6 g/cm^3, while porosity can exceed 40%, depending on the degree of weathering and compaction Friable ore typically shows a steep desaturation curve, indicating rapid loss of water content with a small decrease in suction—reflective of its weak structural fabric and low air entry value [7]. The hysteresis behavior observed during drying and wetting cycles further complicates moisture modeling and needs to be addressed in numerical simulations.

6 Conclusions

This study underscores the critical need to understand the hydro-mechanical behavior of varied soil and rock types namely phyllite clay, siliceous soils, and friable ore to develop effective drainage and slope stability strategies in mining environments. Fine-grained materials like phyllite clay, characterized by low permeability under high suction, demand precise modeling of hydraulic conductivity near saturation for safe and efficient slope design. Siliceous soils, whose stability is strongly influenced by matric suction, necessitate continuous monitoring and integration of unsaturated hydraulic properties into predictive models. Friable ore, due to its high porosity, structural fragility, and susceptibility to moisture-induced collapse, presents unique challenges that highlight the importance of capturing hysteresis behavior and moisture sensitivity in stability assessments.

These insights are essential for numerical modeling and field monitoring of pore water pressure dynamics in open pit mine slopes design, particularly under fluctuating groundwater conditions.

Acknowledgement. I would like to extend my heartfelt thanks to SERB for sponsoring the work. I would also like to express my sincere gratitude to Director, NITK Surathkal, for his invaluable guidance, encouragement, and support throughout the completion of this work.

References

1. Anderson, M.G., Kemp, M.J.: Towards an improved specification of slope hydrology in the analysis of slope instability problems in the tropics. Prog. Phys. Geogr. **15**(1), 29–52 (1991)
2. Reddy, S.K.: Stability assessment and optimal excavated design of a rock slope in an opencast limestone mine. J. Mines Metals Fuels. **71**(2) (2023)
3. Ram, A.S., Goyal, S.P.: Pit slope failure problems in Goan iron ore mines, Goa, India. Int. Conf. Case Hist. Geotech. Eng. **38**, 1–7 (2008)
4. Kumar, A.B., Meena, S.S., Renaldy, T.A., Babu, A.R.: Design and evaluation of stability of slope geometry of an iron ore mine-a case study. In: International Conference on Geotechnical Challenges in Mining, Tunneling and Underground Infrastructures, pp. 755–772. Springer Nature Singapore, Singapore (2021)
5. Perumalla, R., Reddy, S.K., Govinda Raj, M.: Review of slope stability analysis under drawdown conditions for mine slopes. Disast. Adv. **18**(5), 222–227 (2025)
6. Reddy, S.K.: Analysis of fault's effect on the highwall stability of medapalli open pit coal mine. Geotech. Geol. Eng. **41**(5), 2969–2986 (2023)
7. Reddy, S.K.: Monitoring and prediction of slope failure instability in a limestone mine. J. Mines Metals Fuels. **171**(2) (2023)
8. Du, X., Du, C., Radolinski, J., Wang, Q., Jian, J.: Metropolis-hastings Markov Chain Monte Carlo approach to simulate van genuchten model parameters for soil water retention curve. Water. **14**(12), 1968 (2022)
9. Rahardjo, H., Ong, T.H., Rezaur, R.B., Leong, E.C.: Factors controlling instability of homogeneous soil slopes under rainfall. J. Geotech. Geoenviron. **133**(12), 1532–1543 (2007)
10. Ng, C.W.W., Shi, Q.: A numerical investigation of the stability of unsaturated soil slopes subjected to transient seepage. Comput. Geotech. **22**(1), 1–28 (1998)

11. Ng, C.W., Pang, Y.W.: Influence of stress state on soil-water characteristics and slope stability. J. Geotech. Geoenviron. Eng. **126**(2), 157–166 (2000)
12. Fredlund, D.G., Rahardjo, H.: An overview of unsaturated soil behaviour. Geotech. Spec. Publ. **24**(1) (1993)
13. Onyelowe, K.C., Mojtahedi, F.F., Azizi, S., Mahdi, H.A., Sujatha, E.R., Ebid, A.M., Darzi, A.G., Aneke, F.I.: Innovative overview of SWRC application in modeling geotechnical engineering problems. Designs. **6**(5), 69 (2022)
14. Todd, D.: Groundwater Hydrology, p. 103. John Wiley & Sons Inc., New York, NY, USA (1980)
15. Lin, Z.Q., Qian, J.G., Shi, Z.H.: Modeling unimodal/bimodal soil-water retention curves considering the influence of void ratio under capillarity and adsorption. Rock Soil Mech. **42**(9), 7 (2021)
16. Zheng, W., Shen, C., Wang, L.P., Jin, Y.: An empirical soil water retention model based on probability laws for pore-size distribution. Vadose Zone J. **19**(1), e20065 (2020)
17. Yeh, H.F., Tsai, Y.J.: Analyzing the effect of soil hydraulic conductivity anisotropy on slope stability using a coupled hydromechanical framework. Water. **10**(7), 905 (2018)
18. Yeh, H.F., Wang, J., Shen, K.L., Lee, C.H.: Rainfall characteristics for anisotropic conductivity of unsaturated soil slopes. Environ. Earth Sci. **73**, 8669–8681 (2015)
19. Ng, C.W., Pang, Y.W.: Experimental investigations of the soil-water characteristics of a volcanic soil. Can. Geotech. J. **37**(6), 1252–1264 (2000)
20. Kristo, C., Rahardjo, H., Satyanaga, A.: Effect of variations in rainfall intensity on slope stability in Singapore. Int. Soil Water Conserv. Res. **5**(4), 258–264 (2017)
21. Lense, G.H.E., Parreiras, T.C., Moreira, R.S., Avanzi, J.C., Mincato, R.L.: Estimates of soil losses by the erosion potential method in tropical latosols. Ciênc. Agrotecnol. **43**, e012719 (2019)
22. Seki, K.: SWRC fit–a nonlinear fitting program with a water retention curve for soils having unimodal and bimodal pore structure. Hydrol. Earth Syst. Sci. Discuss. **4**(1), 407–437 (2007)
23. Dong, J.J., Tu, C.H., Lee, W.R., Jheng, Y.J.: Effects of hydraulic conductivity/strength anisotropy on the stability of stratified, poorly cemented rock slopes. Comput. Geotech. **1**(40), 147–159 (2012)
24. Fredlund, D.G., Rahardjo, H.: An overview of unsaturated soil behaviour. Geotech. Spec. Publ. **24**(1) (1993)
25. Zienkiewicz, O.C., Taylor, R.L., Zhu, J.Z.: The Finite Element Method: Its Basis and Fundamentals. Elsevier (2005)
26. Van Genuchten, M.T.: A closed-form equation for predicting the hydraulic conductivity of unsaturated soils. Soil Sci. Soc. Am. J. **44**(5), 892–898 (1980)
27. Malik, M.K., Karim, I.R.: Seepage, and slope stability analysis of Haditha dam using geo-studio software. IOP Conf. Series Mater. Sci. Eng. **928**(2), 022074 (2020)
28. Fredlund, D.G., Rahardjo, H.: An overview of unsaturated soil behaviour. Geotech. Spec. Publ. **24**(1) (1993)
29. Fredlund, D.G., Vanapalli, S.K.: Shear strength of unsaturated soils. Meth. Soil Analy. 4 Phys. Meth. **5**, 329–361 (2002)
30. Fredlund, D.G., Morgenstern, N.R., Widger, R.A.: The shear strength of unsaturated soils. Can. Geotech. J. **15**(3), 313–321 (1978)
31. Ng, C.W.W., Shi, Q.: A numerical investigation of the stability of unsaturated soil slopes subjected to transient seepage. Comput. Geotech. **22**(1), 1–28 (1998)

32. Hoek, E., Bray, J.W.: Rock Slope Engineering. Institute of Mining and Metallurgy (1981)
33. Fredlund, D.G., Xing, A.: Equations for the soil-water characteristic curve. Can. Geotech. J. **31**(4), 521–532 (1994)

Smart Helmet for Underground Miners

S. Kumar Reddy(✉) and H. Guruprasad Bhat

Department of Mining Engineering, National Institute of Technology Karnataka, Surathkal, Mangaluru 575025, Karnataka, India
skreddy@nitk.edu.in

Abstract. Mining is the primary industry, the backbone of a nation's economy. A nation's development depends on the availability of its primary resources, including its mineral resources. However, Mining is considered one of the hazardous job roles, where workers are constantly exposed to toxic gases, low visibility, and hazardous conditions that can occur at any time. In this respect, the smart miner's helmet is designed around protective gear at the centre of a miner's welfare. This helmet was developed using the Internet of Things & sensors that monitor the ambient environment for temperature, humidity, methane and carbon monoxide in real time. When conditions in the mine become too hazardous, it warns both the miner and the control room instantly, so vital minutes can be used to take action. Safety involves not only detecting hazards but also reporting them to the supervisor. This helmet uses the latest technology to withstand harsh mining conditions and is reliable, durable, energy efficient and low cost. Therefore, a smart miner's helmet can make a hazardous job much safer by using real-time monitoring data of the mining environment. It also provides peace of mind to miners and their families. Thus, increase worker efficiency.

Keywords: Smart Helmet · Internet of Things · Sensors · Underground Mine Hazards

1 Introduction

Mining has long been the backbone of industrial development, providing essential raw materials ranging from simple building construction to advanced high-end technologies by extracting valuable minerals from the earth's crust [1]. Although mining is crucial to modern life, it remains one of the most hazardous occupations, exposing men and women to challenging and dangerous conditions. Underground mines often contain highly toxic gases, extreme temperatures, and constant threats of cave-ins or equipment failures [2]. Faulty equipment, negligence of labour towards explosions, structure failure, and gas accumulation are the most common causes of underground mine accidents [3]. These dangers pose severe risks to miners' health and safety and impact the overall efficiency and sustainability of mining operations [4, 5].

Significant improvements in mining safety have been made in recent years, such as enhanced ventilation systems and stricter safety protocols. However, traditional safety

H. P. Dasari et al. (Eds.): NITK-CREST 2025, PTT, pp. 43–57, 2026.
https://doi.org/10.1007/978-981-95-3218-6_4

gear such as helmets and gas masks often works reactively rather than proactively. These tools typically require miners to recognise hazards and respond accordingly, which can be too slow during life-threatening situations. Communication is another difficult problem in mine tunnels because electromagnetic waves are absorbed and echoed in a coal tube. Because of the many corners in the tunnel, the Wave cannot cross these corners easily [6]. It is not an uncommon occurrence that the difference between a successful close call or a tragedy can be the time used to identify the threat and respond to it [3].

The smart helmet is a promising device created using the network of sensors and modern technologies, such as Internet of Things (IoT). The smart helmet is special since it is operational in the extreme environment characteristic of mines and as such can be applied in routine mining activities, search and rescue, repairs, inspection and environmental monitoring [7].

Mining risks result because of human error as well as equipment failure. The smart helmet helps mitigate of these risks, as it actively monitors all elements of the environment by various helmets worn by the miners. Outliers that pertain to data can be easily detected. sensor readings can be averaged to help in reducing minor errors. As an example, the new tragic Kenda coal mine disaster, in which the coal miners were stuck in such a hazardous atmosphere in the presence of carbon monoxide and unsuccessful rescue, might have been rather avoided with helmets that have sensors on them.

Smart miner helmet makes use of advanced technology to meet safeguard. Equipment to fill in important safety loopholes. It is IoT enabled sensors, which monitor environmental factors like, temperature, humidity and toxic gases [4, 8–10]. In the case of poisonous levels, the helmet alerts the miner instantly and the control room which allowed swift response since the response time was greatly minimized. which could result in saving lives.

The smart miner helmet combined advanced technology with practical design, represent a quantum leap in mining safety and productivity. It offers a holistic solution for enhancing miner protection and improving efficiency across mining operations.

2 Literature Survey

Different researchers have studied the literature review on different areas of mine hazards and environmental risks. Sago Mine Disaster (2006): West Virginia, USA, a methane explosion trapped 13 miners. Miscommunication about the status of the workers led to a delay in the rescue operation in reaching the miners, and only one survived [11].

Behr et al. considered three major hazards: air quality, helmet removal, and collision (object impact on miners) while developing their smart helmet for the mining sector [4]. Their system monitors air quality and identifies dangerous events, with emphasis on dangerous gases such as carbon monoxide (CO), sulfur dioxide (SO_2), nitrogen dioxide (NO_2), and particulate matter (PM), all of which have a major impact on miners' health.

In a similar initiative, researchers designed a smart helmet to detect dangerous conditions. Through multiple calibrated sensors, the system detects gases like CO, CH_4, liquefied petroleum gas (LPG), and other natural gases. When gas levels surpass set threshold values, the helmet triggers a buzzer and lights LEDS to warn the miner and their overseers [12].

Effective communication systems are essential for transmitting environmental data and alerts in underground mining. Song et al. developed a coal mine safety monitoring system using a wireless sensor network [5]. A microcontroller collects environmental data such as temperature and humidity from various sensors and sends it via wireless communication modules for real-time monitoring.

Wu, Y et al. suggested that Bluetooth wireless transmission be applied in underground coal mine monitoring [13]. Bluetooth is a low-power, inexpensive short-range communication technology that offers real-time monitoring. Its weak points are the limited transmission range and cable installation difficulties, which are unfavourable to underground mining conditions.

Reddy and others developed wireless monitoring of environmental parameters for underground mines using the Internet of Things with a LoRa Transceiver Module [14]. The proposed system uses the IoT with Long Range (LoRa) wireless communication technology in underground mining to build an effective mine safety system. The system can extend to raise an early warning of any hazardous events by integrating gas sensors. Also, it can connect the LoRa module gateway to a cloud database to monitor data from anywhere and at any time. The developed LoRa-based system can be deployed in an underground mining area by supervisors to monitor the real-time environmental parameters from the surface of the mine site to avoid any hazards.

Naik and others proposed a methodology for adopting IoT in underground mines and a systematic review of the interdisciplinary research on using WSN and IoT with ZigBee or LoRa for real-time monitoring of environmental parameters in underground mines [15, 16]. The author also implemented environmental parameters monitoring and an alert system for underground mining using IoT with LoRa Technology, where they used IoT-based continuous monitoring systems needed for underground mine environment monitoring to avoid dangerous occurrences in underground mines.

Despite all the progress, several gaps still exist. The products are limited in localisation, and many systems today are not well-suited for the Indian underground mine conditions because of the high cost [6]. The developed helmet seeks to overcome these challenges by suggesting a more affordable, localised smart helmet solution for Indian mining conditions. It incorporates essential safety features while keeping it affordable and easy to deploy with a low cost.

3 Methodology

The methodology primarily involves real-time hazard alerts using a helmet embedded with IoT and sensors. The integrated sensors detect and warn miners of possible hazards such as toxic gases and temperature fluctuations, improving overall safety and allowing time for intervention. An LED warning system also alerts machinery operators in dark or low-light environments to prevent accidents.

The development process of the smart helmet followed a systematic workflow, as shown in Fig. 1. It started with the requirement phase, identifying primary functionalities like gas sensing, obstacle detection, and real-time monitoring. This was followed by the design phase, where the system architecture and component selection, such as Arduino Uno, MQ-series gas sensors, and ESP32 modules, were finalised.

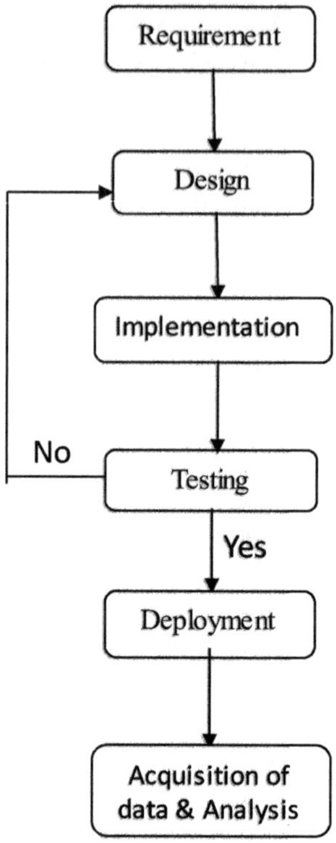

Fig. 1. Flowchart of the development of the model

During the implementation phase, the selected hardware was integrated and programmed. The system underwent thorough testing for functionality, accuracy, and reliability. Iterative changes were made by revisiting the design and implementation phases as needed. Once the prototype met performance criteria, it moved to deployment for operational use in underground mines.

The deployed system continuously acquires environmental data, providing real-time monitoring to enhance miner safety. The gathered data can be further analyzed to detect trends in gas accumulation and environmental risks, supporting proactive hazard response and preventive maintenance practices. This strategy guarantees an aggressive reaction to hazards to safety and upholds preventive maintenance practices. The description of the sensors used is as follows.

3.1 DTH Temperature and Humidity Sensor

The mining conditions pose huge challenges, especially temperature and humidity, which change depending on the working conditions and ambient weather. Water is sprayed to damp down dust, changing the humidity inside the mines. To measure these environmental factors accurately, a DTH Temperature and Humidity sensor is used and incorporated into miners' helmets to monitor in real-time. This sensor works within a voltage of 3.5 V to 5.5 V and drains only 0.3 mA when measuring, but as low as 60 μA in standby. It can measure temperature between 0 °C and 50 °C and humidity from 20% and above, with a high resolution of 16-bit for both parameters. The sensor offers serial data output and features a one-second response time, equivalent to a 1 Hz sampling rate, for timely and precise environmental measurement to provide greater miner safety (Fig. 2).

Fig. 2. DTH temperature and humidity sensor [19

3.2 MQ2 Gas Sensor

The MQ2 gas sensor is an electronic module intended to sense the concentration of gases in the air, such as LPG, propane, hydrogen, alcohol, smoke, and carbon monoxide. It is a chemiresistor that uses a sensing material whose electrical resistance varies based on the presence and concentration of targeted gases. This resistance change is utilized to measure the gas levels and hence the sensor finds application in real-time monitoring in safety-critical conditions like mines. According to TTL logic, the MQ2 sensor works at an operating voltage of +5 V, giving analog output from 0 V to 5 V and digital output at 0 V or 5 V. A preheat and warm-up period of approximately 20 s is needed to stabilise the output reading. The sensor can sense gas concentration between 200 ppm and 10,000 ppm and be adjusted for sensitivity using a potentiometer for digital output calibration [17]. The MQ2 has a functioning current of 150 mA and an accuracy of ±10% and can be effectively used for detecting dangerous gases and miner safety through prompt notice and system integration (Fig. 3).

Fig. 3. MQ2 Gas Sensor [20]

3.3 MQ4 Gas Sensor

In coal mines, gas emission is one of the most important challenges, specifically methane gas emission, resulting from the spontaneous heating of coal and other factors. Methane, though nontoxic, increases in concentration and reduces oxygen levels, creating severe health and safety hazards. If the methane concentration is more than 5.4% and the oxygen level is more than 12%, there is a high possibility of an explosion [18]. Thus, there is a need for a specialized sensor to monitor methane levels continuously for the safety of miners. The sensor works at a +5 V voltage, giving analog output from 0 V to 5 V and digital output of either 0 V or 5 V depending on TTL logic. It needs around 20 s of preheat and warm-up time to give stable data. The sensor can sense LPG, alcohol, propane, hydrogen, carbon monoxide, and methane gases, and the sensitivity is adjustable through a potentiometer for digital output. It can sense gas concentrations from 200 ppm to 10,000 ppm and has a working current of 150 mA, providing an accuracy of ±10% to ensure proper operation in hostile underground conditions, as shown in Fig. 4.

Fig. 4. MQ4 Gas Sensor [21]

3.4 Ultrasonic Sensor

Ultrasonic sensors play a vital role in the mining sector, greatly improving safety and operational effectiveness. They are widely applied for proximity sensing, collision prevention, and silo or tank level measurement. In underground mining, ultrasonic sensors prevent accidents by sensing obstructions and keeping machinery safe from workers and structural components. Their reliability in dusty, low-lit, and aggressive environments renders them very effective for real-time monitoring and automation in contemporary mining conditions. The sensor runs on 5 V DC power, consumes a current of about 15 mA, and operates at a working frequency of 40 kHz. It can measure distances between 2 cm and 400 cm (0.02 m and 4 m) with ±3 mm accuracy and an angle of less than 15°. The sensor takes a 10 µs TTL pulse as a trigger input and delivers a TTL echo output signal proportional to the distance measured. It has a 4-pin interface—VCC, Trig, Echo, and GND—with a quick response time of less than 40 milliseconds. The small size of around 45 mm × 20 mm × 15 mm makes it ideal for incorporating different mining equipment and monitoring systems in Fig. 5.

Fig. 5. Ultrasonic sensor [22]

3.5 ESP8266 Wi-Fi Module

The module features a boot ROM, 64 KB of user data RAM, and 112 KB of instruction RAM, which includes an additional 32 KB segment. It supports 802.11 b/g/n Wi-Fi standards operating at a frequency of 2.4 GHz and integrates a 10-bit ADC, making it suitable for various sensing and communication tasks[19]. Designed for ease of use, it connects to a microcontroller via a simple serial port interface, and an external voltage converter is required only if the operating voltage exceeds 3.6 V. Due to its cost-effectiveness and versatility, this module is widely adopted in robotics and IoT applications. In this project, it functions as a wireless data communication interface. The module operates within a voltage range of 3.0 V to 3.6 V. It exhibits varying current consumption depending on the mode: less than 10 µA in standby, less than 20 µA in deep sleep, 2 mA in modem sleep, 0.9 mA in light sleep, 70 mA during active operation, and up to 300 mA at peak performance. It is equipped with flash memory ranging from 512 KB to 4 MB, depending on the module version, a 32-bit RISC Tensilica Xtensa L106 CPU running at 80 MHz (overclockable to 160 MHz). It offers 16 GPIO pins, providing ample flexibility for interfacing and control, as shown in Fig. 6.

Fig. 6. ESP8266 Wi-Fi module [23]

3.6 Arduino UNO

Arduino Uno is an open-source microcontroller board using the ATmega328P chip that is well known for its ease of use, strong community backing, and support for beginners and hobbyists. It has 14 digital input/output pins, with 6 out of them being capable of supporting PWM output, and 6 analog input pins, all working at a voltage of 5 V with a clock speed of 16 MHz. An integrated USB interface provides easy programming and communication with PCs, making it perfect for home electronics, embedded system prototypes, and educational electronics. In this setup, the Arduino Uno is the central processing unit, the "brain" that makes decisions and performs actions based on the programmed logic. It runs on a recommended input voltage range of 7–12 V, with an acceptable limit of 6–20 V. Each I/O pin can support up to 20 mA of current, and the 3.3 V pin can supply a maximum of 50 mA. The board contains 32 KB of flash memory (0.5 KB reserved for the bootloader), 2 KB of SRAM, and 1 KB of EEPROM to store data. It also provides USB connectivity through a USB-B connector, and its small package dimensions of 68.6 mm × 53.4 mm and weight of around 25 g make it very portable and convenient to integrate into numerous applications, as shown in Fig. 7.

Fig. 7. Arduino Uno [24]

3.7 LCD Display

LCD Display (16x2) is a commonly encountered alphanumeric display module that can display two lines with a maximum of 16 characters per line, making it suitable for real-time display of sensor readings, system messages, and user interfaces in embedded systems and DIY electronics projects. Compatible with the HD44780 controller, the module would normally run on 5 V and can readily be interfaced with microcontrollers like the Arduino via 4-bit or 8-bit parallel communication or using an I2C interface module to enjoy easy wiring and serial communication. The display has a working range of 4.7 V to 5.3 V and consumes around 1 mA of current without backlight, with the backlight consuming around 15 mA depending on the setup. A character is drawn on a 5 × 8 pixel matrix, and the display is TTL logic compatible, usually at 5 V. The backlighting is typically LED, commonly white or green, and the display contrast can be regulated via a potentiometer attached to the V0 pin. Measuring around 80 mm × 36 mm × 12 mm with a typical viewing angle at 6 o'clock, the 16x2 LCD is a space-saving, dependable solution for textual displays in electronic devices, as shown in Fig. 8.

Fig. 8. LCD Display

4 Implementation

Implementing the Smart Helmet system involves hardware and software integration of IoT and sensors. The sensors are interfaced with a microcontroller, which processes and transmits the data using a wireless communication module [17, 18]. A centralized monitoring system receives and displays the data. We use two lights: one is a headlamp, and the other is a warning lamp. It is recommended that these be connected in parallel, as the failure of one should not disturb the other. It's better to have a common circuit breaker to have a warning light turned on when a miner turns his cap lamp on and off when it is unnecessary, like when he is on the surface. Thus, there is no chance of forgetting to turn the warning lights on. The battery is used to supply the power to all the sensors and the aurdino.

The Arduino is programmed according to specific requirements using the Arduino IDE software. The code is written in the C language, checked for errors, and implemented using a data cable for transmission, as shown in Fig. 9.

Fig. 9. Circuit connections of the sensor [25]

5 Result

The model is tested in normal conditions and under the respective sources. The results show that it senses and feeds the data to the selected computer. If any unfavourable incidents occur, the alert is immediately given and the buzzer is activated, indicating that the threshold level has been crossed. The programmed Arduino and interconnected sensors like MQ2, MQ4, and others make it work as a system.

The obtained readings can be plotted as graphs, which makes understanding easier. It also explains when or why the level increases based on the other operations. Again, this serves as the record for future use. The small data makes storage more cost-effective and convenient, as shown in Fig. 10.

Fig. 10. Developed IoT-helmet model [25]

The results achieved through 24-h continuous measurement of the smart helmet under various environmental conditions exhibit its reliability and multi-parameter sensing. Throughout the test duration, gas concentration data (CO, CH_4, and CO_2), temperature, humidity, and proximity data were recorded in real-time and graphed continuously. The plots show evident fluctuations that are responsive to simulated changes in the environment, confirming the accuracy and responsivity of sensors in the helmet. Significant gas level spikes were recorded during controlled exposure tests, whereas temperature and humidity fluctuated according to ambient changes. The proximity sensor also performed well in identifying the proximity of surrounding objects or people, proving useful in collision avoidance or navigating narrow spaces. Overall, the results confirm the helmet's potential to offer complete real-time monitoring in varied and dynamic environments, thus serving as a useful safety tool for miners in Figs. 11, 12, 13 and 14.

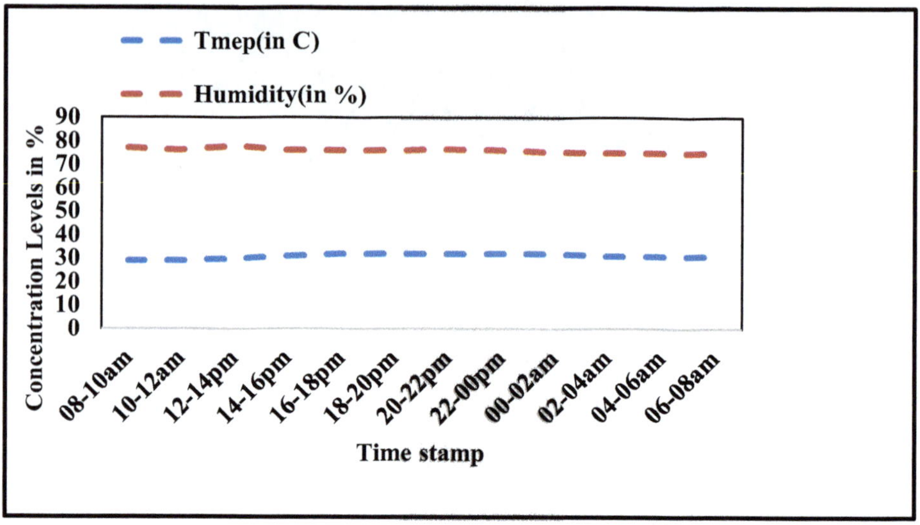

Fig. 11. DTH sensor temperature and humidity readings [25]

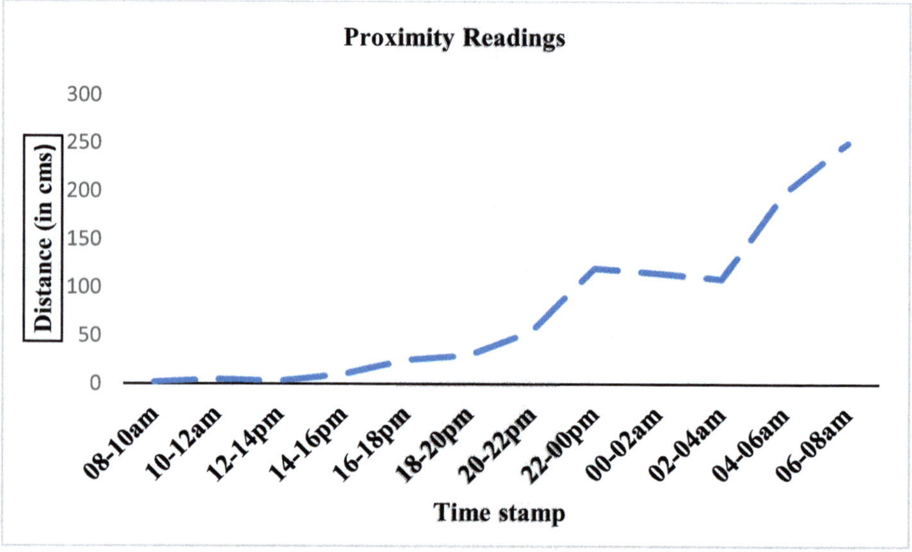

Fig. 12. Ultrasonic sensor data recordings [25]

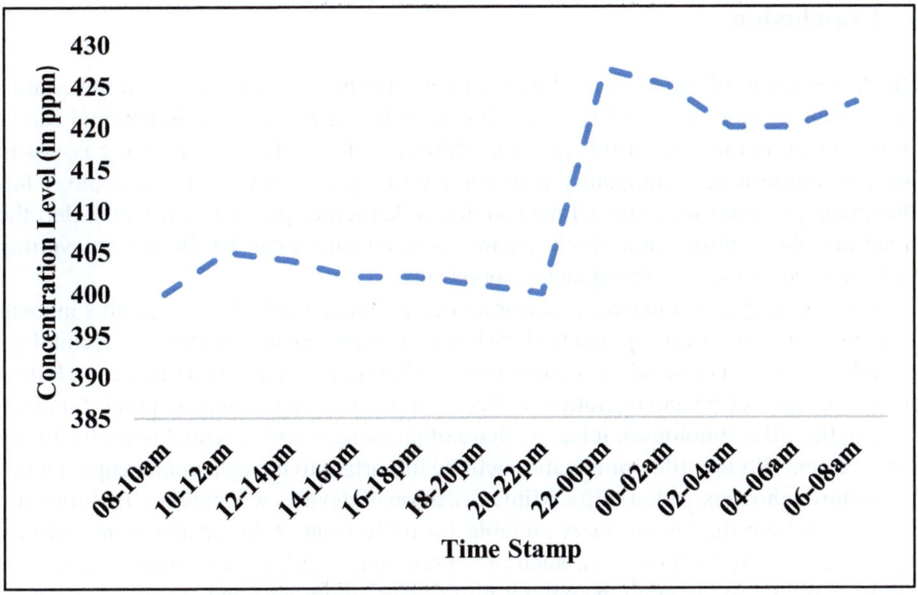

Fig. 13. CO sensor data recordings [25]

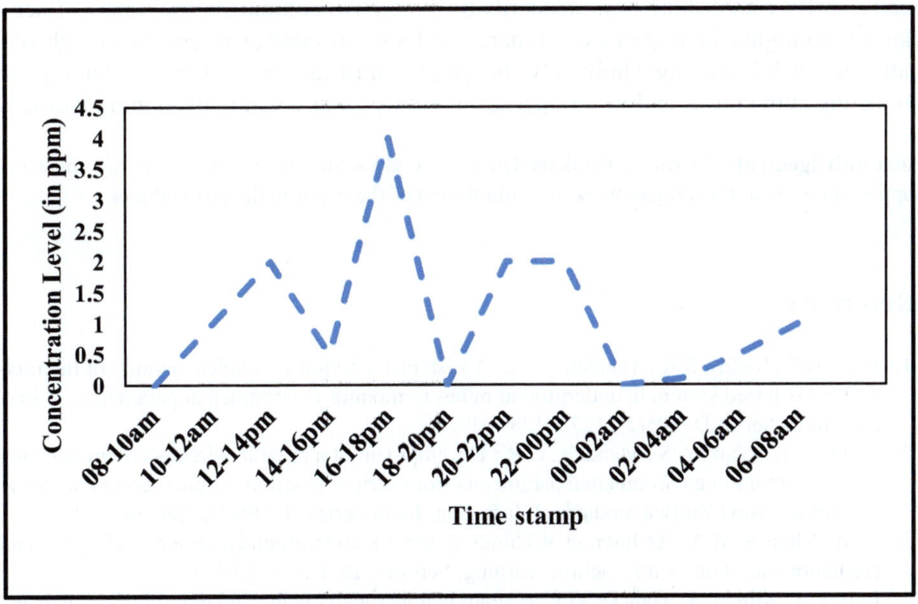

Fig. 14. Methane sensor data recordings [25]

6 Conclusion

The development of the low-cost Smart Miner's Helmet is a major step in the continued pursuit of increasing safety in mining activities. Real-time environmental monitoring, which is the core of this project, addresses one of the most pressing issues in underground mining: anticipatory and reactive safety assurance. This technology has life-saving potential since the helmet can detect dangerous gases and instantly alert the miner and the control room. Such instant communication can be the key to averting accidents and enhancing emergency response times.

With the help of cutting-edge sensor and communication technologies, this innovation raises the bar for safety in a high-risk sector. It proves that continuous technological advancement is crucial in maintaining worker safety. The Smart Miner's Helmet maximizes protection and operational effectiveness, assuring miners and their families.

Yet, like all technologies, it has its share of challenges and potential areas for future improvement. Battery life optimization will be important to provide uninterrupted monitoring throughout extended shifts. Miniaturization of devices will enhance mobility and comfort, making the helmet more suitable for daily wear. Additionally, compatibility with other mining technologies, such as autonomous vehicles and smart ventilation, will be essential to integrate the helmet into a more interconnected and efficient safety system.

This technology has the potential to spur future breakthroughs in industrial safety headgear. By establishing new standards for protective equipment, the Smart Miner's Helmet highlights the importance of increased focus on safety, particularly in high-risk industries such as mining. Ultimately, this project illustrates the power of technology in protecting human lives and fostering a strong safety culture within the mining industry.

Acknowledgement. I cordially thank the Director of NITK Surathkal for his encouragement and support throughout this project work and submission of the paper in the proceedings.

References

1. Naik, A.S., Reddy, S.K., Mandela, G.R.: A systematic review on implementation of Internet-of-Things-based system in underground mines to monitor environmental parameters. J. Inst. Eng. India Series D. **105**(2), 1273–1289 (2024)
2. Reddy, S.K., Naik, A.S., Mandela, G.R.: Development of a reliable wireless communication system to monitor environmental parameters from various positions of underground mines to the surface using ZigBee modules. J. Inst. Eng. India Series D. **105**(1), 359–383 (2024)
3. Jo, B., Khan, R.M.A.: An Internet of Things system for underground mine air quality pollutant prediction based on azure machine learning. Sensors. **18**(4), 930 (2018)
4. Behr, C.J., Kumar, A., Hancke, G.P.: A smart helmet for air quality and hazardous event detection for the mining industry. In: 2016 IEEE International Conference on Industrial Technology (ICIT) 2026–2031. IEEE (2016)
5. Song, J., Zhu, Y., Dong, F.: Automatic monitoring system for coal mine safety based on wireless sensor network. In: Proceedings of 2011 Cross Strait Quad-Regional Radio Science and Wireless Technology Conference IEEE, vol. 2, pp. 933–936 (2011)

6. Gao, J., Gao, X., Zhu, J., Zhu, W., Wei, B., Wang, S.: Coal mine detect and rescue robot technique research. In: 2009 International Conference on Information and Automation, pp. 1068–1073. IEEE (2009)
7. Prabu, A.V., Gayatri, N., Madhuri, J., Rajasoundaran, S., Shaik, M.F.: Smart coal mine safety & monitoring system. In: 2022, the 6th International Conference on Trends in Electronics and Informatics (ICOEI), pp. 408–412 (2022)
8. Bo, C., Xin, C., Zhongyi, Z., Chengwen, Z., Junliang, C.: Web of things-based remote monitoring system for coal mine safety using wireless sensor network. International Journal of Distributed Sensor Networks. **10**(8), 323127 (2014)
9. Kumar Reddy, S., Naik, A.S., Govindaraj, M.: Internet of Things based Data acquisition system for monitoring air quality in underground mines. J. Mines Metals Fuels. **70**(12A), 21–27 (2023)
10. Kumar Reddy, S., Naik, A.S., Govindaraj, M.: Development of a novel real-time environmental parameters monitoring system based on the internet of things with LoRa modules in underground mines. Wirel. Pers. Commun. **133**, 1517–1546 (2023)
11. Sago Report, The United mine workers of America Sago int.indd (wv.gov) (2016).
12. Paul, R., Karar, A., Datta, A., Jindal, S.K.: Smart helmet for coal miners. In: 2022 1st International Conference on the Paradigm Shifts in Communication, Embedded Systems, Machine Learning and Signal Processing (PCEMS), pp. 37–42. IEEE (2022)
13. Wu, Y., Feng, G., Meng, Z.: The study on coal mine using the Bluetooth wireless transmission. In: 2014 IEEE Workshop on Electronics, Computer and Applications, pp. 1016–1018. IEEE (2014)
14. Reddy, S.K., Naik, A.S., Mandela, G.R.: Wireless monitoring of environmental parameters for underground mining using Internet of Things with LoRa transceiver module. In: 2022 IEEE 7th International Conference on Recent Advances and Innovations in Engineering (ICRAIE), vol. 7, pp. 224–229. IEEE (2022)
15. Naik, A.S., Reddy, S.K., Govindaraj, M.: Real-time environmental parameters monitoring system using IoT-based LoRa 868-MHz wireless communication technology in underground mines. IEEE Access. **12**, 7430–7455 (2024)
16. Naik, A., Reddy, S.K., Raj, M.G.: Implementation of environmental parameters monitoring and alert system for underground mining using Internet of Things with LoRa technology. J. Disast. Adv. **18**(5), 1–9 (2025)
17. Elprocus, MQ2 gas sensor working principle and its applications. https://www.elprocus.com/an-introduction-to-mq2-gas-sensor/ (2025)
18. Misra, G.B.: My environment and ventilation. Oxford University Press (1986)
19. Moreware Blog homepage. https://sl1nk.com/ujeTWv. Accesed 09 Oct 2025
20. Rajguruelectronics. https://sl1nk.com/r2fmO. Accessed 09 Oct 2025
21. Philips Electronics. https://l1nq.com/aL0L9. Accessed 09 Oct 2025
22. Tinysine. https://l1nq.com/YpOCm. Accessed 09 Oct 2025
23. Robotics DNA. https://l1nq.com/gwjWd. Accessed 09 Oct 2025
24. Techiesms.com. https://l1nq.com/UurLR. Accessed 09 Oct 2025
25. Reddy, S.K., Guruprasad, B.H.: Project report on design of smart helmet for underground miners (2025)

Effect of Dump Surcharge Loading on Pit Slope Stability and Optimal Safe Distance: A Review

S. Kumar Reddy(✉), G. Dinesh, and K. Venkateshwarlu

Department of Mining Engineering, National Institute of Technology Karnataka Surathkal, Mangalore 575025, India
skreddy@nitk.edu.in

Abstract. The increasing depth of open-pit mines has led to the accumulation of vast quantities of waste material, posing significant waste disposal challenges due to limited space, stringent environmental regulations, and safety concerns for planning and designing open-pit mines. Improper placement of external dumps can impose excessive surcharge loads, altering stress distribution and triggering failure mechanisms such as excessive settlement, rotational sliding, and shear failure. This review paper examines how surcharge loading affects pit slope stability and emphasizes determining an optimal safe distance from the pit crest to mitigate instability. It evaluates critical geotechnical parameters such as material density, grain size, friction angle, cohesion, void ratio, porosity, and compaction along with geometric factors such as bench height, slope angle, and dump height. The paper also explores analytical design approaches, including empirical methods, physical modeling, and numerical techniques like Finite Element Method, Limit Equilibrium Method, and Strength Reduction Method, to analyze slope behavior under varying surcharge conditions. This problem is more than just strategic dump placement, which is crucial for ensuring slope stability, reducing failure risks, and promoting sustainable waste management in open-pit mining operations.

Keywords: Open-pit mining · Slope stability · Surcharge loading · External dumps · Numerical modeling · Physical modelling

1 Introduction

As open-pit mining progresses beyond 500 m in depth, ensuring slope stability becomes increasingly vital. Accident records from Indian mines (2007–2017) indicate that many failures were due to inadequate slope design and monitoring. Key influencing factors include slope geometry, geotechnical properties, and groundwater conditions. Effective stability management requires integrated approaches involving real-time monitoring, physical modeling, and advanced numerical simulations using tools like PLAXIS, Geo Studio, and Slide to predict slope performance under site-specific conditions [1]. Managing overburden (OB) in deep open-pit mining is challenging due to increased stripping ratios and waste disposal needs [2]. In India, external OB dumping is common but poses geotechnical risks due to inadequate investigations and design. Stability is influenced

H. P. Dasari et al. (Eds.): NITK-CREST 2025, PTT, pp. 58–75, 2026.
https://doi.org/10.1007/978-981-95-3218-6_5

by dump height, slope angle, material properties, and foundation shear strength. Low-strength foundations (<25 kPa) can lead to failures, while OB materials like shale and clay often exhibit low shear strength, emphasizing the need for conservative design [3].

The rapid growth of mining operations has increased waste volumes, raising concerns over economic, environmental, and regulatory impacts, particularly in land-scarce or ecologically sensitive areas. External dumping exacerbates issues like inefficient land use, higher handling costs, and a larger environmental footprint compared to internal dumping methods [4, 5]. Historical failures, such as the Aberfan disaster (1966), underscore the risks of poor geotechnical practices, highlighting the need for robust design, drainage systems, and continuous monitoring [6]. The heterogeneity of dump and foundation materials complicates stability assessments, making advanced numerical modeling, including unsaturated soil mechanics and hydro-mechanical behavior, essential for optimizing stabilization strategies [7–11]. In Greece, a lignite mine collapse released 40 million m^3 of spoil due to pore pressure buildup from spoil placement over a natural spring [12]. Faults, folds, and shear zones significantly affect slope stability by weakening rock masses and influencing failure geometry. Faults act as low-strength slip surfaces, folds can unfavorably orient bedding planes leading to sliding or toppling, and shear zones form deep-seated weak zones due to intense deformation. These features also alter groundwater flow, further reducing slope stability [13].

Dump failures in Indian mines have caused significant damage, fatalities, and disruptions. In 2009, an OB dump failure at Singareni Collieries Company Limited (SCCL) displaced material by almost 70 m, triggered by mobilized residual shear strength and pore pressure buildup. At Western Coalfields Limited (WCL), Nagpur, a 75 m high OB dump with a 43° slope slipped by 18 m. Investigations identified that the dump's loose materials, including friable sandstone, shale, and clay, contributed to the failure due to their low shear strength and high susceptibility to water infiltration [14]. A tragic incident at Basundhara mines, Odisha, on August 10, 2013, resulted in 14 fatalities and financial losses of ₹30 crores due to excessive rainfall increased pore water pressures within the dump, reducing effective stress and causing slope failure, highlighting the importance of effective drainage systems during the monsoon season [15]. Similarly, at Jayant Opencast Mine (2009), an unstable foundation, surcharge loading, and water seepage led to the displacement of 146,000 m^3 of waste material as shown in Fig. 1 [16]. A weak interface layer and high water table resulted in a pre-failure Factor of Safety (FOS) of 0.94, well below the recommended static conditions [17]. Water-induced failures are a recurring concern in both international and Indian mine dumps.

Comprehensive geotechnical investigations encompassing field studies, lab testing, and numerical analysis are essential for understanding failure mechanisms and developing effective mitigation strategies in open-pit mines. Critical factors such as slope geometry, material strength, groundwater conditions, and external loads must be integrated into the design. This study reviews the impact of surcharge loading on slope stability, focusing on optimizing external dump placement and employing advanced analytical tools to enhance safety, efficiency, and long-term sustainability in mining operations.

Fig. 1. A dump slide shifted toward the high wall, burying five mine - workers (Sharma & Roy, 2015).

2 Influence of Surcharge Load on Pit Slope Stability

Slope stability is a critical consideration in geotechnical engineering and mining engineering, particularly in contexts involving surcharge loads, which can significantly influence the performance of soil slopes, open-pit mines, and retaining structures. Numerous studies both experimental and numerical have been conducted to understand the effects of surcharge loading on slope behavior, employing advanced modeling techniques such as the Finite Element Method (FEM), Finite Difference Method (FDM), and Limit Equilibrium Method (LEM). Research utilizing tools such as PLAXIS-3D and SLOPE/W-2007 has shown that slope angle plays a dominant role in determining stability under surcharge loads. Slopes with lower inclinations (30° and 45°) demonstrated higher stability and load-bearing capacity, whereas steeper slopes (60°) were more susceptible to sudden failures. The cohesive strength of the soil was found to significantly enhance stability, with failure mechanisms typically involving bulging and crest settlements. FEM-based analyses validated these observations, confirming that slope angle, surcharge intensity, and cohesion are the primary factors controlling stability. As a result, maintaining optimal slope angles under high surcharge conditions is recommended to ensure structural safety [18].

A parametric evaluation further highlighted the influence of multiple factors on slope stability, including slope angle, height, unit weight, cohesion, internal friction angle, and groundwater conditions. An increase in slope angle from 16° to 66° resulted in a significant decrease in the FOS, with an average decline of 2.71% per degree. Similarly, increasing the slope height led to a 0.195% decrease in FOS per meter. An increase in unit weight from 5.886 kN/m^2 to 17.658 kN/m^3 caused a 13.89% reduction in FOS, while each degree increase in the internal friction angle improved FOS by 4.23%. Among all the factors studied, the presence of a shallow groundwater table had the most substantial impact, reducing the FOS by up to 32.63%, thereby underlining the necessity of effective drainage systems in slope and pit design [19].

The role of surcharge loading in nailed soil slopes has been comprehensively investigated using FLAC3D numerical simulations. These studies examined various loading types, including distributed, line, and point surcharges. Results indicated that the magnitude and proximity of surcharge loads significantly reduced FOS and increased both nail forces and lateral displacements. Distributed loads placed 2 m from the slope crest were particularly detrimental, with a 100 kPa load, reducing the FOS to 1.27 according to Federal highway administration and exceeding the horizontal displacement limit of 18 mm. Similar destabilizing effects were observed for line loads within 2–3 m of the crest and large point loads directly at the slope edge. These findings reinforce the importance of considering load position and type in nailed slope design [20].

Mechanically Stabilized Earth (MSE) walls under surcharge have also been evaluated at Senai City Airport involving a 9 m high and 400 m long MSE wall analyzed through PLAXIS 2D demonstrated that increasing surcharge intensity caused notable displacement and stress redistribution. Under a 100 kN/m^2 surcharge, peak displacement reached 19.23 mm, with a corresponding FOS of 0.884. This case emphasizes the need for careful surcharge management and highlights the effectiveness of FEM modeling in capturing failure mechanisms and reinforcement behavior in retaining structures [21].

The laboratory experiments and numerical analyses conducted to investigate the behavior of silica sand slopes under static loading for a small-scale slope was constructed in a transparent acrylic box measuring 500 mm in width, 1000 mm in height, and 1500 mm in length. Numerical analysis was carried out using FEM with PLAXIS 3D, employing the Hardening Soil Model (HSM). A comparative analysis was performed between the LEM, using SLOPE/W, and FEM for a silica sand slope subjected to a 6.4 kPa surcharge load. LEM predicted a factor of safety (FOS) of 0.955, suggesting potential failure, whereas FEM produced a significantly higher FOS of 6.321 due to its ability to account for stress–strain behavior. Laboratory model tests showed actual slope failure under the applied surcharge, aligning more closely with LEM predictions. The findings highlight the importance of selecting suitable analysis methods based on the nature of the problem and soil characteristics [22].

Additional investigations into surcharge parameters intensity, width, height, and distance from the slope edge revealed their influence on failure mechanisms. When surcharges were applied near the crest, rotational failures with significant displacement were triggered due to larger moment arms. As surcharges were moved farther from the slope edge, their effect diminished, eventually becoming negligible. Moreover, increased surcharge width and intensity were found to reduce yield acceleration and increase deformations, underlining the importance of optimal load configurations in design [23].

Notably, excessive surcharges rapidly led to destabilization, even in slopes initially deemed stable. This highlights the need for rigorous geotechnical assessments and appropriate load management strategies in rock slope engineering. Anchored rock slopes have similarly been found to be highly sensitive to surcharge loading. As the surcharge increased, FOS consistently decreased, particularly under low initial load conditions. Factors including slope angle, failure plane geometry, shear strength, and depth of tension cracks played a role in the magnitude of instability [24]. Other studies confirmed that surcharge position and intensity considerably affect slope performance. Higher surcharges increased stress concentrations and deformations, especially when placed near the crest. While wider surcharges distributed loads more evenly, reducing localized effects; narrower loads concentrated deformation and led to more severe failures. These parameters must be accounted for during slope design and risk mitigation planning. Further numerical simulations using GEO5 indicated that increasing slope height and angle decreased FOS, with reductions stabilizing beyond a certain height. The maximum plastic strain was typically concentrated near the crest under surcharge loads, reaffirming the significance of careful surcharge placement in enhancing slope stability [25, 26].

2.1 Surcharge Loading in Open Pit Mines

Understanding surcharge effects in open pit mines allows engineers to optimize slope designs, ensuring adequate stability while accommodating external loads. Properly accounting for surcharge loads helps prevent catastrophic failures in mining pits and waste dumps on steep slopes. Dynamic loading scenarios like earthquakes or traffic vibrations are particularly critical. The integration of FEM and LEM provides accurate predictions for slope behavior under varying surcharge conditions, enabling better decision-making for safety measures.

Surcharge loading significantly influences slope stability by directly affecting the factor of safety (Fig. 2a & b). As surcharge loads increase, FOS decreases, making pit slopes more susceptible to failure under both static and dynamic conditions. Studies indicate that the combination of surcharge loads and horizontal seismic forces notably reduces slope stability and increases permanent displacements sometimes up to twice the values estimated by traditional methods, especially when tensile strength thresholds are exceeded. Excessive surcharge can trigger landslides, particularly when accompanied by elevated pore pressure or seismic activity. However, relocating the surcharge away from the slope crest improves stability. In open-pit mining, careful management of surcharge loads during excavation helps maintain slope integrity, while the post-extraction disposal of waste as external dumps within the confined mine area ensures long-term stability and efficient land use.

Increased surcharge loads can lead to slope failure by reducing the resisting forces against sliding. Studies show that slopes reach critical failure thresholds (FOS $= 1$) as surcharge intensity increases beyond permissible limits. Surcharge effects are often intertwined with other factors like pore pressure changes, seismic forces, and soil cohesion alterations. This complexity makes it challenging to isolate the impact of surcharge loading on stability. Dynamic surcharges from earthquakes or blasting activities introduce transient stresses that complicate stability analyses. Permanent displacements under these conditions may be underestimated without considering tensile strength cut-off effects. The stability of dump slopes in open-pit mines is governed by a combination of geometric configuration, geotechnical and hydrological properties, and external disturbances. Key geometric parameters include the overall slope angle and dump height; research confirms that greater heights and steeper angles increase instability risks [6].

As per the Coal Mines Regulations (CMR), 2017, the overall slope angle of a spoil dump must not exceed the natural angle of repose (typically $\leq 37.5°$), and individual dump decks should be limited to 30 m in height [27]. Shear strength, a crucial factor

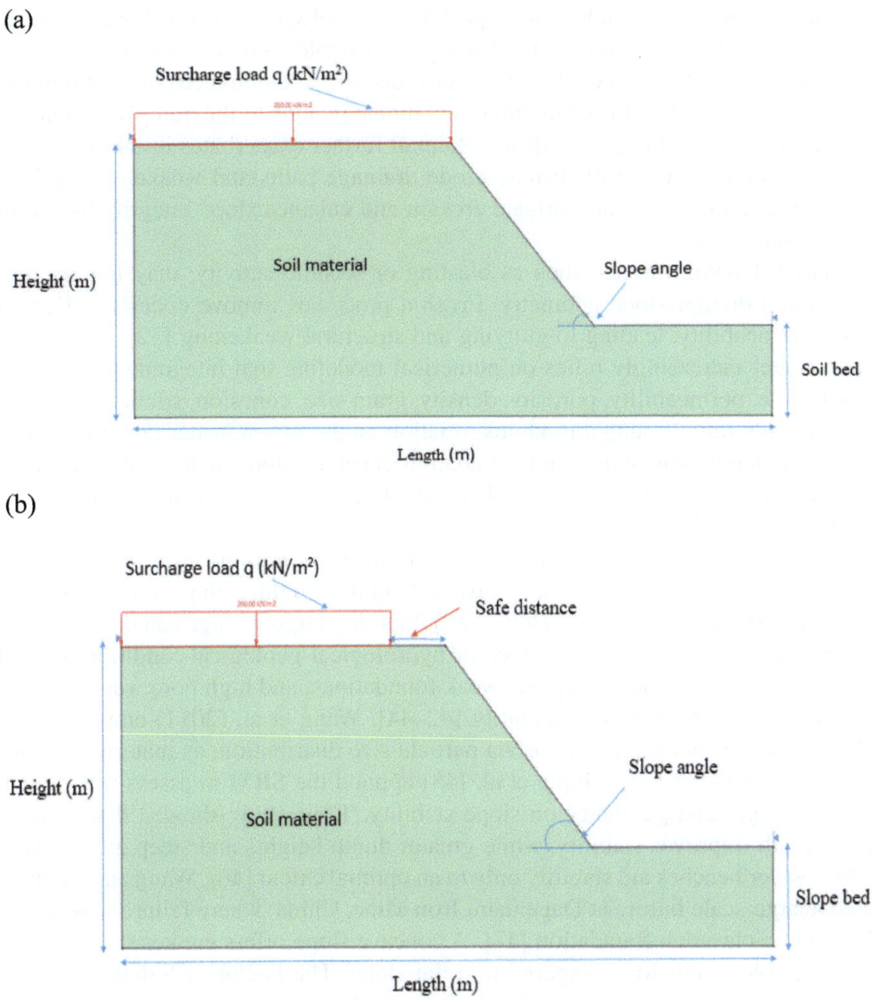

Fig. 2. (a) Slope stability analysis with surcharge load on pit slope. (b) Slope stability analysis with surcharge load and its safe distance from pit crest

in slope stability, depends on cohesion and internal friction angle. Cohesion arises from intermolecular bonding, while the internal friction angle reflects a material's resistance to shear stress. Enhanced shear strength improves slope resistance to failure, thereby increasing safety and operational efficiency [28, 29]. Failure modes depend on geological conditions: slope failure occurs above the toe when the foundation is stronger than the overlying spoil; toe failure results from uniform material strength; and base failure occurs at shallow angles where weak foundations underlie stronger overburden [30].

Hydrological conditions significantly affect stability, though often overlooked. Overburden material, due to its fragmentary and porous nature, allows rapid water infiltration. Without proper drainage, this leads to water accumulation at the dump base, raising hydrostatic pressures and inducing slope failure, as observed in the Jayant Opencast Mine, Singrauli. Porosity and permeability play key roles; interconnected pores facilitate drainage, while disconnected pores retain moisture. Accumulated pore water reduces cohesion and shear strength, while increased moisture adds to the dump's weight, raising shear stress and failure possibilities. Rainfall further exacerbates instability through infiltration and surface runoff. It may erode drainage paths and weaken dump bases, whereas vegetation cover can mitigate erosion and enhance slope integrity by binding loose materials [31].

External dynamic forces, such as blasting or seismic activity, may liquefy dump materials and disrupt slope geometry. Erosion processes remove cohesive fillers and enhance permeability, leading to gullying and structural weakening [32]. Slope stability assessment increasingly relies on numerical modeling that integrates geotechnical parameters i.e. permeability, porosity, density, grain size, cohesion, friction angle, void ratio, poisson's ratio, young's modulus, dilation angle, groundwater conditions, compaction, surcharge, seismicity, and foundation capacity along with geometric factors like bench height, width, number of benches, total height, slope angle, and surface topography [3, 23, 33–38].

Cho and Song (2014) categorized instability triggers into intrinsic (e.g., low shear strength) and external (e.g., rainfall, seismic loading), highlighting failure risks when these interact on already unstable slopes [39]. Stability largely hinges on the mechanical properties of OB materials and site-specific hydrological-geological conditions [40, 41]. Steep benches, high angles of repose, weak foundations, and high pore water pressures are prominent contributors to instability [42–44]. Wang et al. (2011) emphasized the need to assess mechanical behavior via particle size distribution, as material gradation affects slope behavior [45]. Gupta et al. [48] applied the SRM to assess the influence of shear strength and geometry on slope stability. Their study showed that increased shear strength improves stability, while greater dump heights and steep angles reduce the FOS; wider benches aid stability only to an optimal extent [46]. Wang and Chen [49] studied a large-scale failure at Dagenham Iron Mine, China, where failure was initiated at the toe of a clay-rich foundation [47]. A concave slope offers greater stability and is less susceptible to erosion compared to a flat slope. The Factor of Safety varies with elevation, with the lower portion of the slope typically exhibiting a lower FOS, while the upper section demonstrates a relatively higher FOS. To ensure uniform stability across the entire slope, the lower segment should be designed with a steeper inclination than the upper portion. This configuration helps balance the FOS and enhances overall slope stability [48, 49]. Large variations in the dump's mass can weaken the strength of the OB dump [50, 51]. This study evaluates the stability of rock slopes in an opencast limestone mine using the Limit Equilibrium Method in Slope/W, with a focus on optimizing slope angles in geologically complex areas. A steeper 45° slope was successfully implemented in zones with structural discontinuities such as joints, faults, and folds. Geotechnical investigations, including field mapping and laboratory testing of core and block samples, informed the analysis. The study also assessed the impact of surcharge loading from nearby waste dumps on a 110 m deep pit slope. Based on the findings, relocating dumps at least 25 m from the crest and avoiding surface activities near the slope edge are recommended for safer future operations [52].

The foundation soil of the overburden dump may fail to withstand the stress exerted by the overlying material, leading to various types of failures such as surface slides, shallow flow slides, base failures, block translations, and circular failures within the foundation. Analyzing overburden dumps is particularly challenging due to a range of influencing factors, including self-weight, material heterogeneity (ranging from fragmented rock to soil), geotechnical properties, foundation stability, seepage forces, and the presence of tension cracks. Conventional analytical methods often become complex, resulting in dump designs that heavily rely on historical data and experience. Slope stability issues are commonly associated with weak soils and foundations, with failures occurring frequently during the monsoon season. Additionally, structural discontinuities such as strike faults, which align with the slope orientation, exacerbate instability. In some cases, external loading from overburden dumps further intensifies slope failures, worsening overall stability concerns.

3 Methodology

3.1 Field Investigation and Laboratory Test

Comprehensive field investigations and laboratory testing are essential for accurately assessing the geotechnical behavior of mine pit and dumps under various geo-mining conditions. Field investigations involve sampling from different lithologies, in-situ testing, and site characterization to capture the variability in soil and rock properties influenced by regional geology and mining operations. These samples are then subjected to laboratory tests to determine key parameters such as moisture content, specific gravity, grain size distribution, atterberg limits, shear strength, cohesion, internal friction angle, and compaction characteristics. Understanding these parameters is crucial for evaluating slope stability, especially under surcharge loading. The collected data serve as a foundation for physical modeling and numerical simulations, enabling the development of site-specific design guidelines and stability assessments tailored to diverse mining environments.

3.2 Physical Modeling of Slope Stability Using Surcharge Loading

Physical modeling has emerged as a crucial experimental approach to understanding the complex mechanisms governing slope stability under various loading conditions. Among the prominent physical modeling tools, geotechnical centrifuges and shaking tables are extensively utilized for simulating real-world stress conditions and dynamic responses, respectively Physical modeling effectively captures the impact of surcharge loading on slope stability, especially in complex, layered slopes. Applying surcharge at the slope crest in scaled models allows for direct observation of failure initiation and progression. These tests demonstrate that surcharges can trigger both shallow surface slips and deeper failures influenced by subsurface material contrasts. Such insights are essential for understanding deformation behavior under loading and for validating numerical models used in geotechnical design (Fig. 3a & b) [53, 54].

(a)

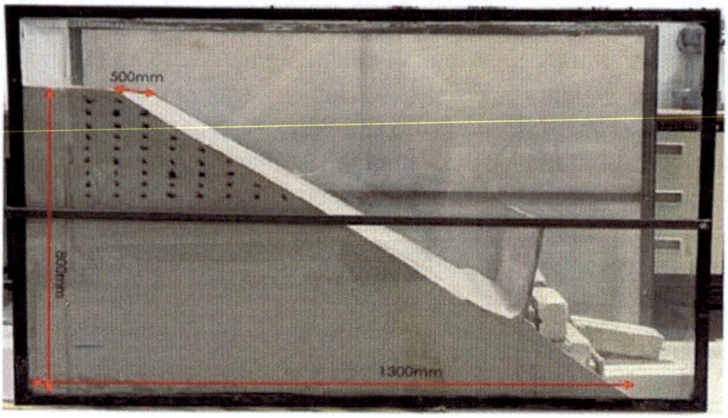

(b)

Fig. 3. (a) Preparation of physical model for slope stability analysis (Ishak et al., 2024). (b) Preparation of physical models for slope stability analysis considering surcharge loading (Ishak et al., 2024)

3.2.1 Geotechnical Centrifuge Modelling

Centrifuge modeling is a powerful technique for investigating slope stability under realistic stress conditions, particularly when analyzing the effects of seismic loading, rainfall infiltration, and vegetation. Liang and Knappett [56] showed that vegetated slopes exhibited increased acceleration near the crest and varying failure modes based on root depth. Reinforced this by demonstrating seismic amplification near the crest and introducing an analytical model for localized safety assessment [55]. Ling et al. (2009) emphasized the role of matric suction loss in rainfall-induced failures [56], while Hung et al. (2020) highlighted that among multiple hazards, seismic shaking caused the most severe slope deformations. These studies collectively demonstrate the versatility of centrifuge modeling in capturing complex, multi-trigger slope failure mechanisms [57].

3.2.2 Shaking Table Modelling

Shake table experiments are a valuable method for analyzing seismic slope stability by simulating earthquake ground motions on scaled models under controlled conditions Wartman et al. (2005) observed deep-seated rotational failures in compacted clayey slopes and validated the Newmark sliding block method through 1 g tests [58]. Similarly, Lin and Wang (2006) demonstrated that sandy slopes remained stable up to 0.4 g but exhibited shallow sliding at higher accelerations, emphasizing the role of proper model construction and boundary conditions. These studies signify the effectiveness of shake table testing in capturing realistic failure mechanisms and supporting analytical validations [59].

3.3 Slope Stability Analysis Techniques

The deterministic approach includes various techniques such as Limit Equilibrium Methods (LEMs), Finite Element Analysis (FEM), and Empirical equations.

3.3.1 Empirical Equation

Empirical equations are designed to estimate the (FOS) for slope stability analysis, offering a simplified alternative to conventional computational methods. The proposed correlations incorporate critical geotechnical parameters, including unit weight (γ), cohesion (c), internal friction angle (φ), slope angle (α), and slope height (H). These equations have been developed for widely recognized limit equilibrium methods such as Spencer's, Morgenstern-Price, Janbu's simplified, and Bishop's simplified techniques, enabling efficient preliminary assessments of slope stability [60].

$$(FS)_{SM} = -0.083\gamma + 0.028C + 2.478\varnothing - 0.864\alpha - 0.110H + 3.625 \quad (8)$$

$$(FS)_{MP} = -0.079\gamma + 0.027C + 2.458\varnothing - 0.927\alpha - 0.106H + 3.569 \quad (9)$$

$$(FS)_{JS} = -0.096\gamma + 0.027C + 2.266\varnothing - 1.013\alpha - 3.432\ \log\ H + 6.313 \quad (10)$$

$$(FS)_{BM} = -0.072\gamma + 0.026C + 2.482\varnothing - 1.103\alpha - 0.10H + 3.507 \quad (11)$$

where, γ = unit weight of soil (kN/m^2), C = Cohesion (kPa), Ø = Friction angle of soil (radians), α = Slope angle (radians), H = Height of the slope (m).

3.3.2 Limit Equilibrium Methods (LEM)

In LEMs, the stability of a soil mass susceptible to sliding under gravitational forces is evaluated by analyzing its equilibrium conditions. Failure is considered to occur when the diving forces or moments exceed the resisting forces or moments as shown in Fig. 4(a & b) [61]. The LEM is a widely used approach for assessing dump slope stability in both natural and engineered terrains [3]. It involves dividing the slope into slices and calculating the FOS by applying the Mohr-Coulomb failure criterion and satisfying force and moment equilibrium. Common LEM techniques include Bishop's Simplified Method (1955), Jambu's Method (1954), Morgenstern-Price Method (1965), Spencer's Method (1967), and the Generalized Limit Equilibrium (GLE) Method, each differing in the treatment of interslice forces [3]. Among these, Bishop's, Janbu's, GLE, and Morgenstern-Price methods are preferred for their efficiency in FOS estimation [62, 63].

$$F = \frac{\tau f}{\tau} = \left\{C^1 + \left(\sigma^1 \tan \theta^1/\tau\right)\right\} \tag{1}$$

where 'τ' represents the mobilized shear stress, τf denotes the shear strength, 'C^{1}' corresponds to cohesion, 'θ^1' signifies the friction angle, and 'σ^1' refers to the normal stress generated at the sliding surface.

Bishop's Simplified Method (BSM)

The Bishop's Simplified Method is widely employed for assessing slope stability involving circular slip surfaces. It considers interslice normal forces but neglects interslice shear forces. The method is based on satisfying vertical force equilibrium to calculate the effective normal force at the base of each slice, as represented by Eqs. (2) and (3) [64].

$$N^1 = \left(\frac{1}{m\alpha} / \Sigma \left(W - \frac{\acute{c}sin\alpha}{F} \text{-} \mu l cos\alpha\right)\right) \tag{2}$$

$$\text{Where, } m\alpha = \left\{ \cos\alpha\left(1 + \frac{tan\alpha tan\acute{\emptyset}}{F}\right)\right\}$$

$$F = [\Sigma\ (\acute{c}L + N^1 tan\acute{\emptyset})] / (\Sigma W sin\alpha)] \tag{3}$$

'N^1' represents the effective force acting on the base, while 'W' denotes the weight of the slice. '$\acute{C}$' corresponds to cohesion, and '$\acute{\emptyset}$' refers to the base friction angle. 'μ' signifies the pore pressure, 'l' represents the slice length, and 'α' indicates the base inclination angle of the slice. Lastly, 'F' stands for the factor of safety.

Fellenius Method

Assumes no interslice forces, providing a simplified solution for slope stability by treating the entire failure mass as a single rigid body. This method calculates the factor of safety based on the moment equilibrium of each slice, where only the weight of the slice and its resistance along the failure surface are considered.

Bishop's Method

This method improves upon the Fellenius method by incorporating both vertical force equilibrium and moment equilibrium, thereby enhancing accuracy. The corresponding force equilibrium is represented in Eq. (4), developed by Bishop (1955). However, it simplifies the analysis by neglecting horizontal interslice forces. It is particularly suited for circular failure surfaces and provides the most accurate results when applied to homogeneous slopes with relatively uniform material properties throughout the depth [65].

$$S = \left[C^1 + (\sigma - u) \tan\phi^1\right] \tag{4}$$

where, S = shear strength, C^1 = Effective cohesion, $\emptyset'$ = Effective angle of shear resistance σ = Total normal stress, μ = Pore water pressure.

Janbu's Method

Janbu's method considers both vertical and horizontal force equilibrium, providing a more comprehensive analysis than simpler techniques. However, it satisfy the moment equilibrium, which limits its accuracy in capturing complex inter-slice interactions. By focusing on force equilibrium, Janbu's method is well suited for analyzing slopes with non-circular failure surfaces, offering greater versatility compared to methods like Bishop's. This is demonstrated in Eq. (5), developed by Janbu (1954) [66, 67].

$$F_{f=} \left\{\frac{[\Sigma(\acute{c}L+(N-\mu)tan\acute{\emptyset})sec\alpha)}{[\Sigma w tan\alpha+\Sigma\Delta E]}\right\} \tag{5}$$

where $\Delta E = E_2 - E_1$ = Net inters slice normal force, N = Normal force, $F_{f=}$ Factor of safety.

Spencer Method

These methods offer a comprehensive and rigorous slope stability analysis by satisfying all three equilibrium conditions: horizontal force equilibrium, vertical force equilibrium, and moment equilibrium. The force equilibrium is represented by Eq. (6), developed by Spencer (1954). This makes the methods significantly more robust and reliable than simplified approaches that may neglect one or more of these conditions [66, 68].

$$F = \frac{\Sigma[c' b sec\alpha + tan\phi'(W cos\alpha - \mu b sec\alpha)]}{\Sigma[W sin\alpha]} \tag{6}$$

where, F = Factor of safety, W = Weight of the soil slice or wedge, α = Inclination angle of the failure surface or slope, c' = Effective cohesion, $\emptyset'$ = Effective angle of internal friction, b = Width of the slice, μ=Pore water pressure.

Morgenstern-Price Method

These methods deliver a comprehensive and rigorous slope stability analysis by satisfying all three equilibrium conditions: horizontal force equilibrium, vertical force equilibrium, and moment equilibrium. The force equilibrium is represented in Eq. (7), developed by Morgenstern and Price (1965). This makes these methods significantly more robust and reliable than simplified approaches that may neglect one or more of these conditions [66, 69]

$$F = \left[\frac{\Sigma(c'l+(N-ul)\tan\phi')\cos\alpha}{\Sigma W \sin\alpha}\right] \tag{7}$$

where, F = Factor of safety, N = Normal force, u = pore water pressure, W = Weight of the slide, l = length of slice, c' = Effective cohesion and $\emptyset'$ = Effective angle of internal friction, α = Inclination angle of the failure surface or slope.

3.3.3 Finite Element Method (FEM)

Finite Element Method is an advanced numerical approach where the slope is discretized into small elements, and a stress-strain relationship is defined. This method naturally

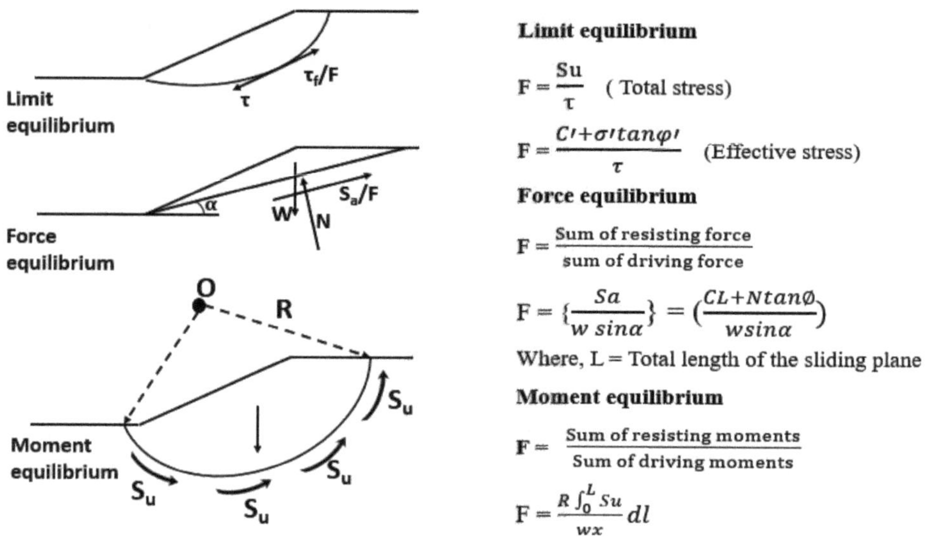

Fig. 4. Limit force equilibrium, force equilibrium, and moment equilibrium (Abramson et al., 2002).

determines the factor of safety and does not require assumptions about failure mechanisms, as in LEM [32]. Beyond traditional limit equilibrium methods, the shear strength reduction method (SSRM) applies a factor to reduce soil cohesion and friction angle until failure, defining the FOS. A continuous plastic zone, non-convergence, and abrupt strain or displacement increases and mark failure. Finite element and finite difference models based on SRM automatically identify the critical slip surface while maintaining equilibrium [11, 12, 70, and]. Meng et al. (2019) enhanced SRM for multiscale analysis by incorporating friction angle, normal cohesion, and tensile cohesion, and Yang et al. (2019) improved soil–rock slope assessments by adjusting Young's modulus and Poisson's ratio during FOS evaluation [37].

Where, The Strength Reduction Factor (SRF) is applied to obtain the factored shear strength parameters$c'f$ and $\emptyset' f$. Equations (8) and (9) are utilized in the analysis to determine the factor of safety.

$$C'f = \frac{C'}{SRF} \tag{8}$$

$$\emptyset'\text{f} = \{\text{arc tan } (\frac{tan\emptyset'}{SRF})\ \} \tag{9}$$

3.3.4 Strength Reduction Method (SRM)

Beyond traditional limit equilibrium methods, the shear strength reduction method (SSRM) applies a factor to reduce soil cohesion and friction angle until failure, defining the FOS. A continuous plastic zone, non-convergence, and abrupt strain or displacement increase mark failure. Finite element and finite difference models based on SRM automatically identify the critical slip surface while maintaining equilibrium. Meng et al.

(2019) enhanced SRM for multiscale analysis by incorporating friction angle, normal cohesion, and tensile cohesion [62], and Yang et al. (2019) improved soil–rock slope assessments by adjusting Young's modulus and Poisson's ratio during FOS evaluation [71].

Where, The Strength Reduction Factor (SRF) is applied to obtain the factored shear strength parameters $c'f$ and $\emptyset'f$. Equations (10) and (11) are utilized in the analysis to determine the factor of safety, as referenced in [72].

$$C'f = \frac{C'}{SRF} \tag{10}$$

$$\emptyset'\mathrm{f} = \{\mathrm{arc\ tan}\ (\frac{tan\emptyset\prime}{SRF})\ \} \tag{11}$$

4 Conclusion

This review paper identifies a key research gap regarding the effect of surcharge loading on slope stability in mining engineering, particularly for overburden dumps in open-pit mines under varied geo-mining conditions. While civil engineering studies have addressed surcharge impacts on structures like embankments, similar research in mining contexts remains limited. Findings highlight that improperly placed or excessive surcharge loads can significantly disrupt dump slope stability, safety problems, triggering failures such as rotational sliding and settlement. The phenomenon of failure mechanism and type need to understand for optimal design in open pit mines and critical influencing factors include slope geometry, bench configuration, surcharge intensity and position, cohesion and internal friction angle of materials, and groundwater conditions of the area. To address this, the proposed research review highlights geotechnical characterization, lab testing, physical modeling (centrifuge and shake table), empirical analysis, and numerical simulations using FEM and LEM tools. The research area is to quantify surcharge effects under realistic mining conditions and develop practical design guidelines to guide safe dumping practices. The outcomes will support improved dump design, enhanced slope stability, safer and sustainable open-pit mining operations globally.

Acknowledgments. I extend my heartfelt appreciation to the NITK Surathkal, and the Science and Engineering Research Board (SERB) project for their support and resources, which have been vital to this research.

References

1. Qudrat-Ullah, H., Panthallor, P.N.: Operational Sustainability in the Mining Industry: The Case of Large-Scale Open-Pit Mining (LSOPM) Operations. Springer Nature (2020)
2. Poulsen, B., Khanal, M., Rao, A.M., Adhikary, D., Balusu, R.: Mine overburden dump failure: a case study. Geotech. Geol. Eng. **32**(2), 297–309 (2014)

3. Agboola, O., Babatunde, D.E., Fayomi, O.S.I., Sadiku, R., Popoola, P., Moropeng, L., Yahaya, A., Mamudu, A.: A review on the impact of mining operation: monitoring, assessment and management. Results Eng. **8**, 100181 (2020)
4. Blight, G.: Slope failures in municipal solid waste dumps and landfills: a review. Waste Manag. Res. **26**(5), 448–463 (2008)
5. Gupta, G., Sharma, S.K., Singh, G.S.P., Kishore, N.: Numerical modelling-based stability analysis of waste dump slope structures in open-pit mines-a review. J. Inst. Eng. India Series D. **102**(2), 589–601 (2021)
6. Eberhardt, E., Stead, D., Coggan, J.S.: Numerical analysis of initiation and progressive failure in natural rock slopes—the 1991 Randa rockslide. Int. J. Rock Mech. Min. Sci. **41**(1), 69–87 (2004)
7. Griffiths, D.V., Lane, P.A.: Slope stability analysis by finite elements. Geotechnique. **49**(3), 387–403 (1999)
8. Jing, L., Hudson, J.A.: Numerical methods in rock mechanics. Int. J. Rock Mech. Min. Sci. **39**(4), 409–427 (2002)
9. Tang, C.A., Yang, W.T., Fu, Y.F., Xu, X.H.: A new approach to numerical method of modelling geological processes and rock engineering problems—continuum to discontinuum and linearity to nonlinearity. Eng. Geol. **49**(3–4), 207–214 (1998)
10. Teixeira, M.B., Nogueira, C.D.L., Oliveira Filho, W.L.D.: Numerical simulation of hillside mine waste dump construction. REM Rev. Esc. Minas. **65**, 553–559 (2012)
11. Kainthola, A., Verma, D., Gupte, S.S., Singh, T.N.: A coal mine dump stability analysis—a case study. Geomaterials. **1**(01), 1 (2011)
12. Steiakakis, E., Kavouridis, K., Monopolis, D.: Large scale failure of the external waste dump at the "South Field" lignite mine, Northern Greece. Eng. Geol. **104**(3–4), 269–279 (2009)
13. Stead, D., Wolter, A.: A critical review of rock slope failure mechanisms – the importance of structural geology. J. Struct. Geol. **74**, 1–23 (2015)
14. Behera, P.K., Sarkar, K., Singh, A.K., Verma, A.K., Singh, T.N.: Dump slope stability analysis–a case study. J. Geol. Soc. India. **88**, 725–735 (2016)
15. Sharma, S., Roy, I.: Slope failure of waste rock dump at Jayant opencast mine, India: a case study. Int. J. Appl. Eng. Res. **10**(13), 33006–33012 (2015)
16. Manna, B., Rawat, S., Zodinpuii, R., Sharma, K.G.: Effect of surcharge load on stability of slopes-testing and analysis. Electron. J. Geotech. Eng. **19**, 3397–3410 (2014)
17. Satyanarayana, I., Budi, G.: Analytical and numerical approach for analysis of factors affecting pit slope stability at Dorli OCP-II, India. J. Min. Sci. **55**, 376–382 (2019)
18. Imani Kalehsar, R., Khodaei, M., Dehghan, A.N., Najafi, N.: Numerical modeling of effect of surcharge load on the stability of nailed soil slopes. Model. Earth Syst. Environ. **1-12** (2021)
19. Dan, M.F.M., Alias, M.A.N.: Effect of surcharges to the stability of mechanically stabilized earth wall using finite element method: Back analysis study case in Senai City Airport. J. Sustain. Undergr. Explor. **2**(1), 20–27 (2022)
20. Ishak, A.S.A., Jelani, J., Wong, S.Y., Suif, Z., Mazuki, A.L.A.: Experimental and numerical study on the behaviour of silica sand slope model subjected to surcharge load. Jurnal Teknologi (Sci. Eng.). **86**(1), 165–173 (2024)
21. Yang, Y., Sun, G., Zheng, H., Qi, Y.: Investigation of the sequential excavation of a soil-rock-mixture slope using the numerical manifold method. Eng. Geol. **256**, 93–109 (2019)
22. Shukla, S.K., Khandelwal, S., Verma, V.N., Sivakugan, N.: Effect of surcharge on the stability of anchored rock slope with water filled tension crack under seismic loading condition. Geotech. Geol. Eng. **27**, 529–538 (2009)
23. Xu, C., Xue, L., Cui, Y., Zhai, M.: Numerical analysis of surcharge effect on stability and interaction mechanism of slope-pile-footing system. J. Earth Sci. **35**(3), 955–969 (2024)

24. Sazzad, M.M., Haque, M.F.: Effect of surcharge on the stability of slope in a homogeneous soil by FEM. In: 2nd International Conference on Advances in Civil Engineering, vol. 1, pp. 314–318 (2014)
25. Chaturvedi, A., Singh, G.S.P.: Influence of interface and induced seismicity on overburden dump slope stability. Indian Academy of Sciences. (2022)
26. Fuentes, W., Gil, M., Duque, J.: Dynamic simulation of the sudden settlement of a mine waste dump under earthquake loading. Int. J. Min. Reclam. Environ. **33**(6), 425–443 (2019)
27. Narwal, S., Mandal, J., Gupte, S.S.: Optimization of overburden dump capacity using limit equilibrium and finite element techniques. World. **8**, 1079–1089 (2017)
28. Coal Mine Regulations. Ministry of labour and employment notification Coal Mines Regulation 2017.pdf (dgms.net), (2017).
29. Pradhan, S.P., Vishal, V., Singh, T.N.: Landslides: Theory, Practice and Modelling. Springer International Publishing (2019)
30. Rai, R., Kalita, S., Gupta, T., Shrivastva, B.: Sensitivity analysis of internal dragline dump stability: finite element analysis. Geotech. Geol. Eng. **30**, 1397–1404 (2012)
31. Bao, Y., Han, X., Chen, J., Zhang, W., Zhan, J., Sun, X., Chen, M.: Numerical assessment of failure potential of a large mine waste dump in Panzhihua City, China. Eng. Geol. **253**, 171–183 (2019)
32. Li, Q.M., Yuan, H.N., Zhong, M.H.: Safety assessment of waste rock dump built on existing tailings ponds. J. Cent. South Univ. **22**(7), 2707–2718 (2015)
33. Yang, Y., Sun, G., Zheng, H.: Stability analysis of soil-rock-mixture slopes using the numerical manifold method. Eng. Anal. Bound. Elem. **109**, 153–160 (2019)
34. Yubonchit, S., Chinkulkijniwat, A., Horpibulsuk, S., Jothityangkoon, C., Arulrajah, A., Suddeepong, A.: Influence factors involving rainfall-induced shallow slope failure: numerical study. Int. J. Geomech. **17**(7), 04016158 (2017)
35. Zou, P., Zhao, X., Meng, Z., Li, A., Liu, Z., Hu, W.: Sample rocks tests and slope stability analysis of a mine waste dump. Adv. Civil Eng. **1**, 6835709 (2018)
36. Cho, Y.C., Song, Y.S.: Deformation measurements and a stability analysis of the slope at a coal mine waste dump. Ecol. Eng. **68**, 189–199 (2014)
37. Amarsaikhan, T., Shimada, H., Wahyudi, S., Sasaoka, T., Hamanaka, A.: Optimization of dump bench configuration to improve waste dump capacity of Narynsukhait open pit coal mine. Int. J. Geosci. **9**(6), 379–396 (2018)
38. Gens, A., Alonso, E.E.: Aznalcóllar dam failure. Part 2: Stability conditions and failure mechanism. Géotechnique. **56**(3), 185–201 (2006)
39. Han, L., Shu, J., Cai, Q., Jing, H., Tian, H.: Mechanical characteristics of dip basement effects on dump stability in the Shengli open pit mine in Inner Mongolia, China. Arab. J. Geosci. **9**, 1–13 (2016)
40. Igwe, O., Chukwu, C.: Slope stability analysis of mine waste dumps at a mine site in Southeastern Nigeria. Bull. Eng. Geol. Environ. **78**, 2503–2517 (2019)
41. Peng, C., Ji, D., Zhao, L., Ren, F.H.: Study on limit height and its stability of open-pit dump based on basement bearing mechanism. Appl. Mech. Mater. **405**, 177–181 (2013)
42. Zhang, D., Inoue, N., Sasaoka, T., Shimada, H., Hamanaka, A., Matsui, K.: Study on formation mechanism of dumping piles on dumping area stability. Open J. Geol. **4**(04), 161 (2014)
43. Jahanfar, A., Dubey, B., Gharabaghi, B., Movahed, S.B.: Landfill failure mobility analysis: a probabilistic approach. Int. J. Environ. Chem. Ecol. Geol. Geophys. Eng. **10**, 476–484 (2016)
44. Lavigne, F., Wassmer, P., Gomez, C., Davies, T.A., Hadmoko, D., Iskandarsyah, T.Y.W., Pratomo, I.: The 21 February 2005, catastrophic waste avalanche at Leuwigajah dumpsite, Bandung, Indonesia. Geoenviron. Disast. **1**, 1–12 (2014)
45. Su, W.X., Miller, H.D.S.: Waste pile stability and debris flow formation. In: ARMA US Rock Mechanics/Geomechanics Symposium, ARMA-95 (1995)

46. Sun, S., Fan, C., Li, Z., Lu, H., Kan, S.: Stress fields evolution law in the piling-up process of waste dump of huge soft hill base. Jinshu Kuangshan/Metal Mine. **8**, 36–38 (2010)
47. Wang, G., Kong, X., Gu, Y., Yang, C.: Research on slope stability analysis of super-high dumping site based on cellular automation. Proc. Eng. **12**, 248–253 (2011)
48. Gupta, G., Sharma, S.K., Singh, G.S.P.: Assessment of dump structure stability via parametric study using numerical modeling. Proc. Recent Chall. Min. Indus. RCMI. **2018**, 103–112 (2018)
49. Wang, J., Chen, C.: Stability analysis of slope at a disused waste dump by two-wedge model. Int. J. Min. Reclam. Environ. **31**(8), 575–588 (2017)
50. Gray, D.H.: Influence of slope morphology on the stability of earthen slopes. In: Geo-Congress 2013: Stability and Performance of Slopes and Embankments III, pp. 1895–1904 (2013)
51. Geete, S.S., Singh, H.K., Singh, T.N., Dahale, P.P.: Stability analysis of overburden dump slope of coal/mineral mines: a review. J. Mines Met. Fuels, 1215–1223 (2023)
52. Li, M., Zhang, J., Sun, K., Zhang, S.: Influence of lateral loading on compaction characteristics of crushed waste rock used for backfilling. Minerals. **8**(12), 552 (2018)
53. Reddy, S.K.: Stability assessment and optimal excavated Design of a Rock Slope in an opencast limestone mine. J. Mines Metals Fuels. **71**(2) (2023)
54. Ni, P., Wang, S., Zhang, S., Mei, L.: Response of heterogeneous slopes to increased surcharge load. Comput. Geotech. **78**, 99–109 (2016)
55. Haundi, T., Okonta, F.: A systematic review of physical modelling techniques, developments and applications in slope stability analyses. Indian Geotech. J., 1–23 (2024)
56. Liang, T., Knappett, J.A.: Centrifuge modelling of the influence of slope height on the seismic performance of rooted slopes. Géotechnique. **67**(10), 855–869 (2017)
57. Ling, H.I., Wu, M.H., Leshchinsky, D., Leshchinsky, B.: Centrifuge modeling of slope instability. J. Geotech. Geoenviron. **135**(6), 758–767 (2009)
58. Hung, W.Y., Tran, M.C., Yeh, F.H., Lu, C.W., Ge, L.: Centrifuge modeling of failure behaviors of sandy slopes caused by gravity, rainfall, and base shaking. Eng. Geol. **271**, 105609 (2020)
59. Wartman, J., Seed, R.B., Bray, J.D.: Shaking table modeling of seismically induced deformations in slopes. J. Geotech. Geoenviron. **131**(5), 610–622 (2005)
60. Lin, M.L., Wang, K.L.: Seismic slope behavior in a large-scale shaking table model test. Eng. Geol. **86**(2–3), 118–133 (2006)
61. Moawwez, M.A., Wang, J.P., Hussain, M.A.: Development of empirical correlations for limit equilibrium methods of slope stability analysis. Arab. J. Geosci. **14**(19) (2020)
62. Kaur, A., Sharma, R.K.: Slope stability analysis techniques: a review. Int. J. Eng. Appl. Sci. Technol. **1**(4), 52–57 (2016)
63. Sangeeta, & Maheshwari, B.K.: Slope Stability Evaluation by Different Limit Equilibrium Methods. (2016).
64. Bishop, A.W.: The use of the slip circle in the stability analysis of slopes. Geotechnique. **5**(1), 7–17 (1955)
65. Janbu, N.: Stability Analysis of Slopes with Dimensionless Parameters (1954).
66. Spencer, E.: A method of analysis of the stability of embankments assuming parallel inter-slice forces. Geotechnique. **17**(1), 11–26 (1967)
67. Morgenstern, N.R.U., Price, V.E.: The analysis of the stability of general slip surfaces. Géotechnique. **15**(1), 79–93 (1965)
68. Singh, T.N., Chaulya, S.K.: External dumping of overburden in opencast mine. Ind. J. Eng. **22**(1 & 2), 65–73 (1992)
69. Meng, Q.X., Wang, H.L., Xu, W.Y., Cai, M., Xu, J., Zhang, Q.: Multiscale strength reduction method for heterogeneous slope using hierarchical FEM/DEM modeling. Comput. Geotech. **115**, 103164 (2019)
70. Pantelidis, L., Griffiths, D.V.: Stability assessment of slopes using different factoring strategies. J. Geotech. Geoenviron. **138**(9), 1158–1160 (2012)

71. Abramson, L.W., Lee, T.S., Sharma, S., Boyce, G.M.: Slope Stability and Stabilization Methods. John Wiley & Sons (2001)
72. Xiao, S., Dai, T., Li, S.: Review and comparative analysis of factor of safety definitions in slope stability. Geotech. Geol. Eng. **42**(6), 4263–4283 (2024)

Mineralogical Study on Sedimentary Rock Samples from the Ramagundam and Manuguru Areas, Godavari Basin, Telangana State

P. Varalakshmi, S. Kumar Reddy(✉), and Ch. S. N. Murthy

National Institute of Technology Karnataka Surathkal, Mangalore, Karnataka 575025, India
skreddy@nitk.edu.in

Abstract. The purpose of this study is to examine the mineralogical properties of abundance available sandstone rock samples from the Ramagundam and Manuguru areas of the Telangana state of the Godavari Basin formations in order to comprehend the geological processes that have shaped their development. In these areas, sandstone is a crucial reservoir rock that is both geologically and economically significant for mineral excavations. The study compares and determines the cementing materials, grain size distribution, and mineral elements in the sandstones from each location using in-depth petrographic examination using Scanning Electron Microscopy and Energy Dispersive X-ray Analysis. A mature sedimentary environment with significant weathering and reworking is indicated by the results, which show that the Ramagundam area sandstones are primarily quartz-rich with trace amounts of feldspar and clay minerals. However, the sandstones from the Godavari Basin have a higher concentration of feldspar and lithic pieces, indicating a less developed, potentially more proximal depositional context with fast sedimentation. As an example of a multi-faceted depositional history affected the Manuguru area sandstones, however, on the other hand, are built by both fluvial and marine processes which have a complicated mineralogical composition and significant concentration of carbonate, chlorite minerals, kaolinite, and chlorite minerals. This is through enhancing awareness about these areas which by extension helps in the acquisition of sedimentary rocks, reservoir potential, mineralogical comparison provides new valuable information on past, and the reservoir potential. Data of future geological and geophysical research and resource management into the future relate to the mining field applications.

Keywords: Sedimentary Rocks · Sandstone · Mineralogical properties · Petrography · EDX Analysis

1 Introduction

In areas such as the Godavari Basin, where sedimentary rocks are rich in minerals, sustainable mining and resource management is vital. It is also possible to mineralogically compare the Manuguru and Ramagundam sedimentary rocks to understand the composition, quality, and environmental impacts of exploiting the resource in a better manner.

H. P. Dasari et al. (Eds.): NITK-CREST 2025, PTT, pp. 76–87, 2026.
https://doi.org/10.1007/978-981-95-3218-6_6

Sustainable mining techniques can be devised to reduce any effects on the environment by studying mineral distribution, geo-chemical characteristics, and history of the deposits to make sure no harm is caused to nature. By utilizing environmentally friendly methods of extracting the resources, effective waste disposal strategies and reversing the damage done over time, mining companies can attain the desired long-term returns on investments without compromising the integrity of the ecosystem. Mineralogical research of sediment rocks often focuses on the distribution and composition of clay minerals such as kaolinite, Illite and Smectite. Based on this study, it can be said that clay minerals play significant roles in recording the climate, source of sediments as well as history of diagenesis in the kind of paleo-environment. The variability in depositional conditions during the fluvial to marine transition is also unveiled by recent research that places emphasis on the usage of X-ray diffraction (XRD) to define clay mineral assemblages [1].

In the study of sandstone mineralogical composition i.e. percent quartz, feldspar and lithic pieces indicate the tectonic setting and the source of the sediment (provenance). According to the studies published in Sedimentary Geology, heavy minerals such as zircon and tourmaline are significant in re-constructing recovery of sediment journeys and source regions. Mineralogical maturity of sandstones is also related to a level of chemical erosion at the source location [2].

The mineralogy of the carbonate sedimentary rocks, e.g. dolostones, and limestones is often analyzed in order to understand diagenetic processes like dolomitization. The role of early diagenetic processes, where the microbial activity influences the formation of such minerals as aragonite and high-Mg calcite, is the subject of research in marine and petroleum geology. These investigations are necessary to explore the carbonate reservoirs and their possibility of trap of hydrocarbons [3].

These tremendous mineral assemblages of sedimentary rocks of the African basins carry the complicated tectonic and depositional history of the continent. Emphasis on the effects of the Precambrian basement rocks on the mineralogy of overlying sediments has allowed them to use mineralogical analysis to find traces of how sedimentary basins evolved with time. There has been an association of authigenic clays and heavy minerals with certain tectonic processes and climatic changes [4].

The mineralogical composition of the sedimentary rocks is important in understanding the mode of bio-geochemical cycles in ancient ecosystems. The iron-rich sedimentary rocks such as banded iron formations (BIFs) have been used to study the mineralogy of its iron content to determine the redox state of the early earth. Precambrian Oxygenation of the atmosphere and oceans of the earth can be determined by study of iron oxides and silicates in these formations [5].

The interaction of organic matter with the inorganic minerals is the most basic line of research when carrying out mineralogical study on the sedimentary sequences that have coal. The results of the interactions which take place between these minerals and organic materials in the course of diagenesis have an immense impact on the ensuing geotechnical characteristics of coal [6]. Critical understanding on the diagenetic and depositional processes in marine environments has been acquired with relation to mineralogical study of shales. The evolution of the organic-rich shales is also conditioned by the diagenetic alteration of these minerals during burial in particular [7].

Siliceous sedimentary rocks such as Cherts have been valuable in shedding more light about the ancient ocean environment. The findings that the accumulation of siliceous organisms contributed to the sedimentary record in the condition of high biological productivity, and, at the same period, Cherts formed [8]. The distinct mineralogy of evaporitic sedimentary rocks, which includes anhydrite, gypsum, and halite, makes them particularly interesting. Chemical geology studies the formation circumstances of these minerals, especially in semi-arid and arid regions. Evaporite mineralogical study has shed light on the geochemical development of confined marine basins and ancient climates when saline fluids were concentrated due to high evaporation rates [9].

The varied mineralogical makeup of sedimentary rocks in Asian basins reflects the region's tectonic and climatic past. The mineral composition of clastic and carbonate rocks, with an emphasis on the part Himalayan orogeny plays in sediment movement and deposition. These rocks' clay minerals, carbonates, and heavy minerals can be studied to uncover hints about previous tectonic processes and environmental circumstances [10].

Sedimentary rocks' mineralogical composition, particularly the occurrence of refractory minerals like zircon, rutile, and garnet often provides clues about their origin and tectonic location. Detrital mineral analysis has been used in Lithos research to recreate paleotectonic settings and trace sedimentary sources. Particularly in continental margin contexts, the investigation of these minerals in sedimentary basins has uncovered intricate histories of long-distance movement and sediment recycling [11].

The regions varied geological past, which includes the Amazonian craton and Andean orogeny, has an impact on the mineralogy of sedimentary rocks in South American basins. The significance of clay mineral assemblages and heavy minerals. In the study of sediments, clay mineral assemblages and heavy minerals are important in historical epidemiology and geology. The sedimentary processes hindering in these basins are underlined. Knowledge about the tectonic development and sedimentary environments of South America has been made can be done through the mineralogical work on sandstones and shales [12].

The digenesis which alters the original composition and texture of the sediments regularly the subject of mineralogical investigation of sedimentary rocks. Reconstructing the post depositional history of sedimentary basins, such as fluid flow changes an examination of these studies is dependent on these investigations [13].

In special cases, especially regarding the rates of subsidence and sedimentations, the mineralogical sedimentary rocks makeup will provide significant knowledge concerning basin evolution. In in a bid to infer depositional conditions and diagenetic processes, Basin Research has been focused on mineral content of sandstones and mudstones. The chronology and the basin formation techniques, such as the effect of tectonics and climate, can be summarized the analysis of authigenic cements and detrital minerals in these rocks has been cleared up [14].

The importance of mineralogical studies in the understanding of the history of sedimentary rocks in complex depositional systems has been highlighted in previous studies into the sedimentary record. When it comes to chronicle the history of diagenesis, deposit and the movement of the sediments, interactivity in terms of the detrital and the authigenic minerals has been primary focus. Mineralogical composition of these rocks is critical to the interpretation of paleo-environments and tectonic settings especially in basins containing complex stratigraphy [15].

The growing demand for natural resources calls for striking a balance between environmental sustainability and economic growth, especially in areas where sedimentary rocks are essential to resource exploitation, such as Ramagundam, the Godavari Basin, and Manuguru. Degradation of the ecosystem, habitat devastation, and resource depletion can result from unsustainable mining methods. In-depth mineralogical research on these sandstone formations is lacking, which makes it more difficult to create ethical mining practices and effectively manage resources. A thorough understanding of these sedimentary rocks mineral makeup, depositional history, and reservoir potential is essential for applying sustainable mining practices that optimize resource efficiency while reducing ecological effect [16]. By carrying out a comparative mineralogical examination, this study seeks to address these issues and provide crucial information to enable long-term resource management and ecologically conscious extraction.

2 Collection of Sedimentary Rocks

In Goa's iron ore mining regions, slope stability is a critical concern, particularly given the state's significant annual rainfall ranging from 2,500 mm to 3,400 mm. The interaction between intense precipitation and the geotechnical properties of local soils can lead to slope failures, posing risks to both mining operations and environmental safety. Previous studies have shown that prolonged rainfall events with high cumulative volumes are among the most critical factors influencing significant variations in pore-water pressure within slopes. This is primarily due to the extended duration allowing water to infiltrate deeper into the soil profile, gradually saturating previously unsaturated zones. Among the various factors, extended rainfall events with substantial cumulative precipitation most significantly alter pore-water pressure in slopes. As infiltration progresses, matric suction decreases, reducing effective stress and potentially destabilizing slopes [3]

2.1 Preparation of Samples for Optical Microscope

Sedimentary rocks from Ramagundam, the Godavari basin, and Manuguru were used to prepare the samples for optical microscope examination. Initially, an appropriate adhesive was used to mount the samples on the glass slides. After that, the mounted samples are polished and ground to a thickness of around 30 μm. The polished surface is covered with a coverslip, and the slide is identified by its label. Through the creation of thin slices, the mineral content, texture, and sedimentary features of the rock may be closely examined under a microscope.

3 Results and Discussion

3.1 Microstructural Analysis of Sedimentary Rocks Collected from Ramagundam

Microstructural analysis of sedimentary rocks from Ramagundam, involves examining the rock's internal structure under a microscope as shown in Fig. 1. Key features analyzed include grain size, sorting, shape, roundness, fabric, cementation, sedimentary structures, and fossils as shown in Fig. 2. The EDX analysis of sedimentary rocks from Ramagundam revealed the elemental composition, highlighting the presence of silica, alumina, and trace minerals as represented in Fig. 3. The analysis aids in understanding the mineralogical characteristics and depositional environment of the region. By studying these features, geologists can infer the depositional environment, diagenetic history, and potential reservoir properties of the rocks. The study of these features provides valuable insights into the depositional environment, diagenetic processes, and reservoir properties of the rocks. The fine-grained, well-sorted rocks with rounded grains and abundant fossils may indicate a calm, shallow marine environment, while coarser-grained, poorly sorted rocks with angular grains and limited fossil might suggest a higher energy, fluvial setting.

Fig. 1. Microstructural analysis of sedimentary rocks from Ramagundam

Fig. 2. Spectrum analysis of sedimentary rocks from Ramagundam

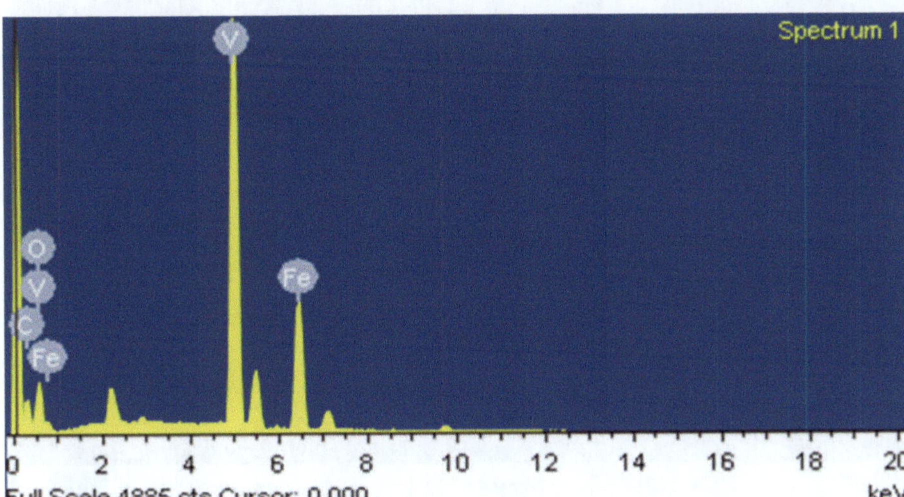

Fig. 3. EDX analysis of sedimentary rocks from Ramagundam

3.2 Microstructural Analysis of Sedimentary Rocks Collected from Godavari Basin

The composition of sedimentary rocks collected from the Godavari basin is primarily determined by the source rocks and depositional processes as shown in Figs. 4 and 5. Common mineral components include quartz, feldspar, clay minerals, calcite, and dolomite is represented in Fig. 6. The relative abundance of these minerals can vary depending on the specific rock type and its formation history. The sandstones are typically composed of quartz and feldspar grains, while shales are rich in clay minerals. The presence of accessory minerals, such as mica, zircon, and garnet, can also provide valuable information about the source rocks and depositional environment.

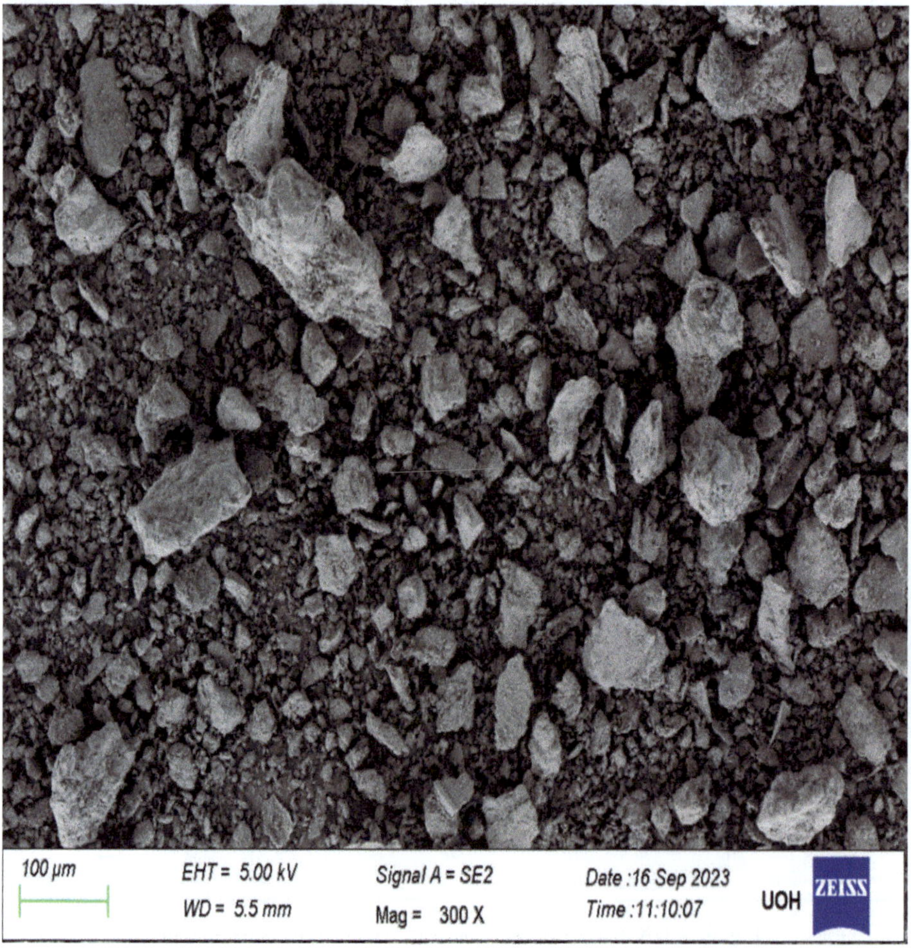

Fig. 4. Microstructural analysis of sedimentary rocks from Godavari basin

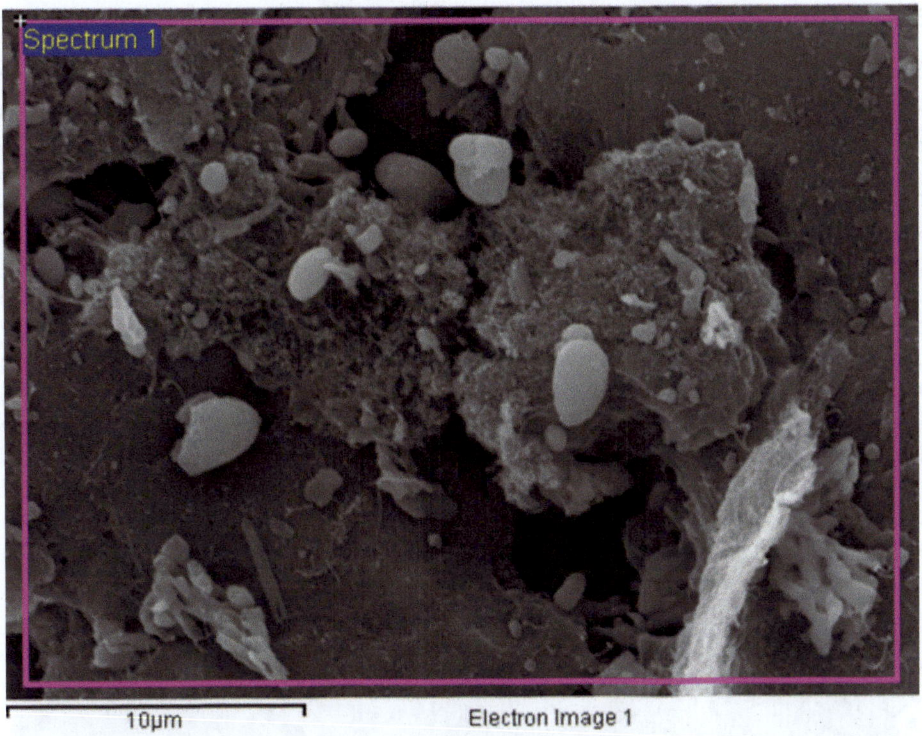

Fig. 5. Spectrum analysis of sedimentary rocks from Godavari basin.

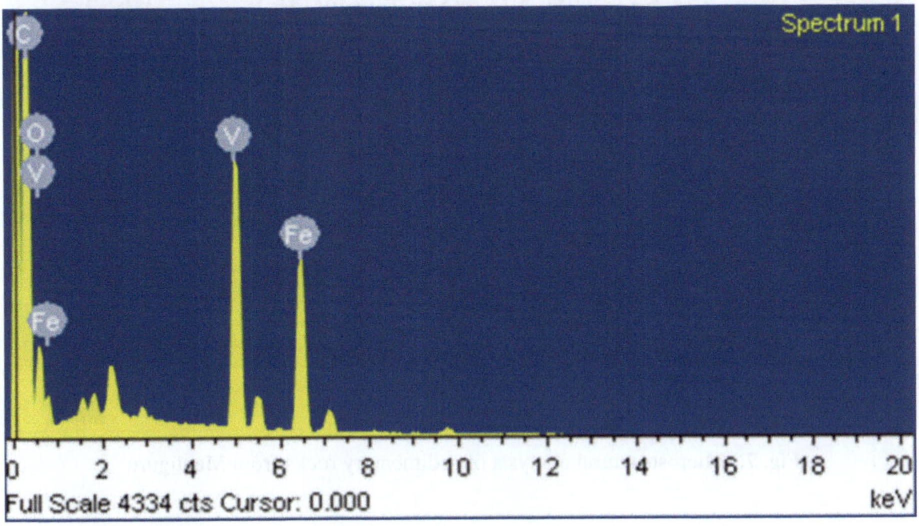

Fig. 6. EDX analysis of sedimentary rocks from Godavari basin

3.3 Microstructural Analysis of Sedimentary Rocks Collected from Manuguru

Microscopic analysis of sedimentary rocks from Manuguru, involves examining the distribution and association of different elements within the rock matrix. By analyzing the elemental composition at a microscopic scale, can clearly observe mineral phases, determine the spatial distribution of elements, and relationships between different components within the rock as shown in Figs. 7 and 8. This information is crucial for understanding the formation conditions, diagenetic processes, and potential economic value of the sedimentary rocks. Figure 9 shows the EDX analysis of sedimentary rocks collected from Manuguru reveals their elemental composition, highlighting the dominance of silica (Si) and oxygen (O), indicating quartz-rich formations typical of sedimentary deposits. The presence of aluminum (Al) and potassium (K) suggests clay minerals like kaolinite or illite, pointing to weathering and low-energy depositional environments. Peaks of calcium (Ca) indicate carbonate minerals such as calcite, hinting at possible marine or lacustrine influences. Additionally, traces of iron (Fe) imply the presence of iron oxides, contributing to the rock's coloration and suggesting oxidative depositional conditions. Overall, the EDX results provide insights into the mineralogical composition and geological history of the Manuguru sedimentary formations.

Fig. 7. Microstructural analysis of sedimentary rocks from Manuguru

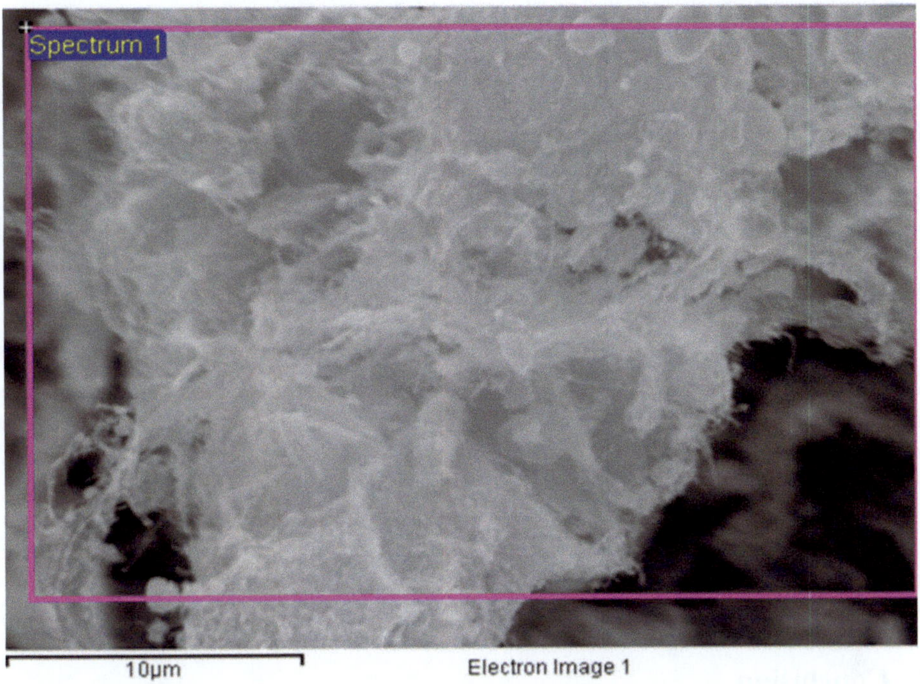

Fig. 8. Spectrum analysis of sedimentary rocks from Manuguru.

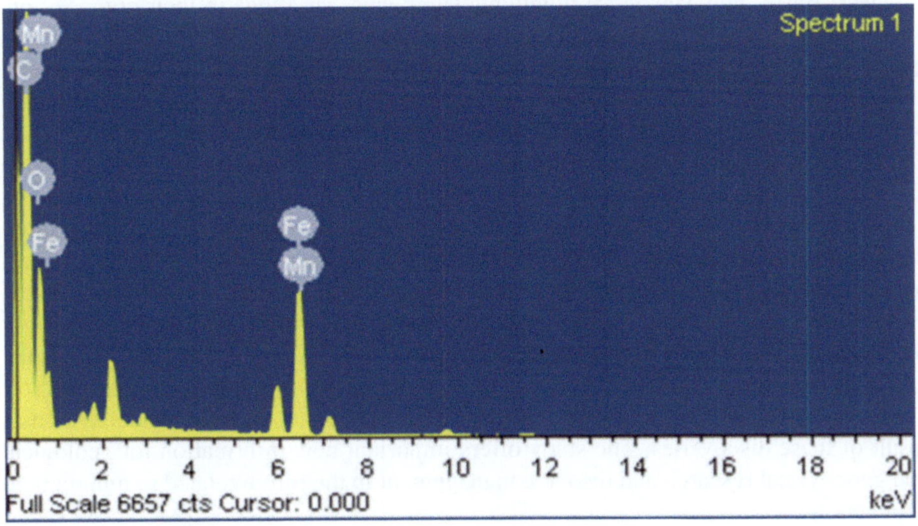

Fig. 9. EDX analysis of sedimentary rocks from Manuguru

3.4 Microstructural Comparative Analysis of Sedimentary Rocks from Three Region

Sedimentary rocks from Ramagundam, the Godavari Basin, and Manuguru exhibit clear differences in grain properties, mineral composition, and depositional conditions when analyzed microstructurally. Coarser, poorly sorted samples point to high-energy river settings, while fine-grained, well-sorted structures from Ramagundam with rounded grains and a wealth of fossils indicate a calm, shallow marine environment. Quartz, feldspar, clay minerals, calcite, and dolomite make up the majority of the rocks in the Godavari Basin; differences in mineral abundance represent various depositional histories.

This region's sandstones are rich in quartz and feldspar, while the shales have a lot of clay minerals. Source rock influences can be inferred from accessory minerals like mica and zircon. On the other hand, microscopic analysis of Manuguru rocks reveals a composition dominated by quartz with significant levels of calcium, iron, potassium, and aluminum, suggesting the existence of carbonate and clay minerals. Weathering effects, oxidative depositional conditions, and possible marine or lacustrine influences are highlighted by the EDX study. Collectively, these discoveries provide important new information about the mineralogical makeup, diagenetic history, and reservoir potential of sedimentary rocks in these three areas.

4 Conclusion

The mineralogical comparative study of sedimentary rocks from Ramagundam, the Godavari Basin, and Manuguru highlights significant variations in their composition, microstructure, and depositional environments. High-energy fluvial conditions are suggested by coarser, poorly sorted samples, while Ramagundam rocks' microstructural examination shows well-sorted, fine-grained textures with rounded grains and a wealth of fossils, suggesting a calm, shallow marine context. Numerous depositional histories and source rock influences are reflected in the mineral abundance of the rocks of the Godavari Basin, which are mostly composed of quartz, feldspar, clay minerals, calcite, and dolomite. The Manuguru rocks, on the other hand, have a quartz-rich composition that includes calcium, iron, potassium, and aluminum. They also contain clay and carbonate minerals, which suggests low-energy depositional settings, weathering impacts, and potential marine or lacustrine influences. Along with trace elements that shed light on diagenetic processes and sedimentary evolution, the EDX study further validates the dominance of silica. The geological past, mineralogical variances, and possible reservoir characteristics of sedimentary strata in these three places are all better understood as a result of these discoveries. The study offers important new information for geological and geophysical research and resource management in the future related to mining field applications.

Acknowledgement. I would like to express my sincere gratitude to my mentor, Dr. Sandi Kumar Reddy, Associate Professor, Department of Mining Engineering, for his invaluable guidance, encouragement, and support throughout the completion of this work. I cordially thank the Director of NITK Surathkal for submission of the paper in the proceedings.

References

1. Wunderlin, C.A., Collo, G., Ezpeleta, M., Carbonell, V.V.R., Nóbile, J.C., Ciccioli, P.L., Faudone, S.: Authigenic and detrital clay minerals as indicators of paleoenvironmental and postdepositional evolution in a Cretaceous–Cenozoic succession from Argentine Central Andes. Sediment. Geol. **437**, 106179 (2022)
2. Garzanti, E.: Petrographic classification of sand and sandstone. Earth Sci. Rev. **192**, 545–563 (2019)
3. Koeshidayatullah, A., Corlett, H., Hollis, C.: An overview of structurally-controlled dolostone-limestone transitions in the stratigraphic record. Earth Sci. Rev. **220**, 103751 (2021)
4. Tankard, A., Welsink, H., Aukes, P., Newton, R., Stettler, E.: Tectonic evolution of the Cape and Karoo basins of South Africa. Mar. Pet. Geol. **26**(8), 1379–1412 (2009)
5. Dauphas, N., Cates, N.L., Mojzsis, S.J., Busigny, V.: Identification of chemical sedimentary protoliths using iron isotopes in the 3750 Ma Nuvvuagittuq supracrustal belt, Canada. Earth Planet. Sci. Lett. **254**(3–4), 358–376 (2007)
6. Ward, C.R.: Analysis and significance of mineral matter in coal seams. Int. J. Coal Geol. **50**(1–4), 135–168 (2002)
7. Rahman, H.M., Kennedy, M., Löhr, S., Dewhurst, D.N., Sherwood, N., Yang, S., Horsfield, B.: The influence of shale depositional fabric on the kinetics of hydrocarbon generation through control of mineral surface contact area on clay catalysis. Geochim. Cosmochim. Acta. **220**, 429–448 (2018)
8. Hoek, E., Bray, J.W.: Rock slope engineering, 3rd edn. Institute of Mining and Metallurgy (1981)
9. Smykatz-Kloss, W., Roy, P.D.: Evaporite mineralogy and major element geochemistry as tools for palaeoclimatic investigations in arid regions: a synthesis. Bol. Soc. Geol. Mex. **62**(3), 379–390 (2010)
10. Jian, X., Guan, P., Zhang, W., Feng, F.: Geochemistry of Mesozoic and Cenozoic sediments in the northern Qaidam basin, northeastern Tibetan Plateau: implications for provenance and weathering. Chem. Geol. **360**, 74–88 (2013)
11. Garzanti, E., Andò, S.: Heavy minerals for junior woodchucks. Minerals. **9**(3), 148 (2019)
12. Oyebanjo, O., Bukalo, N., Ekosse, G.I.: Provenance and paleoenvironmental studies of cretaceous African and South American Kaolins: similarities and differences. Minerals. **11**(10), 1074 (2021)
13. Hashempour, S.S., Maghfouri, S., Rastad, E., González, F.J.: Metallogeny and temporal-spatial distribution of sedimentary phosphorite mineralization in Iran, relation to tethyans oceans evaluation and implications for future exploration. Ore Geol. Rev. **105855** (2023)
14. Dill, H. G. A review of heavy minerals in clastic sediments with case studies from the alluvial-fan through the nearshore-marine environments. Earth Sci. Rev., 45(1–2), 103–132 (1998).
15. Fedo, C.M., Sircombe, K.N., Rainbird, R.H.: Detrital zircon analysis of the sedimentary record. Rev. Mineral. Geochem. **53**(1), 277–303 (2003)
16. Varalakshmi, P., Reddy, S.K., Murthy, C.S.N.: Investigation on estimation and prediction of resistivity of limestone rocks based on physico-mechanical properties of rocks. Disaster Adv. **18**(5), 139–147 (2025)

Fire (energy)

A Comprehensive Analysis of Solar Power in West Bengal

Surjendu Manna[1] and Binoy Krishna Roy[2(✉)]

[1] RCC Institute of Information Technology, Kolkata 700159, West Bengal, India
[2] National Institute of Technology, Silchar 788010, Assam, India
bkr@ee.nits.ac.in

Abstract. This paper analyzes the development of solar power in West Bengal by focusing on the current status, challenges, and prospects of the state. The performance of solar energy is evaluated across different project types, such as ground-mounted and rooftop installations, and then compared with India's overall solar capacity using quantitative data and district-level analysis. It will reflect the regional disparities in the adoption of solar power, point out the key barriers of inadequate policies, constraints due to land, unfavorable climatic conditions, and propose targeted strategies for reducing these gaps. It tries to align the renewable energy track of West Bengal with national goals and aid in the sustainable development of energy.

Keywords: Solar Energy · Renewable Energy · West Bengal · Solar Power · Floating Solar · Rooftop Solar · Hybrid Systems · CUF · Solar Capacities · Energy Policies · Regional Disparities · Energy Storage · Land Acquisition · Grid Infrastructure · Solar Challenges · National Solar Mission

1 Introduction

Solar energy has emerged as a critical pillar in West Bengal's pursuit of a cleaner and more sustainable energy future. This study delves into the current status and untapped potential of solar power in the state, examining its progress in the context of India's broader renewable energy landscape. By drawing comparisons with leading solar-performing states such as Gujarat, Rajasthan, and Karnataka, the paper identifies performance gaps and systemic challenges unique to West Bengal. The analysis goes beyond surface-level metrics to investigate regional disparities in solar adoption across districts within the state. Key factors such as land availability, policy implementation, infrastructure limitations, and socio-economic conditions are explored to understand the causes of uneven solar energy deployment. Based on these insights, the paper proposes targeted strategies to address the disparities and enhance the state's overall solar output. In doing so, the study not only evaluates the effectiveness of existing initiatives—such as rooftop systems, floating solar plants, and canal-top installations—but also outlines prospects and actionable pathways to integrate solar energy more equitably and efficiently across West Bengal. All data, comparisons, and findings presented in this work are derived from peer-reviewed research papers, government publications, and official reports to ensure accuracy and credibility.

H. P. Dasari et al. (Eds.): NITK-CREST 2025, PTT, pp. 91–101, 2026.
https://doi.org/10.1007/978-981-95-3218-6_7

2 Methodology

The methodology followed in this paper adopts a systematic, multi-layered approach to analyzing the present status, challenges, and opportunities for solar power development in West Bengal. The study begins with an extensive literature review using government reports, policy documents, academic papers, and official data from agencies such as WBPDCL, WBSEDCL, WBGEDCL, and MNRE. All information and data used in the analysis reflect the status as of November 2024, including the most recent policy and project developments.

Quantitative data on installed capacities, solar resource classification, and district-wise distribution are collected to identify patterns and trends. These include various configurations such as solar-only systems, solar-diesel hybrids, solar-wind hybrids, and solar-biomass systems. In addition to intra-state analysis, a comparative framework is employed to benchmark West Bengal's solar performance against leading solar energy states in India like Gujarat, Rajasthan, and Karnataka. This helps highlight specific policy, infrastructural, and geographical factors contributing to regional disparities.

The study incorporates both quantitative metrics and qualitative insights to evaluate district-level inequality in solar adoption. Land availability, solar radiation levels, institutional capacity, and public-private participation are also examined. Based on the analysis, targeted strategies are proposed to address performance gaps and promote equitable solar development across the state. Visual tools such as bar graphs, pie charts, and comparative tables are used extensively to summarize key findings and support the analytical narrative of the paper.

All data presented herein are current as of April 30, 2025, unless a specific date or period is explicitly stated.

3 Current Status of Solar Power in West Bengal

West Bengal has been progressively marching forward with incorporating solar energy through multiple initiatives and classifications of solar projects. According to the recent report by the Ministry of New and Renewable Energy, the state's generation of solar power falls into four broad categories: Ground-Mounted Solar, PM-Surya Ghar Yojna (Solar Rooftop), Hybrid Solar Components, and Off-grid Solar/KUSUM Projects [12].

The bar graph in Fig. 1 provides a clear picture of the contributions from these segments, offering insights into West Bengal's evolving solar energy landscape.

3.1 Comparison with India's Solar Power Generation

To begin with, the solar energy performance of West Bengal is benchmarked against the national scenario by comparing the estimated average usable solar power rather than just installed capacities. This approach provides a more realistic assessment by factoring in critical parameters such as the Capacity Utilization Factor (CUF) and mean daily sunshine hours, which directly influence the effective energy output. The comparison spans various deployment types—ground-mounted systems, rooftop installations, hybrid configurations, and off-grid solutions like those under the KUSUM scheme—to evaluate

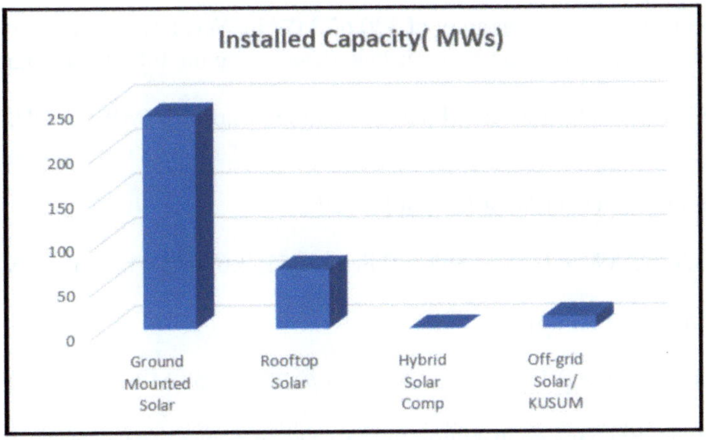

Fig. 1. Deployment of Solar Power in West Bengal [12]

	Installed Capacity(MWs)	
Solar Energy systems	West Bengal	India
Ground Mounted Solar	240.35	82387.48
Rooftop Solar	67.13	17688.44
Hybrid Solar Comp	0	2887
Off-grid Solar/ KUSUM	13.14	4982.69
Total	320.62	107945.1

Fig. 2. Comparison of Installed Solar Power Capacities, West Bengal and India [12]

West Bengal's actual contribution toward meeting national renewable energy targets (Fig. 2).

To evaluate West Bengal's contribution to India's solar power generation, we estimated the yearly average usable solar power output based on the installed capacity, Capacity Utilization Factor (CUF), and average daily sunlight hours.

The Capacity Utilization Factor (CUF) represents the ratio of actual output from a solar plant to the maximum possible output under ideal conditions. For India, an average CUF of 18% is considered, as supported by recent studies [14]. For West Bengal, an average CUF of 18% is considered, as supported by recent studies [15].

Given an installed solar capacity of 320.62 MW in West Bengal. As derived in [30], the yearly average usable solar power is calculated using the following equation:

$$\text{Yearly Avg.Usable Solar Power (MWh)} = \text{Installed Capacity (MW)} \times \text{CUF} \times 365 \times 24\text{h} \quad (1)$$

Substituting the values for West Bengal [12, 27]:

$$= 320.62\text{MW} \times 0.18 \times 365 \times 24\text{h} = 505553.616 \text{ MWh} \approx 506 \text{ GWh}$$

Similarly, for India [12, 26]:

$$= 107945.1 \text{ MW} \times 0.18 \times 365 \times 24 \text{ h} = 170207833.68 \text{ MWh} \approx 170208 \text{ GWh}$$

Therefore, West Bengal's contribution to usable solar power over a year in India

$$= 506 \text{ GWh}/170208 \text{ GWh} \times 100\% = 0.297\%$$

The analysis reveals that despite having a notable installed solar capacity of 320.62 MW, West Bengal contributes only 0.297% to India's total usable solar power output, which stands at approximately 506 GWh annually. This stark disparity underscores a significant underperformance relative to the national benchmark.

3.2 Challenges and Disadvantages of Solar Power in West Bengal

Policy & Incentives Gap: Lack of Subsidies is one of the biggest problems that can be held responsible from April 1, 2023, to March 31, 2024, WBSEDCL did not receive any financial aid or subsidies under the PM Surya Ghar: Muft Bijli Yojana for promoting rooftop solar installations.

This was due to the lack of inspection and approval of rooftop solar installations by WBSEDCL, which is a prerequisite for subsidy disbursement [16].

Poor Net-metering Policies: Before November 2024, the gross metering policy applied to solar systems above 5 kW, requiring all generated power to be exported to the grid, with consumers receiving compensation at a fixed tariff. However, the 5 kW cap limited adoption among larger consumers such as businesses and residential complexes, who were forced to scale down their solar usage or depend on expensive alternatives. Additionally, the low compensation for surplus power further discouraged investment in rooftop solar systems [17].

As per the proposed draft policy update after November 2024 [18], net metering in West Bengal has been extended to cover solar installations up to 500 kW or the consumer's sanctioned load, whichever is lower, across all eligible categories, including residential, commercial, and industrial users. For installations exceeding 500 kW, the policy mandates a gross metering system, wherein the entire solar energy generated is exported to the grid rather than being consumed on-site. In such cases, consumers are compensated at a tariff determined by the West Bengal Electricity Regulatory Commission (WBERC), which is typically lower than the prevailing retail electricity rates.

Land Availability and Solar Potential in West Bengal: West Bengal encompasses a total geographical area of 8,875,000 ha, of which 5,281,000 ha (approximately 59.5%) is designated as net sown area, highlighting the state's heavy dependence on agricultural land [19]. This extensive agricultural usage presents challenges in allocating large tracts of land for utility-scale solar projects. Additionally, in terms of solar potential, regions such as Kolkata receive an average Global Horizontal Irradiance (GHI) of approximately 1684 kWh/m^2 per year, equivalent to about 4.6 kWh/m^2 per day [20]. The city also enjoys around 2,234 sunshine hours annually, averaging 6.12 h of sunlight per day, which, while moderate, limits West Bengal's solar efficiency compared to high-irradiance states [27].

3.3 Future Goals for Solar Power in West Bengal

Several innovation-based large-scale projects by the state have encouraged adoption and brought down a lead in comparison to others. Some of them that deserve mention are Bakreswar Dam's Mega Floating Solar PV project [8], showing the latest trends of use in solar energy; the 10 MW installation near the Teesta Canal Fall Hydro Electric Plant at Uttar Dinajpur in the Canal Bank Solar Power Project [11]. The state is working on a 1200 MW Solar PV Project in Purulia as a source of clean energy supply for pumped storage projects [11]. Its rooftop solar programs include promotion of the installation of 10KWp sets in 200 schools and an urban rooftop solar scheme to 5 MW on 18 districts [11], followed by a proposal for 11 MW with 900 schools and 100 colleges [11]. In addition, the Kalyani Layer Farm Solar Project offers a 1.5 MW grid-connected rooftop system in the Nadia district [12], while there are plans for solar street lighting and high mast systems in Uttar Dinajpur with 121 high masts and 255 streetlights under review [8]. There is also a 24 kWp rooftop solar installation at the Institute of Environmental Studies & Wetland Management, which is now in the tender evaluation stage [8].

Together, WBPDCL, WBSEDCL, and WBGEDCL efforts can be viewed as an integrated effort to move solar energy forward in the state. These projects and proposals are scheduled for the 2024–25 financial year.

4 Statewise Solar Energy Comparison

To put West Bengal's performance in the context of solar power generation, a bar chart is provided that compares the installed capacity of West Bengal with the top 20 Indian states. The visualization brings out stark contrasts where leading states like Rajasthan and Gujarat have much higher capacities due to favorable policies and available land. West Bengal's contribution is very modest in comparison and reflects challenges like limited land resources and policy implementation issues. The following comparisons suggest the need for focused approaches to increase the adoption of solar energy and place West Bengal on par with national leaders in renewable energy development (Fig. 3).

4.1 Factors Driving Better Renewable Energy Performance in Other States

Rajasthan has taken bold and effective steps to position itself as a national leader in solar energy development. A major strength of the state's approach lies in its aggressive

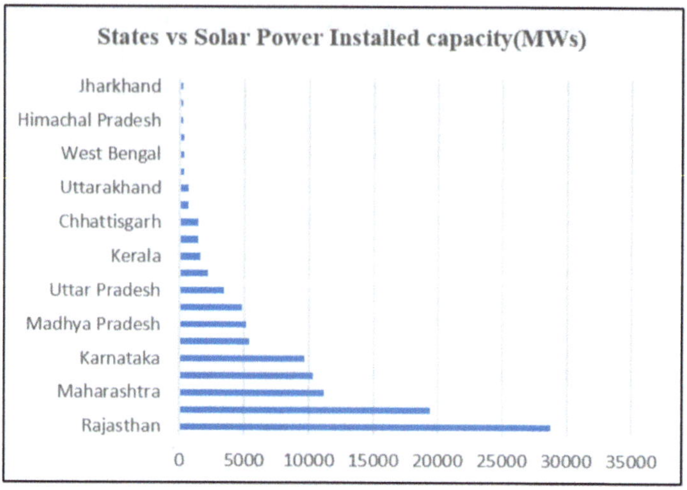

Fig. 3. Solar Power Capacity of Top 11 Indian States (MW) [12]

financial incentives. By offering a 100% exemption from electricity duty for seven years, Rajasthan ensures that solar developers are relieved from a significant operational cost burden. Furthermore, the 75% reduction in stamp duty on land transactions substantially reduces capital expenditure, making the state extremely attractive for solar investment. Rajasthan's effective land policy is also a key driver of its success. The state has allocated vast stretches of barren land specifically for solar park development, transforming otherwise unproductive terrain into high-output energy zones. On the infrastructure front, Rajasthan is proactive, establishing dedicated transmission lines and smart grid projects to ensure that solar energy is efficiently absorbed and distributed through the grid. The implementation of single-window clearance systems greatly simplifies the bureaucratic process, enabling faster execution of projects and encouraging more private sector participation. Beyond infrastructure and policy, Rajasthan also prioritizes awareness and community engagement, running training initiatives and outreach programs aimed at farmers and residents. This holistic approach not only boosts capacity but also ensures local involvement and benefits, making Rajasthan a role model in the integration of solar energy into a state's developmental fabric [7, 21–24].

Gujarat has demonstrated a robust and inclusive model for solar energy adoption by effectively combining grassroots empowerment with large-scale infrastructure development. A flagship initiative, the *Suryashakti Kisan Yojana (SKY)*, provides farmers with a 60% subsidy on solar installation costs, supplemented by an additional 30% as a low-interest loan at rates between 4.5% and 6%. This significantly reduces the financial burden on farmers, making solar-powered irrigation both affordable and sustainable. Gujarat's strategic approach to land utilization is evident in its allocation of vast barren tracts like the 72,600-ha Gujarat Hybrid Renewable Energy Park in the Kutch district, which aims to develop 30 GW of renewable capacity. The state also benefits from excellent solar exposure, with Global Horizontal Irradiance (GHI) levels ranging from 5.5 to

6.0 kWh/m^2/day, ensuring high efficiency and consistent energy yield from solar panels. To support this generation capacity, Gujarat has made major investments in grid infrastructure, including a $360 million solar transmission line initiative and the implementation of the Gujarat Solar and Smart Grid Development Investment Program, aimed at integrating solar PV systems into the grid. Additionally, policies such as net metering encourage prosumer participation by allowing consumers to feed surplus energy back into the grid in exchange for credits. With a streamlined single-window clearance system and community engagement through awareness and training programs, Gujarat has created a conducive environment for both utility-scale and decentralized solar growth, setting an example of policy, infrastructure, and public participation working hand in hand [25–31].

4.2 Barriers to Solar Power Growth in West Bengal

West Bengal's slower progress in solar energy deployment can be attributed to a combination of policy, infrastructural, and administrative challenges. One of the most pressing issues is the lack of aggressive financial incentives. With limited subsidies and less streamlined disbursement mechanisms, solar project developers and individual consumers face higher upfront costs, making investments less attractive. Additionally, the state has not fully capitalized on available land resources. While land scarcity is a known limitation, there has been insufficient identification of underutilized plots or a strong push toward alternatives like rooftop or floating solar installations, which could have bypassed land constraints altogether. On the infrastructure front, the absence of robust, dedicated transmission lines and smart grid systems hampers the seamless integration of solar energy into the existing power grid. The state also suffers from bureaucratic bottlenecks, with slow and complex approval processes discouraging private investment and delaying project execution. Furthermore, the lack of widespread public awareness campaigns and community training programs has left many residents and local stakeholders uninformed or underprepared to engage with solar technologies. Without targeted initiatives for skill development and awareness, the grassroots adoption necessary for decentralized solar growth remains limited. To truly accelerate its renewable transition, West Bengal must address these policy gaps, invest in infrastructure, simplify regulatory processes, and actively involve its communities in the solar journey [6, 13].

5 District-Wise Analysis of Solar Power Potential in West Bengal

This work also presents generational locations and estimated installed capacities across West Bengal. For ease of understanding and clarity, these data points have been grouped by districts, thus providing an analysis that is more localized and actionable. The geographical organization of the data emphasizes district-level contributions and disparities, giving an in-depth view of the distribution of solar energy in the state. This data has been approximated by aggregating the operational solar power projects across districts as of November 2024 to estimate the contribution of solar power generation in each district (Fig. 4).

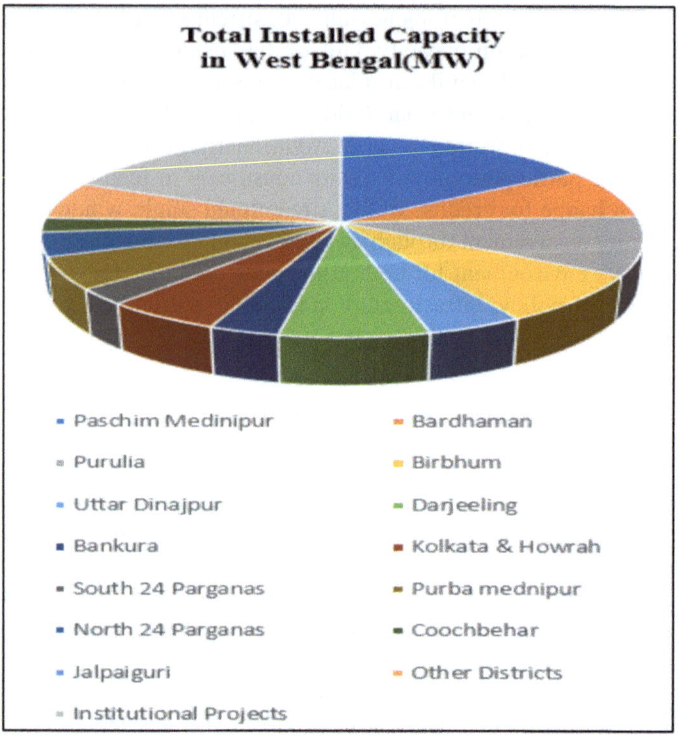

Fig. 4. Solar Power Share Across West Bengal District [1–5, 9, 10, 32].

5.1 Regional Solar Disparities in West Bengal: Uneven Growth and Underlying Barriers

West Bengal's solar energy development is marked by stark regional disparities, with installations heavily concentrated in districts like Paschim Medinipur and Bardhaman—areas that benefit from higher solar insolation and the availability of suitable land. In contrast, many rural districts struggle to keep pace due to multiple structural and logistical challenges. Limited availability of non-agricultural land restricts opportunities for setting up large-scale solar projects, while poor grid connectivity and the high cost of infrastructure development further hinder expansion in these areas. A significant barrier is the widespread lack of awareness and technical expertise in rural communities, where knowledge about solar technology and access to information on existing government incentives remain minimal. This results in low adoption rates despite the potential for decentralized solar solutions. Additionally, a policy bias toward already resource-rich districts exacerbates the problem, as government efforts and subsidies are disproportionately directed to these regions. Land acquisition hurdles, driven by unclear or inadequate land use policies, contribute to delays in project implementation across underserved districts. Compounding the issue, insufficient transmission infrastructure in some areas makes grid integration expensive and technically challenging. These factors collectively lead to an uneven solar development landscape, where the benefits of renewable energy remain inaccessible to large parts of the state's population [19].

5.2 Bridging Regional Solar Gaps Through Inclusive Strategies

To reduce the regional disparities in solar energy adoption across West Bengal, a range of inclusive strategies may prove effective if implemented thoughtfully. Decentralized solutions like rooftop solar, microgrids, and small-scale solar farms could be more suitable for rural districts with limited land availability, while floating solar projects may offer alternatives in areas facing land constraints. Upgrading transmission networks and introducing better energy storage systems might help connect underdeveloped regions and stabilize the solar supply. Targeted incentives, such as district-specific subsidies and low-interest loans, could potentially make small-scale solar adoption more viable for rural communities. Community awareness and skill development programs, especially those involving panchayats, NGOs, and local youth, may also improve adoption and long-term maintenance. Furthermore, reforms in land acquisition policies, including sustainable land-use planning and government-supported leasing, might ease project delays and reduce conflicts. While the impact of these strategies depends on effective execution, they present promising pathways to achieve more balanced solar development across the state.

5.3 Possible Outcomes and Future Prospects

If efforts to expand solar energy across West Bengal prove successful, the state could see a more equitable distribution of solar installations spanning all districts, helping to bridge current regional gaps. Strengthening renewable energy infrastructure—through improved grid connectivity and enhanced stability—would support this growth and enable smoother integration of solar power. The expansion of the solar sector also holds the potential to stimulate economic growth by creating new employment opportunities in installation, maintenance, and related industries. Additionally, increased solar capacity would help West Bengal contribute more significantly toward India's national renewable energy targets, aligning the state's development goals with the country's broader clean energy ambitions. Together, these outcomes could drive a sustainable and inclusive energy future for West Bengal.

6 Conclusion

The solar energy sector is at the heart of West Bengal's transition into a sustainable and renewable energy future. The state has taken commendable strides through innovative initiatives such as floating solar plants, rooftop installations in schools and urban areas, and canal-based solar systems. These projects reflect the state's strong commitment to renewable energy as part of its long-term energy strategy. However, the growth of solar energy is not without its challenges. Regional disparities in adoption, limited land availability, climatic constraints, and infrastructure barriers continue to hinder progress. Urban and developed areas are witnessing rapid growth in solar adoption, whereas rural and underdeveloped districts face challenges in accessing solar technology and infrastructure.

Targeted strategies may be implemented to address these inequalities, including incentivizing investments in solar energy infrastructure, providing localized solutions for rural areas, and fostering awareness about solar benefits. Land acquisition issues will lso need to be addressed, along with optimizing the use of available space for solar installations. This may unlock the true potential of West Bengal for solar power, as long as focused efforts are applied in overcoming these barriers to achieve equitable growth.

References

1. (WBREDA), W. B. Scenario of Renewable Energy in West Bengal. Retrieved from WBREDA: https://cuts-ccier.org/pdf/Scenario_of_RE_in_WB.pptx#:~:text=Solar%20radiation%20is%20abundantly%20available,solar%20energy%20is%20given%20below
2. Department of Power Government of West Bengal. Retrieved from https://wbpower.gov.in/solar-projects/
3. Division, I. E.: Unlocking solar rooftop opportunities in cold storage MSMEs in West Bengal. Retrieved from https://www.sameeeksha.org/pdf/4%20-%20Unlocking%20Solar%20Rooftop%20Opportunities%20in%20Cold%20Storage%20MSMEs%20in%20West%20Bengal.pdf
4. Electricity consumption utilities: West Bengal. Retrieved from https://www.ceicdata.com/en/india/electricity-consumption-utilities/electricity-consumption-utilities-west-bengal
5. Floating solar power plant in West Bengal, India. Retrieved from vikramsolar: www.vikramsolar.com/case-studies/floating-solar-power-plant-in-west-bengal-india/
6. Geological state of West Bengal. (n.d.). Retrieved from https://en.wikipedia.org/wiki/Geography_of_West_Bengal#:~:text=It%20lies%20between%2085%20degree,ethno%2Dlinguistic%20region%20of%20Bengal
7. How to get a Home Solar System Subsidy in Rajasthan? (n.d.). Retrieved from Alpex Power: https://alpexsolar.com/resources/blog/solar-subsidy-in-rajasthan/#:~:text=In%20Rajasthan%2C%20the%20subsidy%20for,capacity%20up%20to%2010%20kW
8. List of works year-wise undertaken by WBGEDCL from 2011 to till date. Retrieved from WEST BENGAL GREEN ENERGY DEVELOPMENT CORPORATION LIMITED: https://www.wbgedcl.in/projects/#:~:text=30%20kWp%20Rooftop%20Solar%20PV,of%20West%20Bengal
9. Power for all West Bengal. Retrieved from https://powermin.gov.in/sites/default/files/uploads/joint_initiative_ofgovt_of_india_and_West_Bengol.pdf
10. Power plant profile. Retrieved from https://www.power-technology.com/data-insights/power-plant-profile-paschim-medinipur-solar-pv-park-india/
11. Solar Program of WBSEDCL. Retrieved from DEPARTMENT OF POWER: https://wbpower.gov.in/solar-projects/#:~:text=Three%20numbers%2010%20MW%20solar,18%20districts%20in%20West%20Bengal
12. State-wise installed capacity of renewable power. Retrieved from Ministry of New and Renewable Energy: https://drive.google.com/drive/folders/19JmoFY3Lo5l_-M48-YHwAMeePs-NZvFA?usp=sharing retrieved from https://mnre.gov.in/en/physical-progress/
13. West Bengal govt must rise above politics to roll out PM Surya Ghar scheme: Josh. Retrieved from Energy World: https://energy.economictimes.indiatimes.com/news/renewable/west-bengal-govt-must-rise-above-politics-to-roll-out-pm-surya-ghar-scheme-joshi/
14. Average CUF for India. https://www.researchgate.net/publication/341766845_Operational_Performance_of_On-grid_Solar_Photovoltaic_System_Integrated_into_Pre-fabricated_Portable_Cabin_Buildings_in_Warm_and_Temperate_Climates

15. Average CUF for West Bengal. https://cea.nic.in/wp-content/uploads/resource_adequacy_st/2024/11/Report_on_Resource_Adequacy_Plan_for_West_Bengal_Up_to_2034_35.pdf
16. Subsidies for PM Surya Ghar Yojna to WBSEDCL https://sansad.in/getFile/annex/266/AS226_m96e2Z.pdf?source=pqars
17. Previous Net metering guidelines before November 2024. https://www.cesc.co.in/roofTopSolarGuide?
18. 500 KW Permission for Net Metering. https://wberc.gov.in/sites/default/files/DraftWBERCGridInteractiveRooftopSolarPVSystemForProsumersRegulations2024_0.pdf?
19. Land Availability in West Bengal. https://desagri.gov.in/wp-content/uploads/2024/09/Final-file-of-LUS-2022-23-for-uploading.pdf?
20. Solar Irradiance in Kolkata. https://globalsolaratlas.info/map?c=22.572646,88.363895,11&m=site&s=22.572646,88.363895global_irridance_rate
21. Land Availability in Rajasthan. https://www.britannica.com/place/Thar-Desert
22. SolarProjectRajasthan https://www.power-technology.com/news/indias-rajasthan-solar-projects/
23. Solar Irradiance rate in Rajasthan. https://rajras.in/rajasthan/geography/sunshine-availability-solar-radiation/
24. Rajasthan Solar Power Policy. https://rising.rajasthan.gov.in/storage/app/public/files/pdf/rips-2022.pdf
25. Gujarat Hybrid Renewable Energy Park. https://en.wikipedia.org/wiki/Gujarat_Hybrid_Renewable_Energy_Park
26. Global Horizontal Irradiance of Gujarat. https://www.nrel.gov/docs/fy14osti/60991.pdf?
27. Gujarat Solar Policy. https://www.tnpscthervupettagam.com/currentaffairs-detail/gujarat-launches-suryashakti-kisan-yojana?
28. Grid in Gujarat. https://www.adb.org/projects/45418-003/main
29. Grid investments in Gujarat. https://www.reuters.com/world/india/adani-energy-spend-360-mln-build-gujarat-power-transmission-line-2023-12-27/?
30. Solar Advisor Model User Guide for Version 2.0 – Publications page 67. https://docs.nrel.gov/docs/fy08osti/43704.pdf#:~:text=The%20Solar%20Advisor%20Model%20is,Research%20Institute%20(%22MRI%22)%20for%20the
31. Gujarat Announces New Solar Policy. (n.d.). Retrieved from The Times of India: https://timesofindia.indiatimes.com/city/ahmedabad/gujarat-announces-new-solar-policy/articleshow/80007550.cms?utm_source=chatgpt.com
32. WEST BENGAL STATE ELECTRICITY DISTRIBUTION COMPANY LIMITED. Retrieved from https://www.wbsedcl.in/irj/go/km/docs/internet/new_website/solar.html

Design and Development of Solar Air Heater for Drying Clothes

Rawal Diganjit[1](✉) and N. Gnanasekaran[2]

[1] Government Polytechnic Kolhapur, University Road, Vidyanagar, Kolhapur, Maharashtra 416004, India
diganjitrawal@gmail.com

[2] Indian Institute of Technology, Tirupati Venkatgiri Road, Yerpedu post, District Tirupati,, Andhra Pradesh 517619, India

Abstract. The use of solar energy is pure and most proven alternative energy source. Solar air heaters are used as a renewable energy source in regions where huge sunlight is accessible. They are useful for drying or heating the product. In this paper, single pass solar air heater using forced convection is studied experimentally and then the theoretical calculation are performed for simple payback period and net present value. This study is performed for National Institute of Technology Karnataka, Surathkal. The outlet temperature and thermal efficiency is obtained experimentally upto 58 °C and 58.33% respectively. The simple payback period and internal rate of return is 12 years and 9.25% respectively considering 10% discount rate.

Keywords: Single pass solar air heater · thermal efficiency · Drying clothes

1 Introduction

Conventionally, clothes are dried in open spaces or under the shade to protect the clothes from direct exposure to the Sun to avoid the color fading and deterioration of the fabric. In urban areas, because of unavailability or lack of open spaces, electrical drying machines are widely used. Such machines consumes lots of energy and cannot be run in the absence or erratic supply of the electricity, which is a common sight in developing countries like India. In such conditions, clothes are hanged in balconies or in closed space like rooms. This practice of drying clothes in the balcony or in rooms not only ruins the aesthetics of the building but also leads to the thermal discomfort because of increase in the moisture. Electrical clothes dryer is mostly used in family buildings and hostels. It consumes a lot of electricity. Drying clothes is a dual action: heat transfer to the clothes from the absorber hot surface and mass transfer of moisture from the clothes to the surface and from the surface to the nearby air [1]. The terrace space of hostels can be used to dry the clothes of students, thus upgrade the terrace utilization (i.e. utilizing space of terrace of each hostel)

These case study considers the space available in the hostel of National Institute of Technology Karnataka (NITK), its environment and climatic conditions. There are

H. P. Dasari et al. (Eds.): NITK-CREST 2025, PTT, pp. 102–109, 2026.
https://doi.org/10.1007/978-981-95-3218-6_8

eleven hostel buildings. The one room is allotted for two or three students. The rooms are smaller [2]. Each floor of buildings has 14 rooms towards right and left sides. One floor has capacity of 42 students. Each room contains the two straight bars which has length of 3 m. It is used for drying clothes. The students can use laundry shop who wants to spend more time for study [2].

The space for drying the clothes is less and moisture, humidity content in the air is more at this location so time required for drying the clothes is more (in natural way of drying clothes). The number of students are more compared to space availability in the hostel. The main disadvantage of laundry service is misplaced of clothes and wastage of time and delivery time is 3 days. In the hostel (apartment type) system, the number of students are more and space for drying the clothes is less.

Hence, the problem statement is to develop solar clothes drying system for the Indian climatic conditions. Here, considering location of Solar drying system at National Institute of Technology Karnataka Surathkal –Hostel block. The National Institute of Technology Karnataka Surathkal (NITK) is situated in southern India in the state of Karnataka at a latitude of $13^\circ\ 0'\ 12''$ & longitude of $74^\circ\ 47'\ 24''$. The one room of hostel is assumed as 3 m × 3 m. One hostel block has 14 rooms for right and left side. So total rooms available on terrace is equal to 28 rooms. Area of one room will be 9 m^2. The area of the 28 rooms is equal to 260 m^2. Considering shadow effect and easy to move through all around solar air heater and clothes dryer, per area of room only one solar air heater and clothes dryer is placed. The total number of rooms in hostel block are 28. Hence, total 28 solar air Heater (SAH) and clothes dryer can be settled on terrace. The assumptions considered during design are the system has proper drainage system and air tight seals. Considering only cotton clothes. Cotton clothes are used they hold the maximum amount of moisture per unit weight.

2 Methodology

The solar air heater (SAH) is single pass with forced convection. The design details and its dimensions are considered for the thermal performance as mentioned in the author previous study of [3–5]. The assumptions for the cloth dryer chamber are considered as - the system has regular and suitable drainage pipe connections and no air leakages. Considering only cotton clothes. Cotton clothes carry the large amount of moisture per unit weight. The design details of SAH is represented in the Table 1 (Figs. 1, 2 and 3).

Table 1. Design details of single pass solar air heater

Name of Component	Material	Dimensions (in meter)	Quantity
Cover Plate	Toughened Glass	2 × 1 × 0.004	1
Absorber Plate	Aluminium	2 × 1 × 0.004	1
Collector box	Polyvinyl chloride board sheet	2.1 × 1.1 × 0.004	1

(continued)

Table 1. (*continued*)

Name of Component	Material	Dimensions (in meter)	Quantity
Insulation	Ceramic Wool	2 × 1 × 0.005-Bottom 1 × 0.005–2 Sides	1
Nut, Bolt Fitting Assembly	MS	As per requirement	25(Approx.)
Black Paint Coating	Blank Paint	200 ml	1

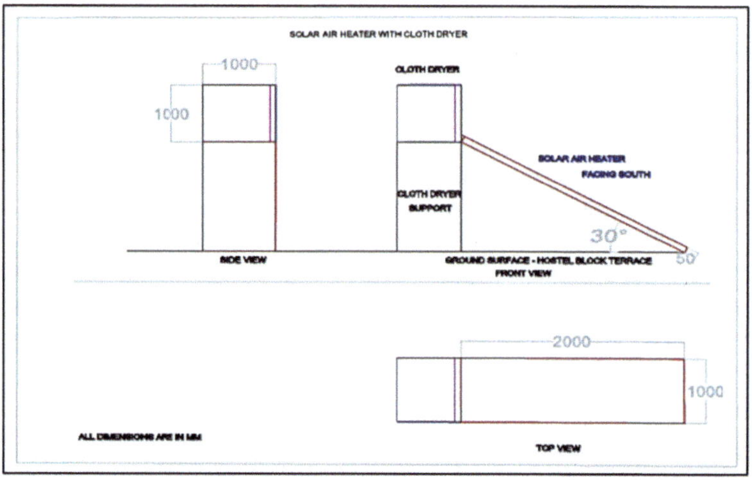

Fig. 1. Layout of the solar air heater with its cloth dryer

Fig. 2. Front view of solar air heater

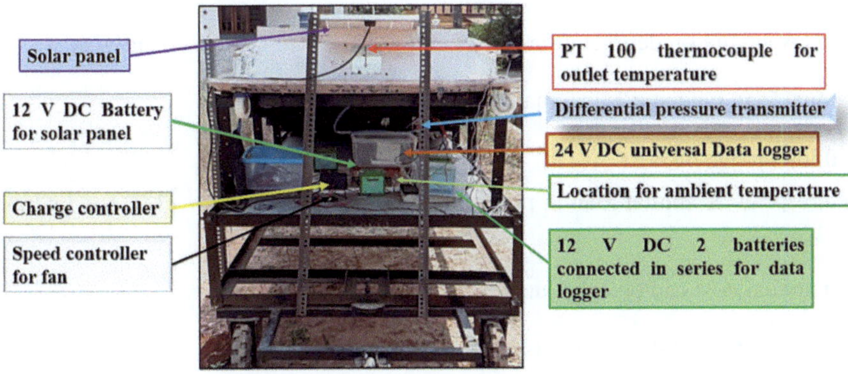

Fig. 3. Backside view of solar air heater

3 Performance Parameters

The ambient, inlet, outlet temperatures, solar intensity, inlet velocity and other surface temperatures of absorber plate and glass plate are calculated and measured as mentioned in the author's previous study of [4]. The detailed calculations of operational parameters are mentioned in reference [4, 5]. The life cycle of the solar air heater and clothes dryer system is assumed 20 years. 0.38 units of electricity are used to dry the 90% clothes using washing machine. Hence, it uses 50% more electricity. If a family washes cloths and uses dryer every day, the total units will come about 40–120 units a month. Hence assuming 500 units per month consumption by students in one Hostel block. The cost per unit electricity is Rs.7.75/unit (Current rate of Mangalore). Consider the electricity saving deduction (or decreases) yearly is 2% of first year. The rate of depreciation is 80% in first year and remaining 20% in second year. The corporate tax rate is 31%. Assuming discount rate is 10% for period of 20 years. The annual loan is considered upto 85% of the initial cost. The interest rate of 5% to be repaid in equal installment for 5 years. The rate of depreciation is 80% in first year and remaining 20% in second year. The corporate tax rate is 31%. Assuming discount rate is 10% for period of 10 years. Assuming no incentives and subsidy from government (MNRE).

The consumption of electricity for one year by students in one hostel block is calculated by Eq. (1), [6, 7]

$$\text{Electricity consumption for one year} = \text{Unit consumed in one month} \times 12 \tag{1}$$

The total cost of electricity saving is calculated by Eq. (2),

$$\begin{aligned}&\text{Total cost for electricity saving in one year}\\&= \text{Electricity consumption for one year} \times \text{per unit rate}\end{aligned} \tag{2}$$

The tax saving on depreciation for first year and second year is calculated by Eqs. (3) and (4),

$$\text{The tax saving on depreciation for first year} = 0.80 \times \text{initial cost} \tag{3}$$

$$\text{The tax saving on depreciation for second year} = 0.20 \times 0.80 \times \text{initial cost} \quad (4)$$

The annual loan payment is calculated by Eq. (5),

$$\text{Annual loan payment} \\ = (\text{interest rate X annual loan in percentage X Initial cost})/\left(\left[1 - \left\{1/((1 + \text{interest rate}))^{\text{year}}\right\}\right.\right. \quad (5)$$

The net annual cost can be calculated by Eq. (6),

$$\text{Net annual cost} = \text{Cost - - Benefits} \\ = (\text{Repayment of loan} + \text{Maintenance cost}) - (\text{Total saving} - \text{Tax saving on depreciation}) \quad (6)$$

The present worth of net annual cost can be calculated by Eq. (7)

$$\text{Present worth of Net annual cost} = (\text{Net annual cost})/((1 + 0.1))\hat{}\,\text{Year} \quad (7)$$

The calculation of IRR involves trial and error process till the Net present value becomes near to zero. For IRR 9.26%, the NPV is Rs. 8.16 and for IRR 9.25%, the NPV is Rs.33. Hence IRR value is 9.25%. The simple payback period for this project is 12 years.

4 Result and Discussions

The rainy season is most challenging to dry the clothes. This location has frequent rain for few hours in day time. This situation happens again and again for more times. Due to this condition, drying clothes in open space is more challenging in this coastal area. The sudden rain may wet the clothes which is half dried. The climatic conditions of Surathkal is most humidity and rain is also more from June to till December month. So, considering these scenarios, the experimental results has considered in the month of September, October and November. The 21-09-2022, 27-10-2022 and 01-11-2022 experimental result are considered for thermal performance of single pass forced convection solar air heater in the present study. The average outlet temperature for 0.05 kg/s mass flow rate is 48, 52 and 50 °C for the 21-09-2022, 27-10-2022 and 01-11-2022 respectively which is more than sufficient to dry the clothes. The solar air heater has average rise is temperature for above dates 15, 16 and 16 °C respectively. The average thermal efficiency for the 21-09-2022, 27-10-2022 and 01-11-2022 experimental result are 55, 54 and 66% respectively. The comparative study with previous literature is shown in the Table 2 (Fig. 4, Tables 3 and 4).

Table 2. Comparative study of outlet operational parameters of solar air heater

Author	Absorber plate area (m^2)	Air gap (m)	Outlet Temperature (°C)	Average Temperature Gain (°C)	Collector Efficiency (%)
Nikhil Jain et al. (2)	1.25	0.015	52	18	23
Mohamad Ramadan (5)	–	–	45	15	–
Present Study (Average)	2	0.12	50	16	58.33

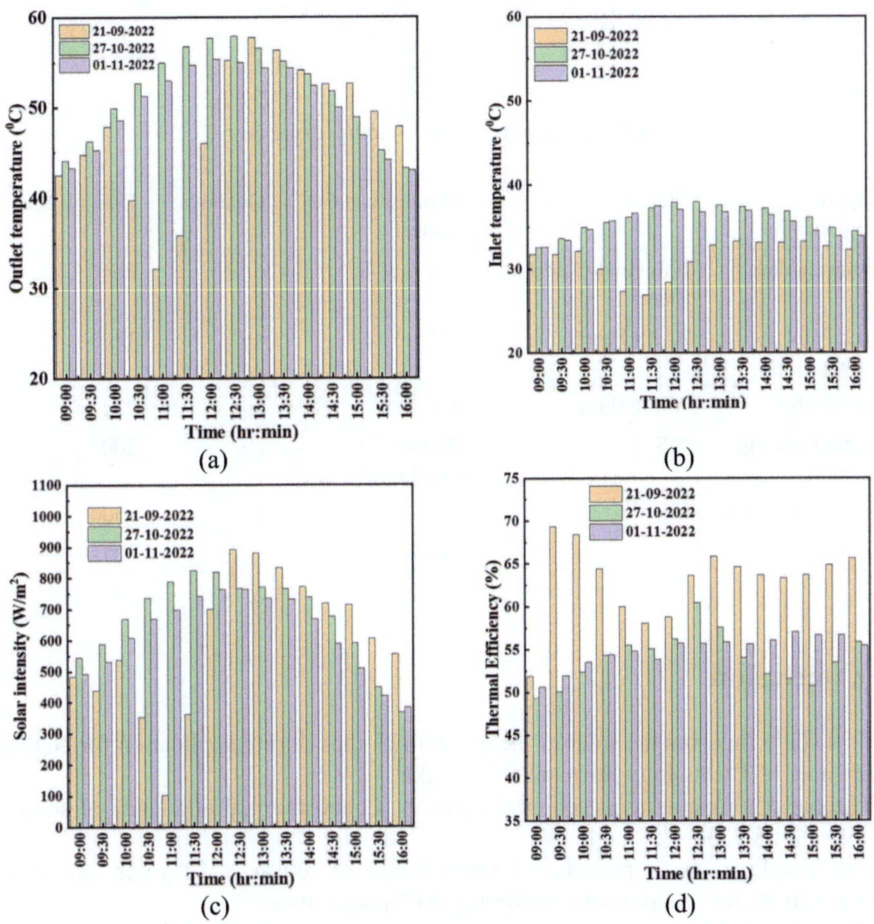

Fig. 4. The different thermal performance parameters (a) Outlet temperature, (b) Inlet temperature, (c) Solar intensity, (d) thermal efficiency for different time interval for 0.05 kg/s mass flow rates

Table 3. Details of single solar air heater cost

Name of Component and its quantity	Material	Dimensions (in meter)	Cost (In Rs.)
Cover Plate,1	Toughened Glass	2 × 1 × 0.004	2905
Absorber Plate, 1	Aluminium	2 × 1 × 0.004	450
Collector box, 1	Polyvinyl Chloride sheet	2.1 × 1.1 × 0.004	8000
Insulation, 1	Ceramic wool	2 × 1 × 0.005-Bottom 1 × 0.005–2 Sides	2400
Nut, Bolt Fitting Assembly, 60 pieces	MS	As per requirement	600
Black Paint Coating 6 bottles	Blank Paint	1200 ml	1038
		Total	15393

Table 4. Details of single clothes dryer cost

Name of Component	Material	Dimensions (in Meter)	Quantity	Cost (in Rs.)
Drying Cabinet	Polyvinyl Chloride sheet	1 × 1 × 1	6	1646
Blank Paint Coating	Black Paint	200 ml	1	173
Wire Mesh	Aluminium	1 × 1	1	732
Nut, bolt, Fitting Assembly	MS	As per requirement	20	200
Supporting Stand	MS	1 × 1 × 1	1	1500
		Total		4251

5 Conclusion

- This study involves the solar air heater with clothes drying application for National Institute of Technology Karnataka, Surathkal.
- The outlet temperature is achieved upto 58 °C and thermal efficiency of solar air heater is 58.33%.
- The simple payback period and internal rate of return is 12 years and 9.25% respectively for this project considering 10% discount rate.
- This solar air heater with clothes drying system is pollution free.

References

1. Jain, N., Garg, V., Mathur, J.: Thermal performance analysis of solar clothes dryer. J. Renew. Sustain. Ener. **5**(4) (2013). https://doi.org/10.1063/1.4816498
2. Geethanjali, H. S. S. B., Kumar, M.: https://thenitklife.weebly.com/boys.html, THENITKLIFE.
3. Diganjit, R., Gnanasekaran, N.: Chapter 2: Performance evaluation of single pass solar air heater with stepped-type arrangement of metal foam by a numerical study. In: Chaari, F., Gherardini, F., Ivanov, V., Haddar, M. (eds.) Conference on Fluid Mechanics and Fluid Power, vol. 7, 1st edn, pp. 17–29. Springer Nature Singapore, Singapore (2022)
4. Diganjit, R., Gnanasekaran, N., Mobedi, M.: Thermohydraulic efficiency of a solar air heater in the presence of graded aluminium wire mesh — a combined experimental – numerical study. Energies (Basel). **16**(15), 1–35 (2023). https://doi.org/10.3390/en16155633
5. Diganjit, R., Gnanasekaran, N., Mobedi, M.: Numerical study for enhancement of heat transfer using discrete metal foam with varying thickness and porosity in solar air heater by LTNE method. Energies (Basel). **15**(23), 8952 (2022). https://doi.org/10.3390/en15238952
6. Sukhatme, S.P., Nayak, J.K.: Solar energy, 4th edn. McGraw Hill (2018)
7. Garg, H.P., Prakash, J.: Solar Energy Fundamentals and Applications. McGraw Hill (2016)

Machine Learning Studies for Pyrolysis of Biomass: A Comprehensive Review

Ritisha Digvijay Kale and Chinta Sankar Rao(✉)

Department of Chemical Engineering, National Institute of Technology Karnataka Surathkal, Mangalore, Karnataka 575025, India
csrao@nitk.edu.in

Abstract. Biomass pyrolysis is a complex thermochemical conversion process influenced by multiple interdependent parameters such as temperature, heating rate, pyrolysis time, composition of feedstock, and type of catalyst. Traditional modeling approaches often struggle to capture these nonlinear interactions effectively. This comprehensive review explores the integration of machine learning (ML) techniques in biomass pyrolysis, highlighting their growing role in optimizing process conditions, predicting product yields, and accelerating research and development. ML algorithms, ranging from decision trees, support vector machines, and neural networks to ensemble and hybrid models, offer promising alternatives by learning patterns from experimental and literature datasets. The review evaluates recent studies on ML applications for predicting bio-oil, biochar, and syngas yields, analyzing feature importance, and model performance metrics like R^2, RMSE, and MAE. Finally, future trends are suggested, including model interpretability, integration with kinetic modeling, and the use of real-time data.

Keywords: Biomass pyrolysis · machine learning · thermochemical conversion

1 Introduction

Energy scarcity has become an economic threat to the growth of nations worldwide, and today's energy consumption is a significant and essential part of the development of a nation [1]. Nowadays, fossil fuels are the world's primary energy source. Fossil fuel consumption and natural resource wasting have risen dramatically in recent decades due to the ongoing development and encouragement of globalization, modernization, urbanization, and industrialization [2]. Concern about climate change, degradation of the environment, and global warming has been growing due to the use of fossil fuels [3]. The management of energy and technological advancement has emerged as key public goals in most industrialized and emerging nations due to the combined problems of environmental deterioration and the depletion of fossil fuels. Renewable energy appears as the most viable and sustainable alternative in this context. Solar and wind energy are great examples of renewable energy sources, but have significant output fluctuations, making them heavily dependent on energy storage systems [4]. With consideration of current energy circumstances, biomass is a promising, ecologically friendly alternative source of renewable energy that is widely accessible worldwide [5]. It includes agricultural residues, forestry waste, organic municipal waste, and energy crops.

H. P. Dasari et al. (Eds.): NITK-CREST 2025, PTT, pp. 110–121, 2026.
https://doi.org/10.1007/978-981-95-3218-6_9

The domestic biomass supply was 57.5 EJ worldwide in 2020. The domestic supply was 86% derived from solid biomass sources, including wood chips, wood pellets, and traditional biomass sources, and liquid biofuels constituted 7%, followed by biogas at 2% and the municipal and industrial waste sectors from 2% to 3%. As shown in Fig. 1, Total biomass supply has increased steadily, growing from about 42 EJ in 2000 to around 65 EJ in 2020 [6]. Currently, about 10% of the world's energy comes from waste and biomass. Nearly ten times the world's current energy needs, or 1.08 x 10^{11} tons of oil equivalent, would be the full estimated potential of annual biomass availability [7, 8]. The abundant availability of biomass, its renewability, CO_2 neutrality, and compatibility with coal-based technologies make it a highly attractive resource for clean energy and materials [7]. As shown in Fig. 2, in 2020, global electricity generation from biomass reached approximately 685 TWh, with solid biomass remaining the principal source of biopower, owing to its wide availability and compatibility with existing combustion technologies. Municipal and industrial waste contribute significantly to biomass-based electricity generation, serving the dual purpose of energy recovery and effective waste management. Biogas represents a smaller portion of biopower generation and is crucial in decentralized energy systems, particularly in regions with abundant organic waste resources [6]. These factors justify the growing research interest in biomass and support its practical utilization as a key component of the global energy transition. With continued innovation and sustainable practices, biomass can play a central role in reducing fossil fuel dependence and combating climate change. Particularly, choosing the right feedstock is essential to selecting the conversion process.

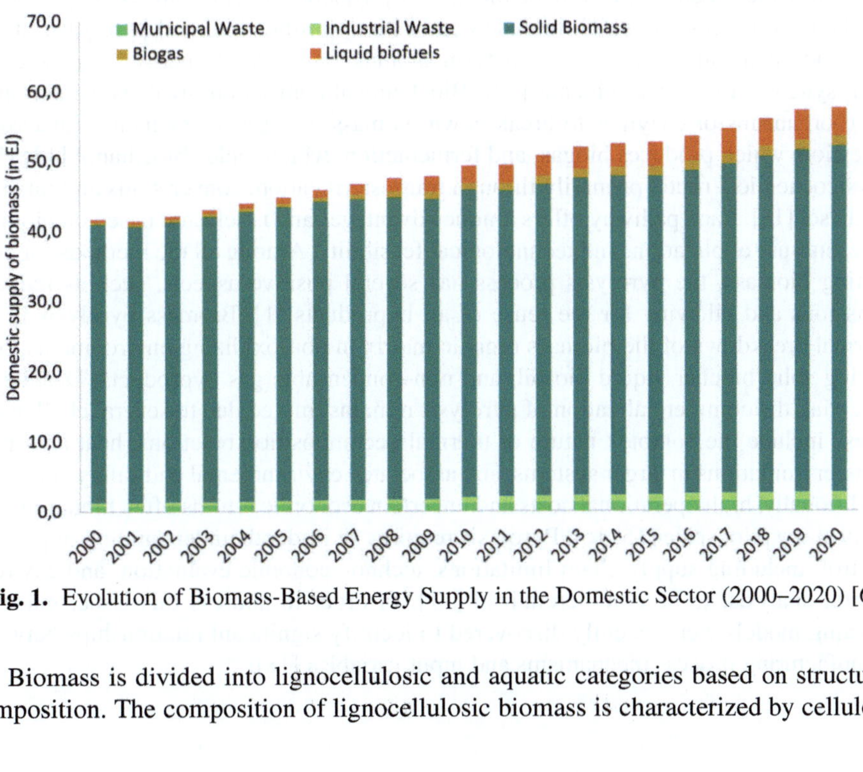

Fig. 1. Evolution of Biomass-Based Energy Supply in the Domestic Sector (2000–2020) [6]

Biomass is divided into lignocellulosic and aquatic categories based on structural composition. The composition of lignocellulosic biomass is characterized by cellulose

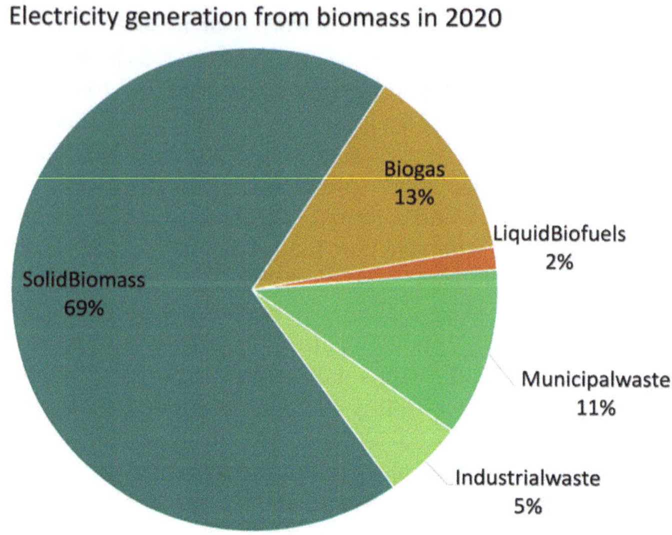

Fig. 2. Biomass Contribution to Global Power Generation in 2020 [6]

(37–50%), lignin (15–26%), and hemicellulose (22%–32%) [9]. Similarly, depending on the species, proteins compose 14–65% of aquatic biomass, carbohydrates 3–30%, and lipids 1–51% [10, 11]. Biomass can be converted into energy through three principal pathways: thermochemical, biochemical, and physicochemical conversion [5]. Thermochemical processes, which include combustion, gasification, and pyrolysis, utilize heat and chemical reactions to transform biomass into useful energy forms such as heat, syngas, bio-oil, and biochar [12]. Biochemical conversion involves the action of microorganisms or enzymes to break down biomass; key processes include anaerobic digestion, which produces biogas, and fermentation, which yields bioethanol [13]. The physicochemical route, primarily through transesterification, converts oils and fats into biodiesel [14]. Each pathway offers unique advantages and is selected based on biomass type, end-use application, and technological feasibility. Among all the methods for converting biomass, the pyrolysis process has several positive aspects, such as reduced emissions and allowing for the reuse of all byproducts [1]. Biomass pyrolysis is the thermal breakdown of the biomass organic matrix in non-oxidizing environments, producing solid biochar, liquid bio-oil, and non-condensable gas byproducts. Despite its potential, the commercialization of pyrolysis remains limited due to several challenges. These include the complex nature of thermal decomposition reactions, heat and mass transfer limitations in larger systems, and associated environmental and safety concerns. Additionally, high operational costs and uncertain economic returns often hinder its viability at the pilot scale [15, 16]. Pyrolysis requires an in-depth understanding of process control, including supply chain limitations, techno-economic evaluation, and environmental analysis, to be commercialized [17]. In order to address this issue, machine learning models were recently discovered to identify significant relationships between manufacturing process mechanisms and input variables [18].

Machine Learning (ML), a subfield of Artificial Intelligence (AI), has emerged as a powerful tool in industrial automation and intelligent decision-making. It involves the development of algorithms and models that can learn from data and make accurate predictions or decisions without being explicitly programmed for every scenario [19]. As industrial systems become increasingly complex and data-intensive, ML will continue to be an indispensable tool for automation, innovation, and intelligent decision-making. Presently, the majority of AI applications in renewable energy focus heavily on solar and wind energy forecasting—specifically, predicting power output based on weather data, irradiance, wind speed, and other environmental parameters [20]. Among various renewable energy sources, bioenergy has garnered significant attention, and one of the most promising pathways for harnessing bioenergy is thermochemical conversion, particularly pyrolysis [21].

Because of the simultaneous transfer of mass and heat as well as the subsequent interacting chemical reactions, pyrolysis is an extremely complex process. The content of biomass feedstocks also varies greatly, and the pyrolysis process is sensitive to a number of variables, such as reactor configuration, residence time, temperature, heating rate, catalyst use, and heating type [22]. Due to the complex and nonlinear nature of biomass pyrolysis, traditional modeling methods struggle to capture its intricacies with limited experimental data. In contrast, machine learning excels at learning from such data, making it highly effective for predicting bio-oil properties [16]. Machine learning accelerates pyrolysis research by enabling efficient prediction and optimization of product yields, process monitoring, and chemical kinetics. By analyzing large datasets, ML helps uncover complex relationships, aiding in process control and improvement [23, 24].

The objective of this review is to provide a comprehensive and critical assessment of recent advancements in the application of machine learning (ML) techniques to the pyrolysis of biomass, a key thermochemical pathway for renewable energy production. Biomass pyrolysis involves intricate interactions among feedstock properties, operating parameters, and catalytic effects, which challenge traditional modeling approaches. This review systematically explores how ML models, ranging from classical algorithms like decision trees and support vector machines to advanced neural networks and ensemble methods, have been employed to predict product yields, estimate kinetic parameters, optimize operational conditions, and assess catalyst performance. By identifying current limitations such as data scarcity, lack of standardization, and issues with model generalization, this review aims to inform future research directions and facilitate the integration of ML into practical and scalable biomass pyrolysis systems.

2 Fundamentals of Biomass Pyrolysis

2.1 Biomass Composition and Its Impact on Pyrolysis

Biomass energy resources are renewable, organic materials that can be converted into energy. Aquatic plants, timber and wood waste, animal waste, municipal waste, agricultural crop residues, and other biomass sources are all readily available [25]. Cellulose, hemicellulose, and lignin are the three major structural components of plant cell walls. These components are organic polymers formed via natural photosynthetic processes, making them biologically derived and abundant in all plant-based material. Lignocellulosic biomass typically consists of cellulose (40–60%), hemicellulose (20–40%), and lignin (10–24%) [26], though exact proportions vary with biomass type.

Cellulose, a crystalline polymer of β-linked glucose $(C_6H_{10}O_5)_n$, is the main structural component of biomass. Hemicellulose, a branched, amorphous polysaccharide with lower molecular weight, supports the structure. Though similar to cellulose, it differs in composition, containing various monosaccharides like xylose, galactose, glucose, arabinose, mannose, and sugar acids [27]. Xylans and glucomannans are the two most common forms of hemicellulose. Hardwood has considerable amounts of xylans. However, glucomannans are primarily found in softwood. Lignin offers rigidity and resistance. Between and around the cellulose and hemicellulose molecules, lignin acts as a binder. Lignin is a cross-linked amorphous copolymer made by randomly combining three distinct phenylpropane monomers: sinapyl, coniferyl, and pcoumaryl alcohols that have zero, one, and two methoxyl groups, respectively [28]. These three key components of lignocellulose material are arranged with acetyl groups, minerals, and phenolic substituents to form an extremely complex structure [27]. Using biomass as the basis for the pyrolysis-bioenergy-biochar platform has significant effects on the economic, environmental, and product yield performance [29]. The decomposition behavior of cellulose, hemicellulose, and lignin significantly affects pyrolysis product distribution. The author mentioned that Cellulose and hemicellulose primarily contribute to bio-oil and gas production, while lignin tends to yield more biochar [30].

2.2 Types of Pyrolysis

Three main types of biomass pyrolysis can be distinguished based on the heating rate and solid residence time: flash pyrolysis, fast pyrolysis, and slow (traditional) pyrolysis. Traditionally, slow pyrolysis, also known as carbonization, has been used to produce charcoal. Because the residence period is prolonged (ranging from hours to days), a broad range of particle sizes (5–50 mm) and a relatively low temperature (~300–700 °C). Biomass, mostly lignocellulosic, is broken down thermally at a very slow rate, allowing time for repolymerization reactions to maximize solid yields. Fast pyrolysis is the process of quickly heating biomass to a high temperature without oxygen. Biomass is often converted into liquids under conditions of low temperature, high heating rate, and brief residence time. The method of fast pyrolysis is gaining enormous traction in the production of various specialty and commodity chemicals, as well as liquid fuels [31]. Flash pyrolysis is a promising method that can produce up to 75% of the bio-oil yield from biomass and produce solid, liquid, and gaseous fuel. This process's features include high reaction temperatures between 450 and 1000 °C, quick devolatilization in an inert atmosphere, instantaneous particle heating rates, and a very brief gas residence time (less than 1 s) [1].

2.3 Pyro-Products

Pyrolysis produces three main products: biochar, bio-oil, and non-condensable gases. Biochar is a carbon-rich solid residue with carbon content ranging from 53–96 wt.%, depending on feedstock and conditions. It serves as a soil amendment for carbon sequestration, fertility enhancement, and pollution control. Biochar also finds use in catalytic processes, energy storage, and environmental applications. Additionally, it acts as a sustainable carbon platform for high-value materials [32]. The dark brown organic liquid known as "bio-oil" is produced when biomass is pyrolyzed. It contains water, oxygenated compounds, coal particles, and dissolved alkali metals. The composition is influenced by biomass moisture and reaction conditions. Bio-oil can serve as a renewable fuel for heating, transport, and power generation. It is also used as a feedstock for chemicals like adhesives, resins, and plasticizers [33]. Biogas from biomass pyrolysis mainly consists of CO_2, CO, H_2, CH_4, and other light gases. CO and CO_2 form from the breakdown of carbonyl and carboxyl groups. From methoxy and methylene linkage decomposition, light hydrocarbon arises. Hydrogen is produced through the breakdown of aromatics and C–H bonds. Syngas, a key component, serves as a sustainable fuel for IC engines and industrial use [34].

3 Machine Learning Application in Biomass Pyrolysis

Machine learning (ML) has emerged as an essential tool in pyrolysis research, allowing researchers to predict product yields, enhance process parameters, and categorize biomass or catalysts of various types with greater precision and efficacy [35]. Regression models, such as gradient boosting algorithms like XGBoost, decision trees, random forests, support vector regression (SVR), and linear regression, are frequently employed to estimate bio-oil, gas, or char yields based on factors such as temperature, heating rate, and biomass composition [36]. Artificial Neural Networks (ANNs), particularly deep neural networks, are adept at identifying intricate, non-linear correlations within the data, especially when dealing with multiple interacting variables. In pyrolysis research, machine learning models are used for a variety of tasks because they have strong prediction and analytical skills [37]. Predicting product yield, such as bio-oil, syngas, and biochar, based on input parameters like temperature, heating rate, biomass type, and catalyst employed, is one of the most common applications. Beyond prediction, machine learning supports the optimization of process conditions to maximize desired outputs or minimize energy consumption [38]. Through these applications, machine learning significantly enhances the ability to design, control, and improve pyrolysis systems with data-driven precision. As an instance, [39] used an ANN model in conjunction with the most successful Rao-2 algorithm to forecast biochar yield. The integrated model achieved low error (RMSE ~ 1.74%) and high accuracy (R^2 ~ 0.93). The investigation revealed that the three most crucial factors for this performance were the pyrolysis temperature (56%) and residence duration (23%), as well as the heating rate (8%) [39]. The motivation behind this study is that, although an extensive number of studies have examined the relationship between bio-oil yield and biomass attributes (Table 1), there are very few studies assessing the concurrent impact of operating circumstances and biomass composition.

In pyrolysis studies, accurate machine learning models rely on high-quality data and relevant features. Data typically comes from three sources: controlled experimental setups, literature (previous studies), and simulations from kinetic or process models.

Table 1. Literature Review

Research Objective	Model Developed	Key Finding	References
Using Pearson feature selection, 13 chosen variables were predicted.	RF, GBDT, XGB, SVR, RBFNN, and LM - BPNN	The gradient boosting decision tree with the Levenberg-Marquardt backpropagation neural network attained a training set R^2 greater than 0.9 and a testing set R^2 above 0.8.	[41]
The study's objective is to evaluate the accuracy of machine learning models in forecasting the yield and makeup of products during microwave pyrolysis.	SVR, RFR, and GBR	As demonstrated by their lower RMSE (below 0.765), lower R^2 value (above 0.822), and lower RMSE (less than 12.38), the models demonstrated superior prediction accuracy.	[42]
To develop a precise model for forecasting the production of nitrogen-enriched biomass pyrolysis products.	RF, GBDT, SVR, and ANN	The optimized GBDT model achieves an MSE of 10.04 and an R^2 of 0.92 in cross-validation.	[43]
To develop and validate a machine learning model using polynomial regression and Leave-One-Out Cross-Validation (LOOCV) to predict the yield of pyrolysis products.	Polynomial Regression machine learning model	All developed polynomial regression models achieved high prediction accuracy with $R^2 > 0.93$	[44]
The study intends to delineate the pyrolysis of wastewater sludge by the application of machine learning algorithms.	MLPNN, SVR, and RFR	The random forest regression model exhibited the highest accuracy among the evaluated models, achieving a correlation value of 0.813 and a relative mean squared error below 12.51.	[22]

Experimental data offers precision, literature data adds diversity, and simulations help when experiments are limited. Feature selection is critical to retain only impactful variables. Common features include biomass properties (moisture, volatile matter, fixed carbon, elemental composition) and process parameters (temperature, heating rate, residence time, particle size, catalyst type/concentration, and microwave power) [24]. In pyrolysis studies, regression models are evaluated using metrics that assess prediction accuracy and generalization. The coefficient of determination (R^2) shows how well the model explains target variability, with values near 1 indicating a strong fit. RMSE measures the square root of average squared errors, emphasizing large deviations. MAE calculates the average absolute error, offering a balanced, outlier-resistant view. MAPE shows errors as a proportion of real values, which makes it easier to compare different variables, though it can be unreliable when actual values are near zero [40].

4 Challenges and Future Directions

One of the primary challenges in applying machine learning (ML) to biomass pyrolysis is the limited availability of high-quality, standardized datasets. Most studies rely on small, lab-scale experimental data with varying feedstocks, operating conditions, and measurement protocols. This heterogeneity leads to inconsistent data structures and missing or poorly labeled information, reducing the reliability of ML model training and validation. Moreover, a lack of open-access databases hinders collaboration and benchmarking across research groups.

Many ML models in the pyrolysis domain are trained on narrow datasets that represent specific biomass types or experimental setups. As a result, these models often struggle to generalize across different feedstocks or scales (e.g., from laboratory to industrial reactors). The transferability of models remains limited due to variations in physicochemical properties, reactor designs, and operating conditions. Without robust generalization, deploying ML in real-world pyrolysis processes remains a significant hurdle. The future of ML in biomass pyrolysis lies in overcoming these limitations through several key advancements. First, creating comprehensive, shared datasets with standardized descriptors and metadata will enhance model robustness. Second, transfer learning and domain adaptation techniques can improve model performance across diverse conditions. Third, there is a growing trend toward hybrid models that integrate ML with mechanistic or kinetic models, combining data-driven predictions with physical interpretability. Ultimately, integrating ML into real-time process optimization and control systems will be a crucial step toward industrial deployment.

5 Conclusion

Machine learning offers significant potential to enhance the understanding, prediction, and optimization of biomass pyrolysis processes. Models such as Artificial Neural Networks (ANN), Support Vector Machines (SVM), Random Forest (RF), and Gradient Boosting (e.g., XGBoost) have shown strong predictive capabilities for outputs like bio-oil yield, gas composition, and char production. Commonly used inputs include process parameters (temperature, time, heating rate), proximate and ultimate analysis data, and catalyst properties. Feature selection significantly influences model accuracy and interpretability. This review has highlighted the various ML models applied in the field, their strengths, and limitations. While notable progress has been made, key challenges related to data availability, model generalization, and transferability still need to be addressed. These efforts will not only improve predictive accuracy but also enable the practical deployment of ML tools in real-time and industrial-scale pyrolysis systems. Continued interdisciplinary collaboration will be essential in driving this transformation and realizing the full potential of machine learning in biomass valorization.

References

1. Uddin, M.N., Techato, K., Taweekun, J., Rahman, M.M., Rasul, M.G., Mahlia, T.M.I., Ashrafur, S.M.: An overview of recent developments in biomass pyrolysis technologies. Energies. **11**(11), 3115 (2018). https://doi.org/10.3390/en11113115
2. Mohanty, A., Ajmera, S., Chinnam, S., Kumar, V., Mishra, R.K., Acharya, B.: Pyrolysis of waste oils for biofuel production: an economic and life cycle assessment. Fuel Commun. **18**, 100108 (2024). https://doi.org/10.1016/j.jfueco.2024.100108
3. Yang, Y., Shahbeik, H., Shafizadeh, A., Masoudnia, N., Rafiee, S., Zhang, Y., Pan, J., Tabatabaei, M., Aghbashlo, M.: Biomass microwave pyrolysis characterization by machine learning for sustainable rural biorefineries. Renew. Energy. **201**, 70–86 (2022). https://doi.org/10.1016/j.renene.2022.11.028
4. Ascher, S., Watson, I., You, S.: Machine learning methods for modelling the gasification and pyrolysis of biomass and waste. Renew. Sustain. Ener. Rev. (2022). https://doi.org/10.1016/j.rser.2021.111902
5. Jahirul, M.I., Rasul, M.G., Chowdhury, A.A., Ashwath, N.: Biofuels production through biomass pyrolysis- A technological review. Energies. **5**(12), 4952–5001 (2012). https://doi.org/10.3390/en5124952
6. WBA GBS: GLOBAL BIOENERGY STATISTICS 2022 World Bioenergy Association. (2022).
7. Kan, T., Strezov, V., Evans, T.J.: Lignocellulosic biomass pyrolysis: A review of product properties and effects of pyrolysis parameters. Renew. Sustain. Ener. Rev. (2016). https://doi.org/10.1016/j.rser.2015.12.185
8. Khan, A.A., de Jong, W., Jansens, P.J., Spliethoff, H.: Biomass combustion in fluidized bed boilers: potential problems and remedies. Fuel Process. Technol. **90**, 21–50 (2009). https://doi.org/10.1016/j.fuproc.2008.07.012
9. Hameed, S., Sharma, A., Pareek, V., Wu, H., Yu, Y.: A review on biomass pyrolysis models: Kinetic, network and mechanistic models. Biomass Bioener. (2019). https://doi.org/10.1016/j.biombioe.2019.02.008
10. Qu, T., Guo, W., Shen, L., Xiao, J., Zhao, K.: Experimental study of biomass pyrolysis based on three major components: hemicellulose, cellulose, and lignin. Ind. Eng. Chem. Res. **50**, 10424–10433 (2011). https://doi.org/10.1021/ie1025453
11. Pattanayak, S., Hauchhum, L., Loha, C., Sailo, L.: Selection criteria of appropriate bamboo based biomass for thermochemical conversion process. Biomass Convers. Biorefinery. **10**, 401–407 (2020). https://doi.org/10.1007/s13399-019-00421-5
12. Mckendry, P.: Energy production from biomass (part 1): overview of biomass. Bioresour. Technol. **83**(1), 37–46
13. Balat, M., Balat, H., Öz, C.: Progress in bioethanol processing. Prog. Ener. Combust. Sci. **34**(5), 551–573 (2008). https://doi.org/10.1016/j.pecs.2007.11.001

14. Kumar, P., Barrett, D.M., Delwiche, M.J., Stroeve, P.: Methods for pretreatment of lignocellulosic biomass for efficient hydrolysis and biofuel production. Indus. Eng. Chem. Res. **48**(8), 3713–3729 (2009). https://doi.org/10.1021/ie801542g
15. Khandelwal, K., Nanda, S., Dalai, A.K.: Machine learning to predict the production of bio-oil, biogas, and biochar by pyrolysis of biomass: a review. Environ. Chem. Lett. **22**, 2669–2698 (2024). https://doi.org/10.1007/s10311-024-01767-7
16. Akinpelu, D.A., Adekoya, O.A., Oladoye, P.O., Ogbaga, C.C., Okolie, J.A.: Machine learning applications in biomass pyrolysis: From biorefinery to end-of-life product management. Digit. Chem. Eng. **8**, 100103 (2023). https://doi.org/10.1016/j.dche.2023.100103
17. Khandelwal, K., Seraj, S., Nanda, S., Azargohar, R., Dalai, A.K.: Subcritical water conversion of biomass to biofuels, chemicals and materials: a review. Environ. Chem. Lett. **22**, 2191–2211 (2024). https://doi.org/10.1007/s10311-024-01750-2
18. He, Y., Liu, G., Li, C., Yan, X.: Reaching the full potential of machine learning in mitigating environmental impacts of functional materials. Rev. Environ. Contam. Toxicol. **260**, 21 (2022). https://doi.org/10.1007/s44169-022-00024-8
19. Tang, Q., Chen, Y., Yang, H., Liu, M., Xiao, H., Wang, S., Chen, H., Raza Naqvi, S.: Machine learning prediction of pyrolytic gas yield and compositions with feature reduction methods: effects of pyrolysis conditions and biomass characteristics. Bioresour. Technol. **339** (2021). https://doi.org/10.1016/j.biortech.2021.125581
20. Lai, J.P., Chang, Y.M., Chen, C.H., Pai, P.F.: A survey of machine learning models in renewable energy predictions. Appl. Sci. **10**(17), 5975 (2020). https://doi.org/10.3390/app10175975
21. Wang, S., Dai, G., Yang, H., Luo, Z.: Lignocellulosic biomass pyrolysis mechanism: A state-of-the-art review. Prog. Ener. Combust. Sci. (2017). https://doi.org/10.1016/j.pecs.2017.05.004
22. Shahbeik, H., Rafiee, S., Shafizadeh, A., Jeddi, D., Jafary, T., Lam, S.S., Pan, J., Tabatabaei, M., Aghbashlo, M.: Characterizing sludge pyrolysis by machine learning: towards sustainable bioenergy production from wastes. Renew. Energy. **199**, 1078–1092 (2022). https://doi.org/10.1016/j.renene.2022.09.022
23. Dong, Z., Bai, X., Xu, D., Li, W.: Machine learning prediction of pyrolytic products of lignocellulosic biomass based on physicochemical characteristics and pyrolysis conditions. Bioresour. Technol. **367** (2023). https://doi.org/10.1016/j.biortech.2022.128182
24. Cheenkachorn, K., Prapainainar, C., Wijakmatee, T.: Machine learning-driven modeling of biomass pyrolysis product distribution through thermal parameter sensitivity. Renew. Energy. **248** (2025). https://doi.org/10.1016/j.renene.2025.123108
25. Banerjee, D., Ghosh, U., Karmakar, S., Das, J.: Biomass energy potential in India: A review. Int. J. Eng. Res. **9**(11), 42–45 (2021)
26. Liu, J., Jia, H., Mairaj Deen, K., Xu, Z., Cheng, C., Zhang, W.: Application of machine learning methods for lignocellulose biomass pyrolysis: activation energy prediction from preliminary analysis and conversion degree. Fuel. **343** (2023). https://doi.org/10.1016/j.fuel.2023.128005
27. Putro, J.N., Soetaredjo, F.E., Lin, S.Y., Ju, Y.H., Ismadji, S.: Pretreatment and conversion of lignocellulose biomass into valuable chemicals. RSC Adv. (2016). https://doi.org/10.1039/c6ra09851g
28. Phanthong, P., Reubroycharoen, P., Hao, X., Xu, G., Abudula, A., Guan, G.: Nanocellulose: Extraction and application. Carb. Res. Conv. **1**(1) (2018). https://doi.org/10.1016/j.crcon.2018.05.004
29. Li, W., Dang, Q., Brown, R.C., Laird, D., Wright, M.M.: The impacts of biomass properties on pyrolysis yields, economic and environmental performance of the pyrolysis-bioenergy-biochar platform to carbon negative energy. Bioresour. Technol. **241**, 959–968 (2017). https://doi.org/10.1016/j.biortech.2017.06.049

30. Vuppaladadiyam, A.K., Vuppaladadiyam, S.S.V., Sahoo, A., Murugavelh, S., Anthony, E., Bhashkar, T., Zheng, Y., Zhao, M., Duan, H., Zhao, Y., Antunes, E., Sarmah, A.K., Leu, S.Y.: Bio-oil and biochar from the pyrolytic conversion of biomass: A current and future perspective on the trade-off between economic, environmental, and technical indicators. Sci. Total Environ. **857**(Pt 1), 159155 (2023). https://doi.org/10.1016/j.scitotenv.2022.159155
31. Demirbaş, A., Arin, G.: An overview of biomass pyrolysis. Energy Sources. **24**, 471–482 (2002)
32. Hu, X., Gholizadeh, M.: Biomass pyrolysis: A review of the process development and challenges from initial researches up to the commercialisation stage. Journal of Energy Chemistry. **39**(1–3) (2019). https://doi.org/10.1016/j.jechem.2019.01.024
33. Neves, D., Thunman, H., Matos, A., Tarelho, L., Gómez-Barea, A.: Characterization and prediction of biomass pyrolysis products. Fuel and Energy Abstracts. **37**(5), 611–630 (2011). https://doi.org/10.1016/j.pecs.2011.01.001
34. Valin, S., Cances, J., Castelli, P., Thiery, S., Dufour, A., Boissonnet, G., Spindler, B.: Upgrading biomass pyrolysis gas by conversion of methane at high temperature: experiments and modelling. Fuel. **88**, 834–842 (2009). https://doi.org/10.1016/j.fuel.2008.11.033
35. Khan, M., Raza Naqvi, S., Ullah, Z., Ali Ammar Taqvi, S., Nouman Aslam Khan, M., Farooq, W., Taqi Mehran, M., Juchelková, D., Štěpanec, L.: Applications of machine learning in thermochemical conversion of biomass-a review. Fuel. **332** (2023). https://doi.org/10.1016/j.fuel.2022.126055
36. Chinenye Divine, D., Hubert, S., Epelle, E.I., Ojo, A.U., Adeleke, A.A., Ogbaga, C.C., Akande, O., Okoye, P.U., Giwa, A., Okolie, J.A.: Enhancing biomass pyrolysis: predictive insights from process simulation integrated with interpretable machine learning models. Fuel. **366** (2024). https://doi.org/10.1016/j.fuel.2024.131346
37. Sunphorka, S., Chalermsinsuwan, B., Piumsomboon, P.: Artificial neural network model for the prediction of kinetic parameters of biomass pyrolysis from its constituents. Fuel. **193**, 142–158 (2017). https://doi.org/10.1016/j.fuel.2016.12.046
38. Sun, Y., Liu, L., Wang, Q., Yang, X., Tu, X.: Pyrolysis products from industrial waste biomass based on a neural network model. J. Anal. Appl. Pyrolysis. **120**, 94–102 (2016). https://doi.org/10.1016/j.jaap.2016.04.013
39. Khan, M., Ullah, Z., Mašek, O., Raza Naqvi, S., Nouman Aslam Khan, M.: Artificial neural networks for the prediction of biochar yield: a comparative study of metaheuristic algorithms. Bioresour. Technol. **355** (2022). https://doi.org/10.1016/j.biortech.2022.127215
40. Sinha, S., Sankar Rao, C., Kumar, A., Venkata Surya, D., Basak, T.: Exploring and understanding the microwave-assisted pyrolysis of waste lignocellulose biomass using gradient boosting regression machine learning model. Renew. Energy. **231** (2024). https://doi.org/10.1016/j.renene.2024.120968
41. Ma, J., Zhang, S., Liu, X., Wang, J.: Machine learning prediction of biochar yield based on biomass characteristics. Bioresour. Technol. **389** (2023). https://doi.org/10.1016/j.biortech.2023.129820
42. Yang, Y., Shahbeik, H., Shafizadeh, A., Masoudnia, N., Rafiee, S., Zhang, Y., Pan, J., Tabatabaei, M., Aghbashlo, M.: Biomass microwave pyrolysis characterization by machine learning for sustainable rural biorefineries. Renew. Energy. **201**, 70–86 (2022). https://doi.org/10.1016/j.renene.2022.11.028

43. Bi, D., Wang, H., Liu, Y., Qin, Z., Song, X., Liu, S.: Machine learning-based prediction model for the yield of nitrogen-enriched biomass pyrolysis products: performance evaluation and interpretability analysis. J. Anal. Appl. Pyrolysis. **182** (2024). https://doi.org/10.1016/j.jaap.2024.106723
44. Potnuri, R., Suriapparao, D.V., Sankar Rao, C., Sridevi, V., Kumar, A.: Effect of dry torrefaction pretreatment of the microwave-assisted catalytic pyrolysis of biomass using the machine learning approach. Renew. Energy. **197**, 798–809 (2022). https://doi.org/10.1016/j.renene.2022.08.006

Machine Learning Studies for Pyrolysis of Plastic: A Comparative Review

P. V. Malavika(✉) and Chinta Sankar Rao

Department of Chemical Engineering, National Institute of Technology Karnataka Surathkal, Mangalore 575025, Karnataka, India
malavikapv7041@gmail.com, csrao@nitk.edu.in

Abstract. This review explores the role of machine learning (ML) in predicting and optimizing product yields from plastic pyrolysis. It begins by outlining the fundamentals of plastic pyrolysis, including the behavior of various plastic types and key process variables such as temperature, catalysts, and residence time. The focus is on regression-based supervised learning models, including Support Vector Regression, Gaussian Process Regression, and ensemble tree methods. Applications of these models in enhancing oil, gas, and char yield predictions are critically discussed. The review concludes with insights into current challenges, data limitations, and recommendations for future research in ML-guided pyrolysis.

Keywords: Plastic pyrolysis · Machine learning · Optimization

1 Introduction

Plastic is a high-molecular-weight synthetic polymer derived from organic compounds such as ethylene and vinyl chloride, primarily sourced from natural gas, coal, and petroleum. Due to its durability, versatility, and low cost, plastic is widely used across sectors—including packaging, construction, electronics, and especially healthcare, where single-use medical devices account for 85% of equipment [1]. Global plastic waste generation has surged with increased reliance on plastics, particularly single-use items, leading to severe environmental concerns. The output of plastic increased from 2 million tons in 1950 to 400.3 million tons in 2022, and by 2050, it is expected to surpass 650 million tons [2]. In India, plastic consumption grew from 5 million tonnes in 2005 to an estimated 24 million tonnes by 2020, with per capita usage expected to double from 11 kg to 20 kg by 2022. A large portion of plastic waste ends up in landfills or is incinerated, the latter releasing harmful emissions that threaten public health and the environment. Additionally, plastic production consumes 7–8% of fossil fuels 4% as feedstock and 3–4% for energy [2]. Post-consumer plastic waste undergoes pre-treatment (e.g., washing, size reduction, extrusion) before chemical recycling, which includes methods such as depolymerization, pyrolysis, gasification, and hydrogenation. Depolymerization recovers monomers for new polymer synthesis but requires purification. Hydrogenation yields saturated liquids for energy use but is costly due to hydrogen demand, while gasification

H. P. Dasari et al. (Eds.): NITK-CREST 2025, PTT, pp. 122–132, 2026.
https://doi.org/10.1007/978-981-95-3218-6_10

emits harmful compounds and is energy-intensive and economically less viable. Pyrolysis offers an efficient alternative for converting plastic waste into energy-rich products at approximately 500 °C under inert conditions. Owing to plastics' high carbon, hydrogen, and volatile content, the process yields valuable fuels like paraffins, olefins, and aromatics. It produces three main outputs: non-condensable gases (2–47%), condensable vapors (31–91%) convertible into oil/wax, and char (0–16%) [3]. Pyrolysis is favored for its low pre-treatment needs, high efficiency, cost-effectiveness, and reduced emissions. It can be thermal,operating at higher temperatures for broader hydrocarbon yields or catalytic, which promotes selective production of highvalue products at lower temperatures via vapor-phase upgrading. Product yield and quality are influenced by feedstock type, particle size, and process conditions such as temperature and heating rate. Catalyst variability adds complexity, making optimization resource-intensive. Large, high-dimensional datasets are used in machine learning to create prediction models, which are becoming more popular in the study of plastic conversion and pyrolysis process optimization. To illustrate the potential of machine learning for in-depth research of pyrolysis behavior, Hooda et al. [4] used random forest (RF) models for predictive modeling of plastic pyrolysis kinetics. These models not only enhance the understanding of complex reaction mechanisms but also enable the prediction of product yields and optimization of process conditions with greater accuracy and efficiency. Consequently, integrating machine learning into pyrolysis research paves the way for data-driven decision-making, reducing experimental costs and accelerating the development of sustainable waste-to-energy technologies. The objective of this review is to systematically examine and synthesize existing research on the application of machine learning (ML) techniques in the pyrolysis of plastic. It focuses on identifying and categorizing various ML models used for predicting product yields, optimizing process parameters, and interpreting complex process behaviours. Methodologies for feature selection, model training, dataset preparation, and performance evaluation are also highlighted in the paper. The scope centres on studies that specifically address plastic pyrolysis using ML approaches, encompassing supervised, unsupervised, and explainable AI techniques. Particular emphasis is placed on the integration of ML with process optimization, design of experiments, and its potential for real-time industrial implementation.

2 Fundamentals of Plastic Pyrolysis

2.1 Types of Plastics and Their Pyrolysis Behaviour

Different types of plastics exhibit distinct pyrolysis behaviours due to variations in their chemical composition, thermal stability, and volatile content. Most early pyrolysis studies predominantly used virgin or resin plastics in small-scale reactors, while recent work has increasingly focused on real municipal plastic waste (MPW) to better represent practical conditions. Studies have also explored the recovery of valuable aromatics like styrene, ethylbenzene, and toluene from MPW-derived oils. Thermogravimetric (TG) and derivative thermogravimetric (DTG) analyses reveal that pyrolysis behaviour varies: polystyrene (PS) and polyethylene (PE) decompose in a single stage, while polyvinyl chloride (PVC) undergoes two stages. During co-pyrolysis, PS decomposition is restrained by PE, whereas PS accelerates the degradation of PVC, indicating synergistic or antagonistic effects based on the plastic mix [5]. The variation in behaviour is also linked to proximate analysis parameters like volatile matter and ash content. High volatile content enhances liquid oil yield, while high ash content leads to increased gas and char formation. For example, polyethylene terephthalate (PET), which has lower volatile content (86.8%), yields less oil (23–40 wt %) and more gases (52–77 wt %) under similar pyrolysis conditions [6]. In contrast, high-density polyethylene (HDPE), due to its high volatile content and linear polymer structure, produces significantly higher oil yields—up to 84.7 wt % under optimized conditions. PVC is chemically different from common thermoplastics like polyethylene and polystyrene due to its high chlorine content (57%), however, pyrolysis of PVC is less explored because it releases hydrogen chloride (HCl) gas at elevated temperatures.

2.2 Mechanism and Pathways of Plastic Pyrolysis

Understanding the thermal degradation pathways and underlying mechanisms is crucial to optimizing pyrolysis processes for different plastic types. Long-chain polymers are broken down into smaller fractions during the plastic pyrolysis process, yielding extremely valuable oils and gases that can be used as fuel and chemical precursors [7]. When polypropylene and polyethylene are pyrolyzed at 500–600 °C without the use of a catalyst, the main product is C20+ hydrocarbons, which at room temperature exist as solid wax. Higher temperatures favour the formation of lighter fractions such as C1–C4 olefins, C5–C20 olefins, and a minor portion of aromatics [73].Thermal cracking of PE produces a range of hydrocarbons: C1–C4 gases, C5–C10 non-aromatics, monoaromatics, and waxes (>C11). Higher temperatures increase gas yield, decrease waxes, and slightly enhance aromatic formation. Pyrolysis involves free radical mechanisms, including chain scission, secondary decomposition, and radical recombination [8]. PS pyrolysis proceeds through similar free radical mechanisms. Its major products include aromatic monomers, dimers, and trimers. In contrast, PVC behaves differently under pyrolysis.

Even without catalysts, it can yield significant amounts of liquid oil rich in polycyclic aromatic hydrocarbons (PAHs), rather than wax. The thermal degradation of PVC follows three stages: (1) dehydrochlorination with internal cyclization, (2) aromatic chain scission, and (3) release of 2–4 ringed aromatic compounds. PET follows yet another pathway. Its thermal degradation primarily involves β -scission and retro-hydroalkoxylation, leading to benzoic acid and vinylic compounds. PET decomposition also involves the breakdown of ester linkages, releasing substantial amounts of CO_2, CO, and ethylene [9].

2.3 Process Types: Thermal, Catalytic, Microwave-Assisted Pyrolysis

Plastic pyrolysis is broadly categorized into slow and fast types based on heating rates. Slow pyrolysis, with longer vapor residence times, is ideal for char production, while fast pyrolysis—subdivided into flash and ultra-rapid—promotes high yields of oil and gas. Typical pyrolysis temperatures for plastics range from 500 °C to 700 °C, depending on plastic type and catalysts. For instance, Miandad et al. [10] achieved oil yields of 54% and 50% from PS, PE, PET, and PP waste at 450 °C using zeolite catalysts, while

Das and Tiwari reported high-octane oil from PP slow pyrolysis at 300–400 °C. PP pyrolysis mainly produces C_3–C_4 gases, and the resulting char, with low moisture and carbon content, has been used as an adsorbent for water treatment. Process parameters such as temperature, reactor type, pressure, residence time, and catalysts are critical for optimizing yield and product composition.

2.3.1 Temperature

Temperature is an important factor in thermal pyrolysis, driving polymer chain scission by enabling cracking reactions once molecular vibrations exceed the enthalpy of C-C bonds. TGA is commonly used to study this behavior, offering TG and DTG curves that reveal weight loss and degradation rates with temperature and time [8]. Thermal degradation profiles differ significantly across plastic types. For instance, PET begins degrading at 400 °C, peaking at 427.7 °C and fully decomposing around 470 °C within 350–520 °C [6]. HDPE degrades between 378–404 °C, peaks at 467 °C, and completes at 517–539 °C. PVC exhibits a two-stage degradation: 260–385 °C (major weight loss) and 385–520 °C (over 83% weight reduction), with minimal changes beyond 800 °C. PP degrades slightly earlier than HDPE, initiating just below 400 °C and peaking at 447 °C. PS shows the lowest onset, with full decomposition into oil by 350 °C and maximum yield at 425 °C, shifting to gas at higher temperatures. Overall, 300–500 °C optimizes liquid yields, while temperatures above 500 °C favor gas and char [11].

2.3.2 Type of Reactors

Reactor selection greatly affects mixing, residence time, heat transfer, and efficiency in plastic pyrolysis. Laboratory-scale setups use batch, semi-batch, and continuous-flow systems like fluidized bed, fixed-bed, CSBR, and microwave-assisted reactors. Batch reactors, while simple and high-conversion, are labor-intensive and hard to scale, whereas semi-batch reactors offer better selectivity with small-scale suitability. Fixed-bed reactors face limitations from irregular feed and catalyst surface, while fluidized beds improve heat/mass transfer and continuous operation. Microwave-assisted pyrolysis enables rapid, efficient heating and high liquid yields, though scale-up is hindered by limited industrial data and material-specific requirements [12].

2.3.3 Catalysts

Catalysts play a vital role in plastic pyrolysis by enhancing reaction rates, reducing activation energy, and lowering energy costs, critical factors for large-scale commercialization. They optimize product yields, selectivity, and quality, producing chemicals like automotive fuels and light olefins. Heterogeneous catalysts, including zeolites (HZSM-5, HUSY), silica-alumina, MCM-41, metal-supported (Ni, Fe, Pt), and basic oxides (MgO, CaO), are preferred for their reusability and efficiency [13]. Zeolites and FCC catalysts improve cracking efficiency, product selectivity, and pyrolysis oil quality to meet fuel standards.

2.4 Key Products: Liquid Oil, Syngas, Solid Residue and Their Applications

Plastic pyrolysis yields three main products: liquid oil (or wax), non-condensable gas (syngas), and solid char. These outputs have diverse applications, including as fuels, chemicals, and materials [14].

2.4.1 Liquid Oil/Wax

Plastic pyrolysis yields liquid oil or wax, with slow pyrolysis favoring oil and fast pyrolysis yielding more wax. This oil, rich in paraffins, olefins, and aromatics, varies in composition based on the plastic feedstock—PS produces high styrene and aromatics, while PE and PP yield aliphatic-rich oils. Plastic oil generally offers a high heating value (45 MJ/kg), low water and ash content, and near-neutral pH [14]. Its applications range from electricity generation and boiler fuel to blending with conventional diesel (up to 50% with satisfactory performance) and as a feedstock for petrochemical processes.

2.4.2 Non-condensable Gas (Syngas)

The non-condensable gas produced during pyrolysis is a valuable co-product with a high calorific value (44 MJ/kg), comparable to natural gas [15]. Its yield is generally enhanced at higher pyrolysis temperatures, with typical components including methane, ethylene, ethane, propylene, butadiene, and butane. This gas can be directly utilized for heat generation, potentially making the pyrolysis process self-sustaining and independent of external energy inputs. Beyond energy recovery, it serves as a precursor for high-value products such as ethylene and propylene through steam reforming, or carbon nanotubes (CNTs) via chemical vapor deposition (CVD) processes [14]. Additionally, non-condensable gas can be reutilized as a fluidizing medium in continuous pyrolysis, which has been shown to increase liquid oil yields.

2.4.3 Solid Residue (Char)

The solid product of pyrolysis primarily consists of coke and ash. Its yield is generally lower compared to bio-based char due to the lower fixed carbon and higher volatile matter content of plastics. Despite its relatively low yield, char adds value to the pyrolysis process by serving as a substitute for fossil fuels, carbon black, and pigments, contributing to a "plastic-closed-loop economy" [14]. Additionally, char can be combusted alongside non-condensable gases to recover thermal energy, enhancing the self-sustainability of the process. It also holds potential as an adsorbent and can be further processed to recover high-value products like multi-walled carbon nanotubes (MWCNTs).

3 Machine Learning Applications in Plastic Pyrolysis

3.1 Overview of ML Techniques in Plastic Pyrolysis

ML has emerged as a powerful tool in plastic pyrolysis, enabling accurate prediction of product yields, identification of key process parameters, and process optimization. Techniques such as ensemble learning (boosting and bagging), Support Vector Machine

(SVM), Neural Networks Models and Gaussian Process Regression (GPR) model oil, gas, and char yields based on feedstock properties, operating conditions, and catalysts. Dataset quality—typically sourced from laboratory experiments or literature—and composition are critical for model accuracy. These datasets include plastic types (HDPE, LDPE, PP, PET), process conditions (e.g., temperature, heating rate), and physicochemical properties (e.g., proximate, ultimate analysis). Target outputs focus on product yields or compositions. With limited data sizes, effective feature selection via domain expertise, correlation analysis, or PCA is essential to enhance model performance and prevent overfitting. Tree-based models and feature engineering (normalization, encoding, derived features) further improve prediction effectiveness.

3.1.1 Performance Evaluation Metrics for ML Models

In plastic pyrolysis studies using machine learning, performance is typically evaluated using metrics score such as coefficient of determination (R^2), mean absolute deviation (MAD), mean squared error (MSE) and mean absolute percentage error (MAPE) [16]. Better performance is indicated by a smaller MAD, which calculates the average absolute difference between predicted and actual data (Eq. 1). Root mean squared error (RMSE) assesses the spread of prediction errors, with lower values suggesting a closer fit to actual data (Eq. 2). MAPE is used to express prediction accuracy in percentage terms; values below 10% are considered highly accurate, 10–20% indicate good accuracy, and values exceeding 50% suggest poor predictive ability (Eq. 3). R^2 is another widely used metric, reflecting how well the model explains the variance in the response variable, values closer to 1 indicate a stronger fit (Eq. 4) [17].

$$\text{MAE} = \frac{1}{n}\sum_{i=1}^{n}|y_i - \hat{y}_i| \tag{1}$$

$$\text{RMSE} = \sqrt{\frac{1}{n}\sum_{i=1}(y_i - \hat{y}_i)^2} \tag{2}$$

$$\text{MAPE} = \frac{100\%}{n}\sum_{i=1}^{n}\left|\frac{y_i - \hat{y}_i}{y_i}\right| \tag{3}$$

$$R^2 = 1 - \frac{\text{Sum of Squares of Residuals}}{\text{Total Sum of Squares}} \tag{4}$$

3.2 Supervised Learning for Plastic Pyrolysis Predictions

Supervised learning, particularly regression-based models, is widely applied in predicting the product yields (oil, gas, and char) from plastic pyrolysis. In supervised learning, the model is trained on labeled datasets, allowing algorithms to learn the relationship between input and output and make reliable predictions, as illustrated in Fig. 1.These models learn the mapping between input features (such as feedstock properties and operating conditions) and output responses (product distribution) (Table 1).

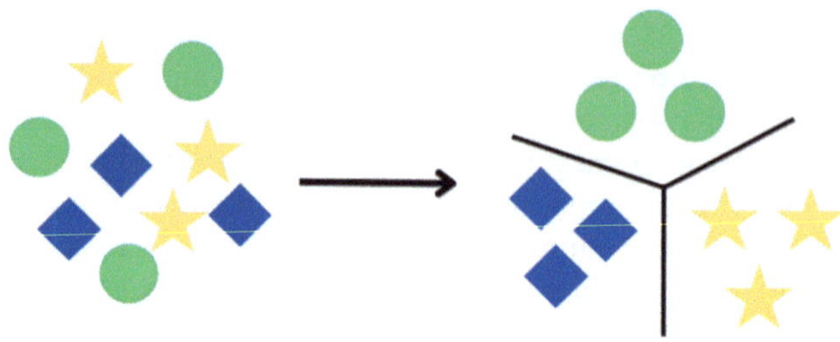

Fig. 1. An outline of supervised learning. Examples of input are grouped into a predetermined set of classes.

3.2.1 Regression Based Models

Regression methods predict plastic pyrolysis yields from feedstock and process parameters. Linear Regression models simple relationships [23], while SVR and GPR handle non-linear patterns, with GPR also estimating uncertainty [24]. Decision Trees offer interpretability, Random Forests improve accuracy via ensembles, and boosting methods like GB and XGBoost refine complex predictions [25]. Common parameters for these regression-based models are listed in Table 2.

Table 2. Common Parameters for Regression-based Machine Learning Models with References [24–26]

Model	Common Parameters
Support Vector Regression (SVR)	Kernel (linear, poly, RBF), C (regularization), Epsilon, Gamma
Gaussian Process Regression (GPR)	Kernel, Alpha, Optimizer, N-restarts optimizer
Decision Tree (DT)	Criterion (gini, entropy), Max depth, Min samples split, Min samples leaf
Random Forest (RF)	N estimators, Criterion, Max depth, Min samples split, Bootstrap
Gradient Boosting (GB)	Learning rate, N estimators, Max depth, Min samples split, Subsample
XGBoost (XGB)	Learning rate (eta), Max depth, N estimators, Subsample, Gamma, Min child weight

4 Machine Learning Prediction of the Effect of Different Parameters on Pyrolysis Product Yield

Machine learning models effectively predict how parameters like temperature, catalyst type, feedstock composition, and inert gas flow rate influence pyrolysis product yields. Explainable AI tools such as SHAP and Partial Dependence Plots help identify key factors affecting oil, gas, and char yields. For example, volatile matter and fixed carbon

Table 1. Overview of machine learning models used in plastic pyrolysis literature

Reference	ML model	Feedstock	Input Features	Predicted Output	Key Findings
[18]	LR, SVR, RFR, GBR, XGBR	Polyolefin plastics (PP, LDPE, HDPE, LLDPE)	Plastic properties, pyrolysis conditions, catalyst loading, acid properties, pore structure	Gas and oil yields	XGBR showed $R^2 \geq 0.91$, RMSE and MAE ≤ 6.21 wt%.
[19]	LR, DT, RF, XG-Boost, ANN	PS, HDPE, LDPE, PP, PVC, PET (325 data points)	Plastic concentration, pyrolysis temp, heating rate, particle size, feed amount, catalyst, reactor type	Oil yield	XGBoost gave highest accuracy; MAE of 9.1%.
[20]	SVR, DT, NN, GP	PP, PE, PVC, PS, PET	Plastic type, ultimate analysis, particle size, feed intake, pyrolysis temp, residence time	Oil, gas, char yields, liquid/gas compositions	DT and ANN had R^2 of 0.984 and 0.926; GP and SVR lower (0.592, 0.583).
[21]	NN	HDPE, LDPE, PP, PS mixture	Ratios of different plastics	Char, oil, gas yields	$R^2 > 0.9$, good agreement between experimental and predicted.
[22]	NN	Mixed plastics	Temperature, heating rate	Weight loss	Very high accuracy; $R > 0.9999$.

strongly predict biochar yield, while moisture content and pyrolysis temperature impact bio-oil and biogas production [27]. Fixed carbon, ash, and carbon content also influence activation energy in large-scale pyrolysis [4]. Studies show oil yield rises with temperature up to 500 °C, then declines due to secondary cracking, and that feed-to-catalyst ratio and inert gas flow optimize liquid recovery [4]. Additionally, reaction temperature, BET surface area, and zeolite Si/Al ratio are critical for liquid yield, while the PE ratio affects C5–C12 outputs [16]. These findings highlight ML's role in optimizing pyrolysis for efficient, sustainable bio-product recovery.

5 Challenges and Future Prospects

Machine learning (ML) in plastic pyrolysis shows promise for optimizing parameters and predicting yields, but challenges remain. Data quality and variability in plastic waste complicate model robustness, while capturing complex non-linear relationships among factors like temperature, catalyst, and feedstock is difficult for traditional regression methods. Approaches such as Gaussian Process Regression, Support Vector Regression, and multiple linear regression offer some effectiveness but are limited by scarce high-quality data. Although more interpretable than black-box models, these methods benefit from explainable AI tools like SHAP and Partial Dependence Plots to improve understanding. Scaling ML models for industrial use introduces further uncertainty, requiring better generalization and adaptability. Future work should focus on data augmentation, enhanced interpretability, and integration with real-time monitoring to improve the reliability and scalability of ML for sustainable pyrolysis optimization.

6 Conclusions

This review highlights the integration of machine learning with plastic pyrolysis, demonstrating its capability to optimize process parameters, improve reaction efficiency, and accurately predict product yields. Regressionbased ML models effectively identify key factors such as temperature, pyrolysis time, and catalyst type, enabling data-driven strategies that enhance oil yield and product quality. Beyond efficiency gains, ML-driven pyrolysis supports sustainable waste management by converting plastics into valuable chemicals and fuels, aligning with circular economy goals. Future research should focus on scalability, real-time industrial integration, improved data availability, and coupling ML with computational fluid dynamics and kinetic modeling to refine process predictions. Overall, machine learning offers a transformative path toward more efficient, sustainable, and economically viable plastic waste recycling.

References

1. Ilyas, M., Ahmad, W., Khan, H., Yousaf, S., Khan, K., Nazir, S.: Plastic waste as a significant threat to environment - a systematic literature review. Rev. Environ. Heal. **33** (2018)
2. Fayshal, M.A.: Current practices of plastic waste management, environmental impacts, and potential alternatives for reducing pollution and improving management. Heliyon. **10**(23), e40838 (2024) issn: 2405-8440
3. Gasim, M.F., Choong, Z.-Y., Koo, P.-L., Low, S.-C., Abdurahman, M.-H., Ho, Y.-C., et al.: Application of biochar as functional material for remediation of organic pollutants in water: an overview. Catalysts. **12**(2), 210 (2022)
4. Hooda, S., Mondal, P.: Predictive modeling of plastic pyrolysis process for the evaluation of activation energy: explainable artificial intelligence based comprehensive insights. J. Environ. Manag. **360**, 121189 (2024)
5. Li, D., Lei, S., Wang, P., Zhong, L., Ma, W., Chen, G.: Study on the pyrolysis behaviors of mixed waste plastics. Renew. Energy. **173**, 662–674 (2021)
6. Cepeliogullar, O., Putun, A.E.: Utilization of two different types of plastic wastes from daily and industrial life. In: Ozdemir, C., Sahinkaya, S., Kalipci, E., Oden, M.K. (eds.) ICOEST Cappadocia 2013, pp. 1–13 (2013)

7. Kunwar, B., Cheng, H., Chandrasekaran, S.R., Sharma, B.: Plastics to fuel: a review. Renew. Sust. Energ. Rev. **54**, 421–428 (2016)
8. Kumar, S., Singh, R.K.: Recovery of hydrocarbon liquid from waste high density polyethylene by thermal pyrolysis. Braz. J. Chem. Eng. **28**, 659–667 (2011)
9. Artetxe, M., Lopez, G., Amutio, M., Elordi, G., Olazar, M., Bilbao, J.: Operating conditions for the pyrolysis of poly-(ethylene terephthalate) in a conical spouted-bed reactor. Ind. Eng. Chem. Res. **49**, 2064–2069 (2010)
10. Miandad, R., Barakat, M.A., Aburiazaiza, A.S., Rehan, M., Ismail, I.M.I., Nizami, A.S.: Effect of plastic waste types on pyrolysis liquid oil. Int. Biodeterior. Biodegrad. **119**, 239–252 (2017) issn: 0964-8305
11. Sharuddin, S.D.A., Abnisa, F., Daud, W.M.A.W., Aroua, M.K.: A review on pyrolysis of plastic wastes. Ener. Convers. Manag. **115**, 308–326 (2016) issn: 0196-8904
12. Fernandez, Y., Arenillas, A., Menendez, J.A.: Microwave heating applied to pyrolysis. In: Advances in Induction and Microwave Heating of Mineral and Organic Materials. InTech, Spain (2011)
13. Stelmachowski, M.: Thermal conversion of waste polyolefins to the mixture of hydrocarbons in the reactor with molten metal bed. Energy Convers. Manag. **51**, 2016–2024 (2010)
14. Peng, Y., Wang, Y., Ke, L., Dai, L., Wu, Q., Cobb, K., Zeng, Y., Zou, R., Liu, Y., Ruan, R.: A review on catalytic pyrolysis of plastic wastes to high-value products. Energy Convers. Manag. **254**, 115243 (2022)
15. Tsuji, T., Hasegawa, K., Masuda, T.: Thermal cracking of oils from waste plastics. J. Mater. Cycles Waste Manag. **5**, 102–106 (2003)
16. Luo, W., Fan, Z., Wan, J., Hu, Q., Dong, H., Zhang, X., Zhou, Z.: Study on the reusability of kaolin as catalysts for catalytic pyrolysis of low-density polyethylene. Fuel. **302**, 121164 (2021) issn: 0016-2361
17. Ahmad, A., Yadav, A.K., Hasan, S.: A novel waste-to-energy conversion plant based on catalytic pyrolytic conditions: Modeling and optimization using supervised machine learning and desirability-driven methodologies. Energy. **317**(2025), 134648. issn: 0360-5442
18. Li, H., Wu, Z., He, H., Feng, S., Zhou, Y., Shi, C., Tu, X., Yan, J., Zhang, H.: Machine learning-driven product prediction and process optimization for catalytic pyrolysis of polyolefin plastics. Energy Convers. Manage. **333**, 119789 (2025)
19. Belden, E.R., Rando, M., Ferrara, O.G., Himebaugh, E.T., Skangos, C.A., Kazantzis, N.K., Paffenroth, R.C., Timko, M.T.: Machine learning predictions of oil yields obtained by plastic pyrolysis and application to thermodynamic analysis. ACS Eng. Au. **2**(4), 254–265 (2022)
20. Cheng, Y., Ekici, E., Yildiz, G., Yang, Y., Coward, B., Wang, J.: Applied machine learning for prediction of waste plastic pyrolysis towards valuable fuel and chemicals production. J. Anal. Appl. Pyrolysis. **169**, 105857 (2023)
21. Abnisa, F., Sharuddin, S.D.A., Zanil, M.F.b., Daud, W.M.A.W., Mahlia, T.M.I.: The yield prediction of synthetic fuel production from pyrolysis of plastic waste by Levenberg–Marquardt approach in feedforward neural networks model. Polymers. **11**(11), 1853 (2019)
22. Dubdub, I.: Pyrolysis study of mixed polymers for non-isothermal TGA: artificial neural networks application. Polymers. **14**(13), 2638 (2022)
23. Kumar, S., Bhatnagar, V.: A review of regression models in machine learning. J. Intell. Syst. Comput. **3**(1), 40–47 (2022)
24. Yapıcı, E., Akgün, H., Özkan, K., et al.: Prediction of gas product yield from packaging waste pyrolysis: support vector and Gaussian process regression models. Int. J. Environ. Sci. Technol. **20**, 461–476 (2023)
25. Shen, J., Yan, M., Fang, M., Gao, X.: Machine learning-based modeling approaches for estimating pyrolysis products of varied biomass and operating conditions. Biores. Technol. Rep. **20**, 101285 (2022)

26. Potnuri, R., Rao, C.S., Surya, D.V., Kumar, A., Basak, T.: Utilizing support vector regression modeling to predict pyro product yields from microwave-assisted catalytic co-pyrolysis of biomass and waste plastics. Energy Convers. Manag. **292**, 117387 (2023)
27. Rajpurohit, N.S., Kamani, P.K., Kamani, P.K., Lenka, M., Rao, C.S., Rao, C.S.: Predictive modeling of product yields in microwave-assisted co-pyrolysis of biomass and plastic with enhanced interpretability using explainable AI approaches. J. Anal. Appl. Pyrolysis. **188**, 107021 (2025)

Design and Simulation of Perovskite Solar Cells: A Numerical Approach Towards Device Structure Optimization

C. Guru Santosh[1], Hemant Verma[2], Shreyang Shekhar Singh[1], T. G. Kanakadas[1], S. B. Lekshmi[1], Anisha Mohapatra[2(✉)], and Mohan Lal Meena[1(✉)]

[1] Department of Chemical Engineering, National Institute of Technology Karnataka Surathkal, Srinivasnagar P.O, Mangalore, Karnataka 575025, India
mohan.meena78@nitk.edu.in
[2] Research Center for Applied Sciences, Academia Sinica, Academia Road 128, Section 2, Nangang, Taipei 11529, Taiwan
anishamohapatra1993@gmail.com

Abstract. Perovskite solar cells have emerged as promising candidates for next-generation renewable energy technologies due to their high absorption coefficients, tunable bandgaps, and low-cost fabrication potential. Methylammonium lead iodide ($MAPbI_3$), one of the most studied perovskite materials, has demonstrated exceptional photovoltaic performance. However, challenges such as lead toxicity and long-term material instability continue to hinder commercial viability. Here, we discuss numerical simulation and optimization studies conducted using the SCAPS-1D program to optimize the architecture of single-junction, hetero-junction, and tandem perovskite solar cells. The device structures were analyzed by varying the parameters of various perovskite absorber layers, hole transport layers, and electron transport layers to improve the power conversion efficiency and other critical performance metrics. To address concerns related to lead toxicity, the study explored lead-free alternatives such as $MASnI_3$, $FASnI_3$, etc., and discussed the reasons associated with their low efficiency and long-term stability as compared to the lead-based counterparts.

Keywords: Perovskite solar cells · Lead-free · HTL · ETL · SCAPS-1D

1 Introduction

Among all the renewable energy sources, solar energy is considered one of the most promising alternatives to address the energy crisis. The primary benefit of solar energy over other power generation methods is the ability to directly convert sunlight into energy using small-sized photovoltaic (PV) solar cells, and the phenomenon is termed the PV effect [1]. Solar cells are classified into three generations based on the materials used, technology involved, and performance characteristics, as shown in Fig. 1 (a). Among these, perovskite solar cells (PSCs) have gained a lot of interest for special properties like tunable bandgaps, low-cost fabrication potential, high efficiency, etc. [2]. These

H. P. Dasari et al. (Eds.): NITK-CREST 2025, PTT, pp. 133–142, 2026.
https://doi.org/10.1007/978-981-95-3218-6_11

third-generation cells have rapidly achieved efficiency levels of first-generation cells, while being more cost-effective than second-generation devices. Several key advantages of perovskites are illustrated in Fig. 1 (c). To enable the successful commercialization of PSCs, it is necessary to address key challenges, particularly enhancing their stability and ensuring environmental sustainability [3, 4].

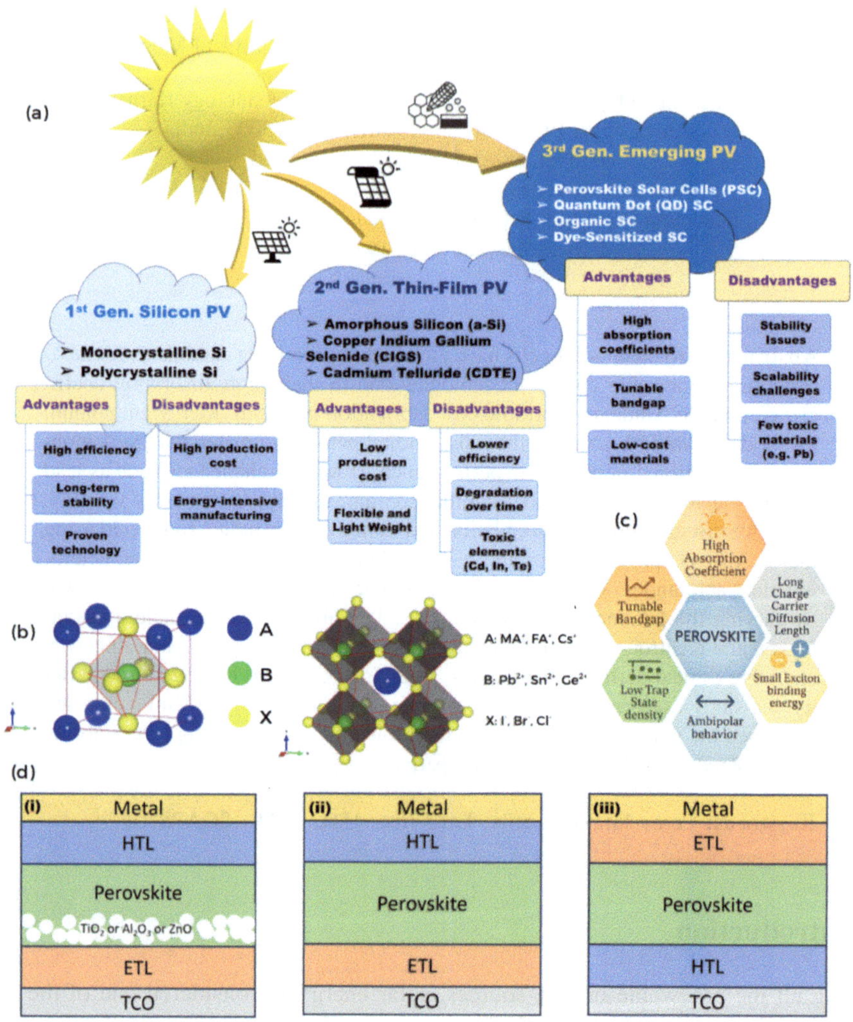

Fig. 1. (a) Classification of PV cells, (b) ABX_3 crystal structure in BX_6 octahedral coordination, (c) advantages of perovskites, (d) Typical PSC device structure in (i) mesoporous (ii) planar n-i-p and (iii) inverted planar p-i-n configuration

Perovskite refers to a class of compounds with a crystal structure similar to that of the mineral calcium titanate ($CaTiO_3$). The structure is described by the general formula ABX_3, where A and B are monovalent and divalent cations, and X is a halide anion [5], shown in Fig. 1 (b). Metal halide perovskites (MHPs) are a common class of perovskites, where the B-site is taken by a divalent metal cation and the X site is occupied by a halogen anion. The A site is typically occupied by methylammonium ($CH_3NH_3^+$/MA^+), or formamidinium ($CH(NH_2)_2^+$, FA^+), etc. [2, 5]. $MAPbI_3$ is one of the most studied MHPs due to its early breakthrough around 2009, which demonstrated significant potential for efficiency improvement, while offering low-temperature, solution-based fabrication [6]. However, lead's proven toxicity [7] is motivating towards lead-free alternatives with Sn, Ge, Ti, Bi, etc. [4]. Furthermore, bandgap tuning in MHPs is achieved via compositional engineering. Cation substitution involves the doping of the A- or B-site with cations that influence the bandgap, like mixing FA^+ decreases the bandgap, while adding Cs^+ increases it. Partial substitution of the B-site with Sn^{2+} or Ge^{2+} reduces the bandgap, due to their different electronic configurations compared to Pb^{2+}. Similarly, anion substitution involves the doping of the X-site halide; substituting I^- with Br^- increases the material's bandgap [8].

Octahedral factor (μ) and Goldschmidt tolerance factor (t) are key empirical parameters for determining the structural stability of perovskites, as defined below [9]:

$$\mu = \frac{r_B}{r_X} \tag{1}$$

$$t = \frac{r_A + r_X}{\sqrt{2}(r_B + r_X)} \tag{2}$$

where r_A, r_B, and r_X are the ionic radii of the A-site, B-site, and X-site ions in the ABX_3 structure, respectively, and are typically measured at room temperature [5, 9]. The octahedral factor quantifies the degree of distortion in the octahedron, and the tolerance factor measures the degree of fit between the A-site cation and the BX_6 octahedron [9]. When $t = 1$, the material adopts an ideal cubic perovskite structure while it remains stable in the ranges of $0.8 < t < 1.1$ and $0.44 < \mu < 0.94$. Within the non-ideal stable ranges, various structural distortions can still emerge and can significantly influence the material's electronic, magnetic, and dielectric properties [5, 9]. The A-site cation plays a critical role in these factors, with larger cations like FA^+ offering lower bandgaps, but also inducing crystallization and phase stability issues. $MAPbI_3$ and $MASnI_3$ have t value around 0.9, while $FAPbI_3$ and $FASnI_3$ typically have $t > 1$ [10].

2 Methodology

The perovskite solar cells (PSCs) are made of an active absorption layer of perovskite sandwiched between thin electron and hole transport layers (ETLs and HTLs), which together are connected to a transparent conducting oxide electrode (front contact) and a metal electrode (back contact) [9] as shown in Fig. 1 (d). The energy band alignment of the materials in these layers depends on their energy level, ionization potential, and electron affinity. The materials with low ionization potentials and electron affinities are used in HTL, while the higher ones serve as ETL materials [5, 11]. Mesoporous PSCs integrate a mesoporous scaffold, commonly TiO_2 or Al_2O_3, to support the absorber layer and increase the surface area for perovskite infiltration, facilitating efficient charge separation and transport [12].

When photons with energy greater than the band gap incident on the PSC, electron-hole pairs are generated in the perovskite layer. These pairs are effectively separated and driven towards the ETL and the HTL by the built-in electric field. The HTL material blocks the passage of electrons and injects the holes into the back electrode [13]. Whereas the ETL material injects the electrons into the electrode in contact [5, 13]. The commonly used HTL materials are Spiro-OMeTAD and Cu_2O, while the typical ETL materials include TiO_2, SnO_2, and PCBM [5]. Moreover, transparent conducting oxides (TCOs) like tin-doped indium oxide (ITO) and fluorine-doped tin oxide (FTO) are commonly used as the front contact (light-facing) electrodes because of their cost-effectiveness, high-temperature tolerance, high optical transparency in the visible region, and high reflectivity in the infrared region. These TCOs extract the carriers from the ETL and transfers them to the external circuit [5, 14]. The back contact is typically a metal electrode, like Au and Ag, to ensure efficient charge carrier conduction [5]. Each of these layers is coated sequentially using spin-coating, beginning with a TCO substrate [15].

2.1 SCAPS Simulation

The commonly used 1D simulation programs include SCAPS-1D, wxAMPS, PC1D, and various 2D and 3D programs, including Silvaco ATLAS, TiberCAD, etc. The methodology used in the SCAPS-1D program is explained, as it is one of the widely used programs for PSC studies.

SCAPS-1D (Solar Cell Capacitance Simulator), developed at ELIS, University of Gent, Belgium, provides extensive control over various features while offering default values and mechanisms suitable for most cases. The primary input parameters include the standard test conditions (Air Mass 1.5 Global spectrum at 1000 W/m^2 at 298 K) [16], the device structure by adding various layers and specific parameters, viz. bandgap (E_g), electron affinity (χ), relative permittivity (ε), donor and acceptor density (N_D and N_A), electron and hole mobility (μ_n and μ_p), density of states in conduction and valence bands (N_C and N_V), defect density (N_t), recombination coefficients, and interface defect states. The simulator then computes various device properties, such as the current-voltage (J-V) characteristics, external quantum efficiency, open-circuit voltage (V_{oc}), short-circuit current density (J_{sc}), fill factor (FF), and power conversion efficiency (PCE). The simulator computes by solving the Poisson and continuity equations. Poisson's equation, given by Eq. (3) establishes the relationship between the electric field (E) and charge carriers [17].

$$\frac{dE}{dx} = \frac{\rho}{\varepsilon} = \frac{q}{\varepsilon}\left(p(x) - n(x) - N_A^- + N_D^+\right) \tag{3}$$

where ρ indicates the charge density, q refers to the elementary charge constant (1.602×10^{-19} C), *p(x)* and *n(x)* indicate the hole and electron concentration at position x, respectively. The Continuity equations [17], given by Eqs. (4) and (5), ensure charge conservation by relating the rate of change of carrier concentration to current flow and generation-recombination processes.

$$\frac{dn}{dt} = \frac{1}{q}\frac{dJ_n}{dx} - (U - G) \tag{4}$$

$$\frac{dp}{dt} = -\frac{1}{q}\frac{dJ_p}{dx} - (U - G) \tag{5}$$

where J_n and J_p represent the electron and hole current densities, respectively, while U and G denote the recombination and generation rates. The Drift (or transport) equations [17], given by Eqs. (6) and (7), describe how the carriers move under the influence of electric fields (drift) and concentration gradients (diffusion).

$$J_n = qn\mu_n E + qD_n\frac{dn}{dx} \tag{6}$$

$$J_p = qp\mu_p E - qD_p\frac{dp}{dx} \tag{7}$$

where D_n and D_p are the diffusion coefficients for electrons and holes, respectively.

3 Results and Discussions

The optimization process of PSCs using the SCAPS-1D program involves the tuning of various parameters, such as the thickness of the perovskite layer, ETL, and HTL; bandgap of the perovskite layer, doping concentration, donor and acceptor densities, defect density of layers and interfaces, etc. A few notable simulation results of single-junction lead-halide PSCs and tin-based lead-free alternatives resulting from the optimization of parameters within the reasonable ranges reported in the literature and potential experimental feasibility are compiled in Table 1. The results primarily focus on simulation and optimization studies addressing the challenges of replacing lead-based PSC devices.

Table 1. Optimized simulation results of notable single-junction lead-halide PSCs and tin-based lead-free alternatives

Sl. No.	Device Architecture	PCE (%)	V_{oc} (V)	J_{sc} (mA/cm^2)	FF (%)	Ref
1	FTO/SnO_2/$MAPbI_3$/CuSCN/Au	26.74	1.19	25.61	87.81	[18]
2	FTO/SnO_2/$MAPbI_3$/Cu_2O/Al	25.97	1.14	25.61	88.59	[18]
3	ITO/$CuAlO_2$/$MASnI_3$/C_{60}/Al (p-i-n)	25.18	0.95	32.54	81.15	[19]
4	ITO/CuSCN/$MASnI_3$/C_{60}/Al (p-i-n)	25.13	0.95	32.49	81.13	[19]
5	FTO/SnO_2/$MAPbI_3$/Spiro-OMeTAD/Au	24.18	1.13	25.60	83.50	[18]
6	FTO/C_{60}/$MASnI_3$/$CuAlO_2$/Pt	24.04	1.08	25.85	85.84	[19]
7	FTO/C_{60}/$MASnI_3$/CuSCN/Pt	23.96	1.08	25.78	85.83	[19]
8	FTO/SnO_2/$FAPbI_3$/Spiro-OMeTAD/Au	21.40	1.20	22.60	77.00	[20]
9	FTO/SnO_2/$FA_{0.85}$ $Cs_{0.15}$ PbI_3/Spiro-OMeTAD/Au	21.00	1.20	21.50	78.00	[20]
10	FTO/TiO_2/$FASnI_3$/PTAA/Au	17.92	0.93	24.81	77.79	[21]
11	FTO/TiO_2/$MAPbBr_3$/Spiro-OMeTAD/Au	17.42	2.10	8.87	93.42	[22]
12	ITO/ZnO/$MAPbCl_3$/CuO/Ni	16.10	0.93	21.99	78.96	[23]

As observed in Table 1, $MAPbI_3$-based PSCs are one of the high-performing devices compatible with most ETL and HTL materials, predominantly due to their favorable bandgap (1.5–1.64 eV), defect tolerance, and strong energy level alignment with common ETL and HTL materials [24]. Also, $MAPbI_3$ devices are more cost-effective due to their use of standard ETL and HTL materials, making the cost proposition more favorable towards lead-based cells.

$FAPbI_3$ offers a slightly lower bandgap (~1.48 eV) and better thermal stability than $MAPbI_3$; however, $FAPbI_3$ tends to have more defects and non-uniform grain boundaries due to slower crystallization kinetics. $FAPbI_3$ also suffers from phase instability of the photoactive black perovskite phase (α-$FAPbI_3$), developed at high temperatures, spontaneously transforming to a non-perovskite yellow δ phase at room temperature [20]. The broader tolerance factor (> 1.0) range [10] also contributes to the acceleration of the formation of non-perovskite yellow impurities. These reasons reflect the relatively lower performance of the $FAPbI_3$-based PSCs (Table 1, Devices 8 and 9). While $MAPbBr_3$ ($E_g \sim 2.3$ eV) and $MAPbCl_3$ ($E_g \sim 3.1$ eV) have bandgaps that are too wide for single-junction PSCs, making them more appropriate for tandem applications [22, 23].

As discussed earlier, tin-based halide perovskites are one of the leading contenders for addressing the lead toxicity issue. $MASnI_3$ benefits from its narrow bandgap (~1.3 eV), which enables the absorption in the near-infrared (NIR) region as well. However, Sn^{2+} is easily oxidized to Sn^{4+} [9], which forms deep trap states that enhance recombination and significantly affect the lifetime of the cell. The use of standard ETLs like SnO_2, and TiO_2 can accelerate Sn oxidation, and thus require the use of organic ETLs like C_{60}, and PCBM to inert the interface properties. The oxidation issue is being addressed by using a mixed-cation and mixed-halide $MASnI_3$ with a trace of Pb and an organic passivation layer [9]. Further, additives like SnF_2 suppress the Sn^{2+} oxidation, reducing the vacancies [4]. The inverted $MASnI_3$ PSC (Table 1, Devices 3 and 4) performs better than the n-i-p one (Devices 6 and 7) due to better transparency of the inorganic HTLs as a window layer than the ETLs [19].

$FASnI_3$ demonstrates poor efficiency due to its low charge carrier mobility compared to $MAPbI_3$ and $MASnI_3$ and suffers from critical device instability. The oxidation of Sn^{2+} to Sn^{4+} persists, but additionally, larger FA^+ weakens the Sn–I bonding, leading to the formation of Sn vacancies that can act as recombination centres [4, 21]. However, $FASnI_3$, having a lower bandgap (~1.4 eV), enables effective absorption of the NIR region, making it a suitable bottom cell in tandem configurations [24] when paired with a higher bandgap top cell.

SnO_2 is widely adopted as an ETL material due to its favorable conduction band alignment with $MAPbI_3$, high electron mobility, and higher UV stability compared to traditional TiO_2 [18]. CuSCN and Cu_2O outperform traditional Spiro-OMeTAD as HTL material owing to their deep valence band alignment, better charge selectivity, and the inorganic nature, enhancing thermal stability and moisture resistance [25].

Table 2 summarizes some notable results from the optimization studies on $MAPbI_3$-based heterojunction and tandem cells. Both heterojunction and tandem configurations enhance the effective light absorption by enabling each absorber layer to capture different parts of the incident light spectrum. The Shockley-Queisser (SQ) limit establishes the maximum theoretical efficiency achievable by a single p-n junction, and arises from fundamental losses, primarily due to the inability to absorb photons with energies below the bandgap and the loss of excess energy as heat from high-energy photons, capping the theoretical efficiency at ~33.7% [30]. Single-junction heterojunction cells are bound by the same limit, but tandem cells can be efficient beyond the theoretical limits of the constituent sub-cells [31].

Table 2. Optimized simulation results of notable lead-based and lead-free heterojunction and tandem (multi-junction) PSCs

Sl. No.	Device Architecture	PCE (%)	V_{oc} (V)	J_{sc} (mA/cm^2)	FF (%)	Ref
1	TiO_2/$MAPbI_3$/Spiro-OMeTAD (top) PCBM/$FASnI_3$/Spiro-OMeTAD (bottom)	38.50	3.08	13.50	77.90	[24]
2	TiO_2/$MAPbI_3$/Cu_2O (top) TiO_2/$MASnI_3$/Cu_2O (bottom)	32.74	1.68	24.60	79.39	[26]
3	FTO/SnO_2/$FASnI_3$/$MASnI_3$/Cu_2O/Ag	28.06	1.07	30.74	85.23	[27]
4	FTO/TiO_2/$PeDAMA_5Pb_6I_{19}$/$MAPbI_3$/Ni	27.42	1.09	32.96	84.20	[28]
5	FTO/$SrTiO_3$/$MAGeI_3$/Au (top) FTO/$SrTiO_3$/$FAMASnGeI_3$/Au (bottom)	26.72	1.07	28.36	84.39	[29]

As discussed earlier, $FASnI_3$ is significantly more suited for heterojunction and tandem configurations, due to its lower bandgap (~1.4 eV), which enables effective absorption in the NIR region. This property is well-utilized in Table 2, Device 1, where $FASnI_3$ is paired with $MAPbI_3$ in a two-terminal tandem configuration, yielding a high efficiency of 38.5%, achieved by optimizing the constituent sub-cells.

Similarly, $FASnI_3$ is coupled with $MASnI_3$ in a lead-free heterojunction configuration (Table 2, Device 3), where the light absorption is maximized [15]. $MASnI_3$, with a slightly lower bandgap (~1.3 eV) and better crystallinity than $FASnI_3$, improves charge transport while maintaining strong absorption in the red-NIR range [27], making it an effective lower absorber in tandem designs (Table 2, Device 2) as well. The use of mixed perovskites like $FAMASnGeI_3$ coupled with $MAGeI_3$ (Table 2, Device 5) enables improved bandgap tunability, leading to the development of efficient entirely lead-free PSC devices. Also, Ge^{4+} helps partially mitigate the Sn^{2+} oxidation and further reduces the bandgap, enhancing NIR absorption [24]. The device 4 in Table 2 highlights the use of 2D Dion-Jacobson (DJ) layered perovskites, such as $PeDAMA_5Pb_6I_{19}$ (E_g ~ 1.65 eV) in heterojunction with $MAPbI_3$. This 2D perovskite contributes towards improved light absorption in the visible range, while also enhancing thermal and structural stability due to its layered configuration and stronger interlayer interactions [28], leading to the development of highly stable PSC devices.

The simulations often exhibit an efficiency error when translated to lab-scale devices due to the inability to simulate complex interfacial phenomena occurring during real-world operation. This can be much higher for tandem cells where thermal, optical, and electrical mismatches at the interconnecting junctions introduce additional loss mechanisms that are difficult to model precisely. Nevertheless, these optimization studies play a critical role in material selection, early development strategies and form a foundational step towards the development of efficient, stable, and cost-effective PV devices for a greener and more sustainable future.

4 Summary

This study emphasized the optimization of perovskite solar cells through comprehensive SCAPS simulations of the PSCs. By studying the effects of varying absorber layers and transport materials alongside other parameters, including layer thickness, bandgap, donor and acceptor density, etc., we discussed the understanding of the underlying physical and optoelectronic mechanisms influencing device efficiency. The $MAPbI_3$ absorber, despite its toxicity concerns, has been shown to yield superior photovoltaic performance. Its well-balanced bandgap and excellent charge transport properties make it a reliable choice for high-efficiency single-junction designs, reaffirming its significance in advancing perovskite solar technologies.

5 Future Perspective

We assertively analyzed both historical and contemporary solar cell technologies, focusing on rigorous numerical evaluations. Perovskite solar cells have emerged as the definitive cost-effective alternative to conventional solar technologies. Ensuring their stability is not just a goal but an essential priority. Recent advancements in passivation, interface engineering, encapsulation, and the strategic mixing of cations and halides have significantly improved stability. Furthermore, perovskite cells represent a robust opportunity for integration with silicon in tandem solar cells, paving the way to exceed the limitations of single-junction efficiency and propelling the swift deployment of next-generation solar energy systems.

References

1. Sharma, S., Jain, K.K., Sharma, A.: Solar cells: In research and applications—A review. Mater. Sci. Appl. **06**, 1145–1155 (2015)
2. Suresh Kumar, N., Chandra Babu Naidu, K.: A review on perovskite solar cells (PSCs), materials and applications. J. Mater. **7**, 940–956 (2021)
3. Meena, M.L., Gupta, K.K., Dutta, S., Kumar, R., Singh, R.K., Lu, C.-H., Lin, S.D., Som, S.: Short review on the instability and potential solutions for perovskite quantum dots. Curr. Res. Green Sustain. Chem. **5**, 100321 (2022)
4. Magdalin, A.E., Nixon, P.D., Jayaseelan, E., Sivakumar, M., Devi, S.K.N., Subathra, M.S.P., Kumar, N.M., Ananthi, N.: Development of lead-free perovskite solar cells: opportunities, challenges, and future technologies. Resul. Eng. **20**, 101438 (2023)
5. Bhattarai, S., Mhamdi, A., Hossain, I., Raoui, Y., Pandey, R., Madan, J., Bouazizi, A., Maiti, M., Gogoi, D., Sharma, A.: A detailed review of perovskite solar cells: introduction, working principle, modelling, fabrication techniques, future challenges. Micro Nanostruct. **172**, 207450 (2022)
6. Kojima, A., Teshima, K., Shirai, Y., Miyasaka, T.: Organometal halide perovskites as visible-light sensitizers for photovoltaic cells. J. Am. Chem. Soc. **131**, 6050–6051 (2009)

7. Ren, M., Qian, X., Chen, Y., Wang, T., Zhao, Y.: Potential lead toxicity and leakage issues on lead halide perovskite photovoltaics. J. Hazard. Mater. **426**, 127848 (2022)
8. Miah, M.H., Khandaker, M.U., Rahman, M.B., Nur-E-Alam, M., Islam, M.A.: Band gap tuning of perovskite solar cells for enhancing the efficiency and stability: issues and prospects. RSC Adv. **14**, 15876–15906 (2024)
9. Afre, R.A., Pugliese, D.: Perovskite solar cells: a review of the latest advances in materials, fabrication techniques, and stability enhancement strategies. Micromachines (Basel). **15**, 192 (2024)
10. Diomandé, I., Bouich, A., Hyacinthe, A.A., Soucasse, B.M., Boko, A.: Comparative performance analysis of MAPbI3 and FAPbI3 perovskites: study of optoelectronic properties and stability. Model. Numer. Simul. Mater. Sci. **13**, 51–67 (2023)
11. Bhatt, P., Pandey, K., Yadav, P., Tripathi, B., P., C.K., Pandey, M.K., Kumar, M.: Investigating the charge carrier transport within the hole-transport material free perovskite solar cell processed in ambient air. Sol. Energy Mater. Sol. Cells. **140**, 320–327 (2015)
12. Di Giacomo, F., Zardetto, V., Lucarelli, G., Cinà, L., Di Carlo, A., Creatore, M., Brown, T.M.: Mesoporous perovskite solar cells and the role of nanoscale compact layers for remarkable all-round high efficiency under both indoor and outdoor illumination. Nano Energy. **30**, 460–469 (2016)
13. Tress, W., Marinova, N., Inganäs, O., Nazeeruddin, M.K., Zakeeruddin, S.M., Graetzel, M.: The role of the hole-transport layer in perovskite solar cells - reducing recombination and increasing absorption. In: 2014 IEEE 40th Photovoltaic Specialist Conference (PVSC), pp. 1563–1566 (2014)
14. Chavan, G.T., Kim, Y., Khokhar, M.Q., Hussain, S.Q., Cho, E.-C., Yi, J., Ahmad, Z., Rosaiah, P., Jeon, C.-W.: A brief review of transparent conducting oxides (TCO): the influence of different deposition techniques on the efficiency of solar cells. Nanomaterials. **13**, 1226 (2023)
15. Mohapatra, A., Singh, A., Abbas, S.A., Lu, Y.-J., Boopathi, K.M., Hanmandlu, C., Kaisar, N., Lee, C.-H., Chu, C.-W.: Bilayer polymer solar cells prepared with transfer printing of active layers from controlled swelling/deswelling of PDMS. Nano Energy. **63**, 103826 (2019)
16. Odeh, S.: Analysis of the performance indicators of the PV power system. J. Power Ener. Eng.. **06**, 59–75 (2018)
17. Kowsar, A., Debnath, S.C., Shafayet-Ul-Islam, M., Hossain, M.J., Hossain, M., Chowdhury, A.F.M.K., Hashmi, G., Farhad, S.F.U.: An overview of solar cell simulation tools. Solar Ener. Adv. **5**, 100077 (2025)
18. Raoui, Y., Ez-Zahraouy, H., Tahiri, N., El Bounagui, O., Ahmad, S., Kazim, S.: Performance analysis of MAPbI3 based perovskite solar cells employing diverse charge selective contacts: simulation study. Sol. Energy. **193**, 948–955 (2019)
19. Afridi, K., Noman, M., Jan, S.T.: Evaluating the influence of novel charge transport materials on the photovoltaic properties of MASnI3 solar cells through SCAPS-1D modelling. R. Soc. Open Sci. **11** (2024)
20. Karthick, S., Velumani, S., Bouclé, J.: Experimental and SCAPS simulated formamidinium perovskite solar cells: a comparison of device performance. Sol. Energy. **205**, 349–357 (2020)
21. Yin, C., Gou, T., Li, J., Li, S., Zhao, M.: Numerical investigation of eco-friendly FASnI3 perovskite solar cells: effects of energy band alignment and interface defect. Micro Nanostruct. **203**, 208147 (2025)
22. Bouazizi, S., Bouich, A., Tlili, W., Kadri, B., Amlouk, M., Omri, A., Soucase, B.M.: Growth and efficiency of MAPbBr3 based perovskite solar cells: insight from experimental and simulation. Ind. J. Phys. **98**, 3341–3353 (2024)
23. Kabir, M.R., Ishraq, M.H., Tarekuzzaman, M., Nahid, N.I., Modak, J.K., Ahmad, S., Arfi, A.M., Rasheduzzaman, M., Hasan, M.Z.: A comprehensive SCAPS-1D simulation study on the photovoltaic properties and efficiency enhancement of CH3NH3PbCl3 perovskite solar cells. Eng. Res. Express. **7**, 015319 (2025)

24. Singh, P., Kumar, A.: Device optimization of FASnI3/MAPbI3 tandem solar cells: evaluating carrier recombination and engineering parameters for high PCE. J. Mater. Sci. Mater. Electron. **35**, 985 (2024)
25. Kim, J., Lee, Y., Gil, B., Yun, A.J., Kim, J., Woo, H., Park, K., Park, B.: A Cu2O–CuSCN nanocomposite as a hole-transport material of perovskite solar cells for enhanced carrier transport and suppressed interfacial degradation. ACS Appl Energy Mater. **3**, 7572–7579 (2020)
26. Ahmed, S., Rahman, N., Dulal Haque, M., Hasan Ali, M., Zafor Md Touhidul Islam, A.: Performance investigation of MAPbI3/MASnI3 based all-perovskite tandem solar cell using SCAPS simulation. Phys. Scr. **99**, 055520 (2024)
27. Panda, P., Kaur, J., Basu, R., Sharma, A.K., Madan, J., Pandey, R.: Toward high-efficiency photovoltaics: MASnI3 and FASnI3 double absorber perovskite solar cells with optimized conversion efficiency of 28%. Phys. B Condens Matter. **710**, 417232 (2025)
28. Ou, M., Qiu, L., Ding, C., Zhou, W., Zheng, C., Wu, Y., Ma, X., Ling, Y., Wu, J., Wang, E., He, Y., Qian, T., Jin, Y.: 2D Dion Jacobson/3D perovskite heterojunction solar cells without hole transport layer: further optimize the performance by SCAPS-1D. Mater Today Commun. **40**, 109575 (2024)
29. Singh, N., Agarwal, A., Agarwal, M.: Numerical simulation of highly efficient lead-free all-perovskite tandem solar cell. Sol. Energy. **208**, 399–410 (2020)
30. Zanatta, A.R.: The Shockley–Queisser limit and the conversion efficiency of silicon-based solar cells. Result. Optic. **9**, 100320 (2022)
31. Futscher, M.H., Ehrler, B.: Efficiency limit of perovskite/Si tandem solar cells. ACS Ener. Lett. **1**, 863–868 (2016)

Space (Climate)

Assessing Agricultural Vulnerability to Floods in Alappuzha District, Kerala

A. Preethi(✉) and Punithraj Gururaj

Department of Civil Engineering, National Institute of Technology Calicut, Kozhikode 673601, Kerala, India
preethi2598@gmail.com

Abstract. The 2018 Kerala flood caused widespread devastation, leaving a lasting impact on agricultural activity in Alappuzha district, which is characterized by its low-lying terrain, extensive water bodies and its vulnerability to extreme weather events exacerbated by climate change. By integrating hydrological and geospatial methods, this research aims to offer a thorough knowledge of the losses associated with the flood's impact on agriculture. A 30-m resolution Digital Elevation Model (DEM) and HEC-RAS hydrodynamic modeling were used to create a flood inundation map that indicated the areas that were submerged during the flood, while a Land Use Land Cover (LULC) map was made to identify agricultural areas throughout the district. By overlaying these maps on LULC map, the extent of the flooded agricultural land was determined. The findings indicate the extent of agricultural damage, highlighting the most vulnerable areas,16% of the total area of the district is covered with agricultural land and in that 97.82% of the agricultural land was affected by the flood. It was determined that both flood depth and flood duration significantly affect the level of agricultural damage the maximum depth of inundation in the agricultural was 13.7 m and the longest duration of inundation is 13 days, with prolonged inundation resulting in greater economic losses. Given the rising frequency and severity of floods linked to climate change, this study emphasizes the urgent need for efficient flood risk management techniques.

Keywords: Flood Damage · Flood Depth · Flood Duration · Flood Inundation Mapping · Remote sensing

1 Introduction

Flooding is one of the most severe natural disasters in India, impacting millions of lives. The country's diverse geography, monsoon-dependent climate, and extensive river systems make it highly susceptible to floods (Pal et al., 2022). Floods in India can be attributed to multiple factors, including excessive monsoon rainfall, poor urban drainage systems, deforestation, and unplanned urbanization (Mohanty et al., 2020). Among the various sectors affected, agriculture is one of the most vulnerable (Shrestha et.al., 2018), Flooding poses a significant threat to agriculture in India, leading to extensive crop destruction, soil erosion, reduced productivity, and disruption of rural livelihoods, especially in low-lying, flood-prone regions (Hussain & Guha, 2021).

H. P. Dasari et al. (Eds.): NITK-CREST 2025, PTT, pp. 145–160, 2026.
https://doi.org/10.1007/978-981-95-3218-6_12

The 2018 Kerala floods were one of the most severe natural disasters in the history of Kerala (Singh et al., 2025), leaving a lasting impact on its people, economy, and environment. Triggered by an unusually high monsoon rainfall, as per Indian Meteorological Department (IMD) data, in contrast to the 1649.5 mm of rainfall that was anticipated, Kerala got 2346.6 mm between June 1 and August 19, 2018, which was around 42% more than the typical amount of rainfall (Sudheer et al., 2023). Further, the rainfall over Kerala during the month of June, July and 1st to 19th of August (Suneela et al., 2023) was 15%, 18% and 164%respectively, above normal (Ministry of Earth Sciences, 2018). This heavy precipitation overwhelmed Kerala's River systems, including the Periyar, Pamba, and Bharathapuzha rivers, leading to widespread flooding across all 14 districts (Walia et al., 2022).

Alappuzha, a district of Kerala known for its extensive network of backwaters, canals, and low-lying terrain. The district faced widespread inundation during 2018 floods from major rivers like Pamba and Manimala merged with the backwaters, submerging homes, agricultural fields, and road networks (Sivan, 2024).

Alappuzha is of strategic importance due to its ecological and economic significance. The district's backwaters play a vital role in Kerala's tourism, while Kuttanad is a key agricultural hub and it is the only part of country where agricultural activities are conducted below the mean sea altitude (Veerakumaran & Santhi, 2019). Along with that Alappuzha's dense population and Vembanand lake ecosystem exposes the land area into flooding conditions. Thus, close monitoring of change in water levels as well as flood management is required. This study emphasizes on developing flood inundation map for 2018 floods affecting agricultural areas of Alappuzha district. These developed maps play a crucial role in future floods to recommend mitigate crops getting affected by flood as well as help decision makers to take action about crop insurance (Shrestha et al., 2015; Win et al., 2018).

2 Material and Methodology

2.1 Study Area

The study area selected is Alappuzha district in Kerala state, India which extends between latitudes 9°05′ and 9°54′ N and longitudes 76°17′ and 76°40′ E having total area of 1,414 km^2. Study area is known for its complexity in hydrological and topographical characteristics consisting of widespread of river systems and lagoons. Alappuzha district consists of four major river basins namely, Pampa, Manimala, Achankovil, and Muvattupuzha. Due to the presence of alluvial soil and abundance river network, Alappuzha is considered has major agricultural hub by growing crops like rice, coconuts, and bananas. Part of Alappuzha district, Kuttanad is called has "Rice bowl of Kerala". The agriculture prominence of this study area shows the importance of the work (Fig. 1).

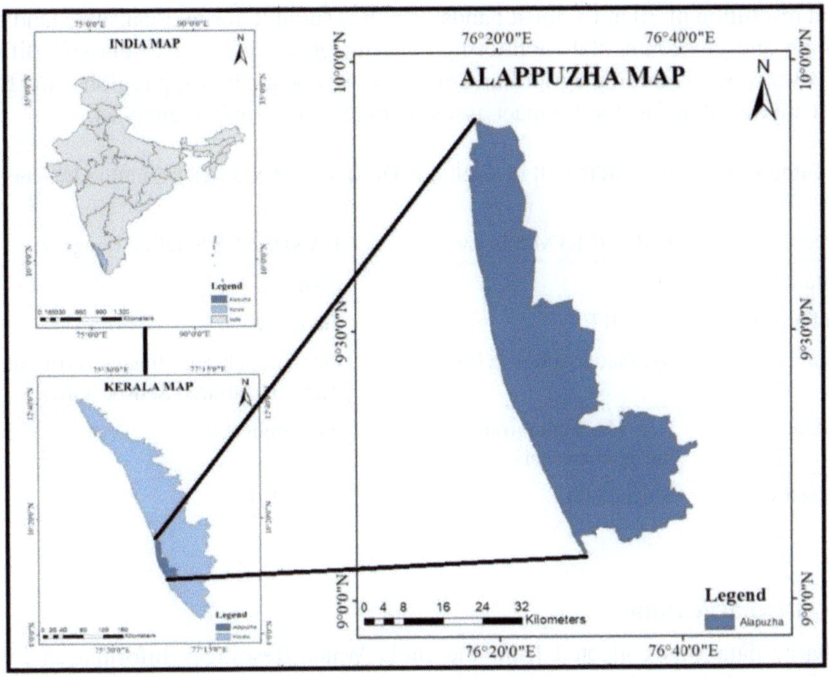

Fig. 1. Study area map. Source: Constructed by authors

2.2 Data Products

Three different set of datasets are utilized in this study to produce the flood inundation map for agricultural risk assessment. Digital Elevation Model (DEM) data was used to understand the topography of the terrain. Discharge data was also used to create the inundation map, and LANDSAT 8 imagery was utilized to produce the LULC map.

2.2.1 Digital Elevation Model (DEM)

Shuttle Radar Terrain Model (SRTM), Digital Elevation Model (DEM) having 30 m resolution was obtained from the United States Geological Survey (USGS) (https://earthexplorer.usgs.gov). DEM data gives the information about the terrain and hydrological characteristics of the study area. The details of SRTM DEM are given in Table 1, these datasets provide essential elevation information for flood modeling, topographic analysis, and land surface characterization.

2.2.2 LANDSAT 8 Imagery

LANDSAT 8 satellite imagery was acquired to analyze Land Use/Land Cover (LULC) information for the study area.

The data was obtained from the United States Geological Survey (USGS) Earth Explorer platform (https://earthexplorer.usgs.gov). The details of LANDSAT 8 Satellite Imagery is given in Table 1 below. LANDSAT 8 provides multispectral imagery with a spatial resolution of 30 m for most bands, which is suitable for regionalscale land cover mapping and environmental monitoring. The imagery was used to classify different land cover types, including agricultural areas, water bodies, built-up regions, and forest, which are essential for flood impact assessment and land surface analysis.

Table 1. Details on SRTM 30 m resolution DEM and LANDSAT 8 Satellite Imagery

Source	USGS (SRTM mission)	LANDSAT 8 Satellite Imagery
Resolution	30 m	30 m
Data Format	GeoTIFF	GeoTIFF
Sensor	Synthetic Aperture Radar	Operational Land Imager (OLI) and the Thermal Infrared Sensor (GIS)
Polarization	HH (Horizontal transmit and horizontal receive)	Not applicable
Released years	Since 2000	Since 2013

2.2.3 Discharge Data

Discharge data was collected from the India Water Resources Information System (WRIS) portal (https://indiawris.gov.in) to model the hydrological behavior of rivers within the study area. Discharge data is available in daily, hourly, weekly, and monthly formats, depending on the monitoring system. Daily data, recorded at gauging stations, is widely used for trend analysis and flood modeling. The daily discharge data was collected for specific time periods corresponding to significant flood events in Alappuzha.

2.3 Methodology

Daily discharge data for the period from 6th August 2018 to 29th August 2018 was obtained from WRIS India to provide the necessary hydrological input. HEC-RAS 6.5 was employed to create the flood inundation maps (Desalegn & Mulu 2021). In the initial setup, the projection coordinate system was configured in RAS Mapper to match the geographical location of the Alappuzha district. The terrain data (SRTM DEM) was prepared and imported into RAS Mapper, where a terrain layer was generated to serve as the foundation for further analysis. The geometry of the study area was defined using the HEC-RAS Geometry Tool. A 2D flow area was drawn to represent the river basin boundary, and a computational mesh with a uniform cell size of 100 m × 100 m was generated within this boundary. Boundary conditions were established for the simulation. A flow hydrograph was specified at the upstream boundary, while a normal depth boundary condition was applied at the downstream boundary (Patel et al., 2016). The daily discharge data was used to simulate an unsteady flow. A 30 m DEM was used to create an inundation map with the help of HEC-RAS software and as a result of simulation flood depth and duration of the study area was extracted (Chung et al., 2019), further these data were imported to Arc -GIS to produce the map (Rutuja & Anant 2023)
.

A Land Use Land Cover Map was generated with supervised classification by Maximum Likelihood Classification (MLC) method with LANDSAT 8 imagery, and accuracy of the map was evaluated by the confusion matrix (Kumar & Daija, 2021; Darem et al. 2023). From the LULC map the agricultural areas of the district was individually extracted from the LULC, mainly concerned on paddy field. To this agriculture area extracted the flood depth and flood duration map was overlayed in the agricultural areas map and the depth and duration value corresponding to the agricultural areas was found. This map created using the HEC-RAS and Arc GIS (Fig. 2).

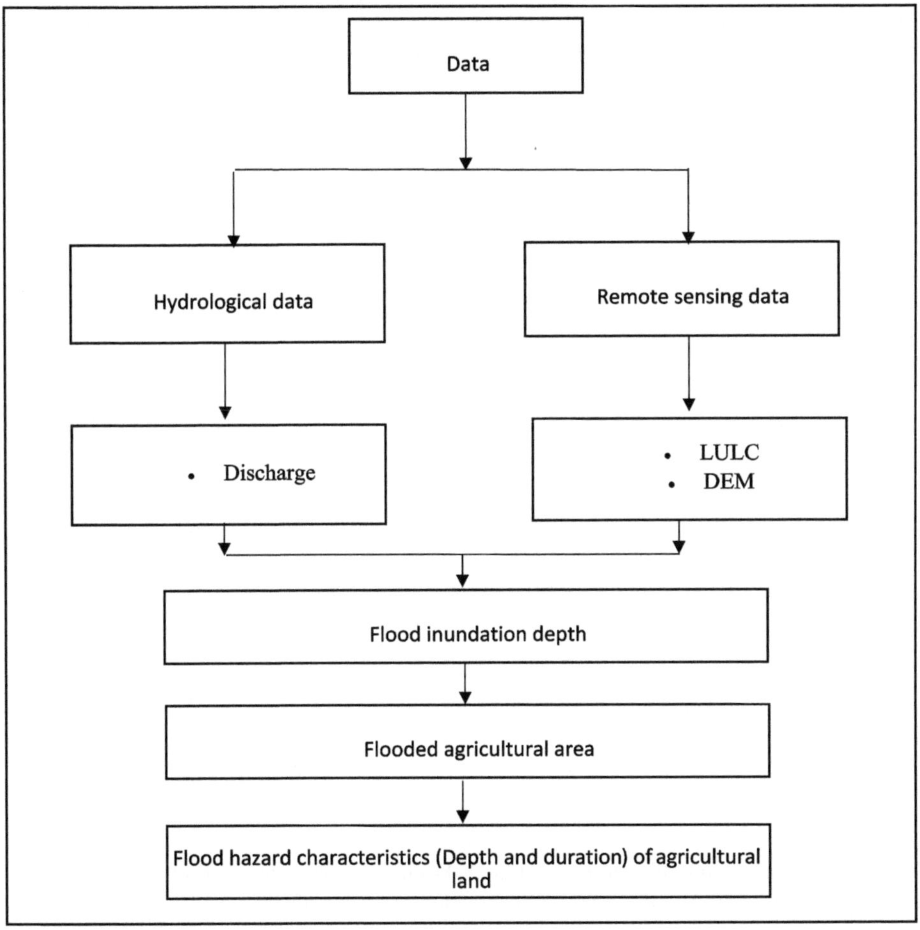

Fig. 2. Flow chart of the methodology. Source: Constructed by authors

3 Results and Discussion

3.1 Graphical Representation of Precipitation and Discharge of Different Basins of Alappuzha District During the Flood Period

The graph illustrates the discharge and precipitation trends in Pampa, Manimala, Achankovil and Muvattupuzha basins of Alappuzha district during mid-July to end of August 2018. In Fig. 3 the blue bars represent precipitation (rainfall), while the orange line indicates river discharge. Peaks in discharge closely follow high rainfall events, particularly around 18th July and 16th August, showing the impact of intense precipitation on river flow. The graph highlights a strong correlation between rainfall and increased discharge, especially during mid-August, suggesting significant runoff from heavy rainfall during that period.

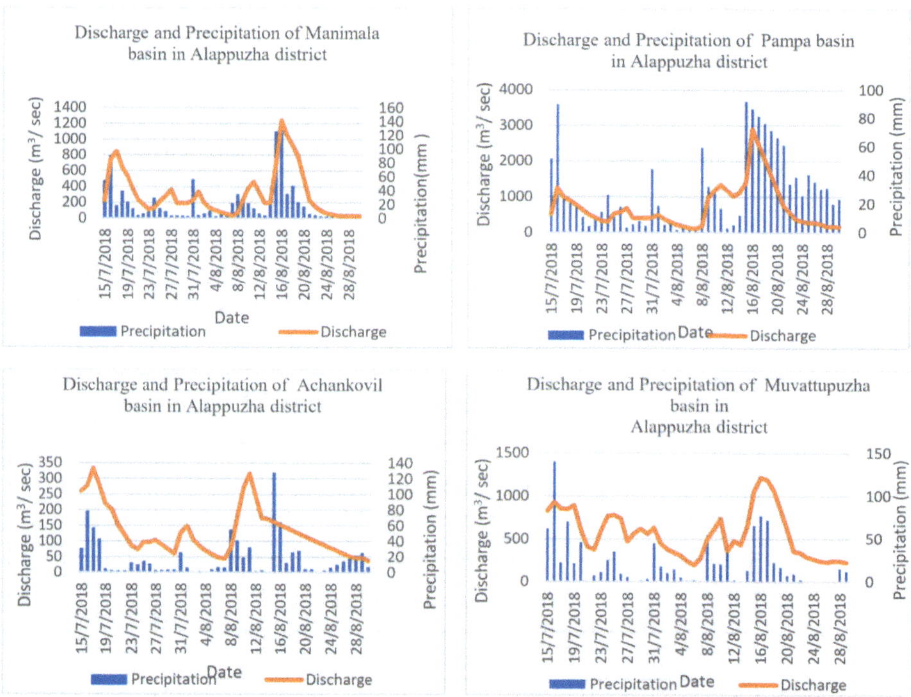

Fig. 3. Precipitation and Discharge of different basins of Alappuzha district during the flood period. Source: Constructed by authors, data downloaded from WRIS.

3.2 Stream Order Map

The stream order map shown in Fig. 4 of district was generated to classify the hierarchy of the drainage network. The map delineates streams from first to fifth order, with first-order streams being the smallest and most numerous, primarily located in the eastern and southern upland regions. As the 1st order stream converges to form higher order stream which is lesser in number and carry more amount of water. Second and third and are widespread throughout the district, fourth and fifth are mainly seen in the central and south part of the district which indicates the major drainage pathways. The presence of higher-order streams suggests well-defined drainage basins and potential floodprone areas. Understanding stream orders helps in assessing watershed characteristics, surface runoff behavior, and planning for flood management. This hierarchical classification provides critical input for hydrological modeling, sediment transport analysis, and water resource planning within the district.

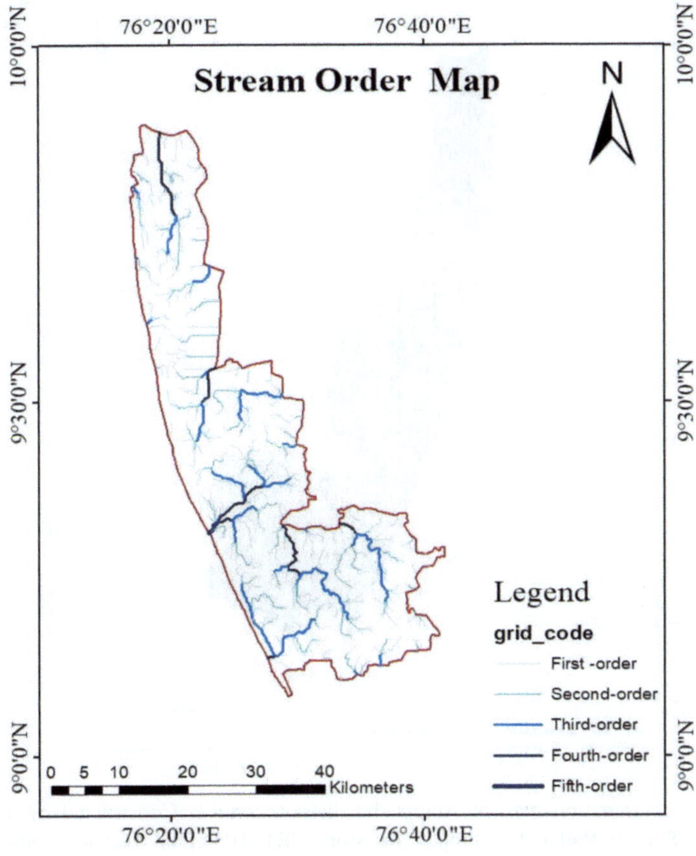

Fig. 4. Stream order map of Alappuzha district. Source: Constructed by authors, data derived from Shuttle Radar Topography Mission (SRTM) Digital Elevation Model (DEM), processed using ArcGIS software.

3.3 Flow Accumulation Map

The flow accumulation map of Alappuzha district shown in Fig. 5 illustrates the cumulative flow of water across the terrain, identifying potential drainage paths and zones of surface runoff concentration. Derived from the DEM, the map highlights areas with high accumulation values, typically representing stream channels, river courses, or zones where water is likely to converge and flow. These high-value areas are crucial indicators of natural drainage lines and flood-prone zones. Higher accumulation numbers indicate strong water flow convergence; the values range from 0 to 223,916. While accumulation is comparatively modest in the western coastal region, it is larger in the midland and eastern zones, where flow concentration is influenced by terrain.

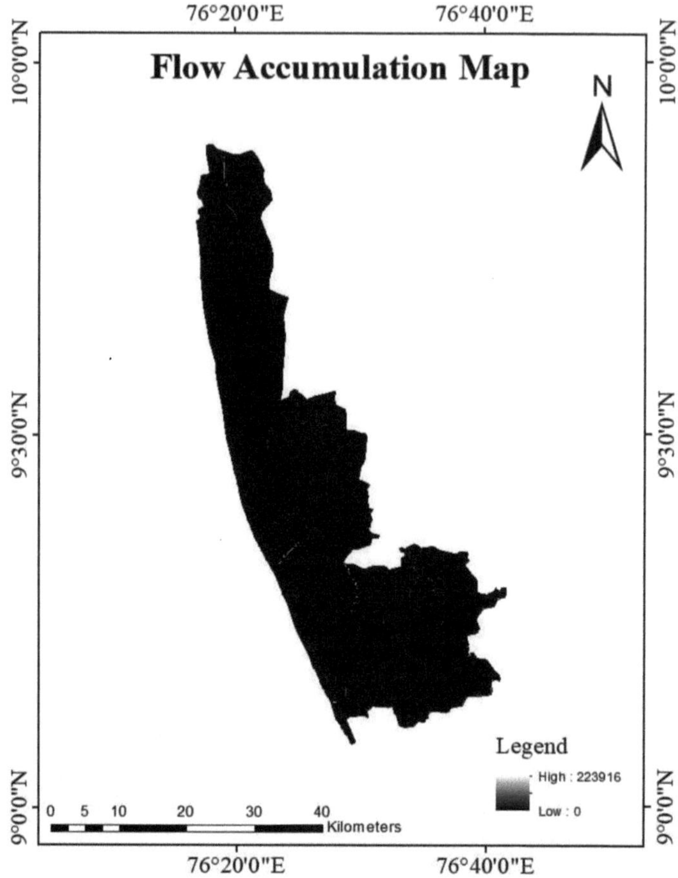

Fig. 5. Flow accumulation map of Alappuzha district. Source: Constructed by authors, data derived from Shuttle Radar Topography Mission (SRTM) Digital Elevation Model (DEM), processed using ArcGIS software.

3.4 Flow Direction Map

The flow direction map of Alappuzha district shown in Fig. 6 was generated, the map reveals spatial variability in flow directions, influenced by the district's topography and drainage patterns. A dominant eastward flow is observed in the upper northern region where elevation is slightly higher, suggesting a gradual slope towards the Vembanad Lake and a dominant southward flow in the Manimala and Pampa basin region, most of the region displays a more diverse pattern, with significant representation of south, southwest, and westward flow directions. This flow direction information is essential for hydrological modeling, watershed delineation, and flood mapping, as it helps determine how surface runoff will travel during rainfall events or flooding conditions across the district.

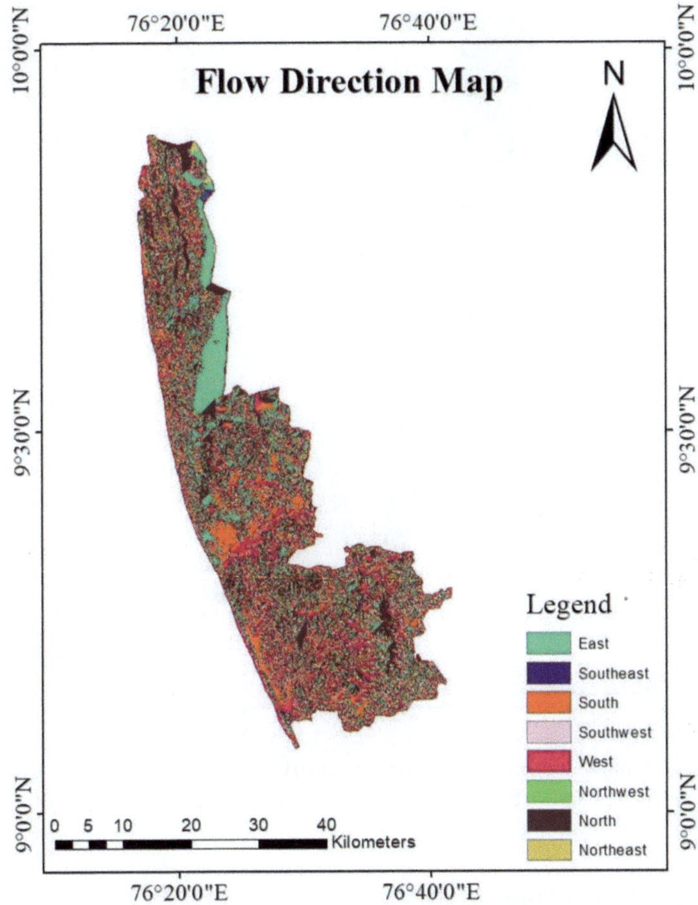

Fig. 6. Flow direction map of Alappuzha district. Source: Constructed by authors, data derived from Shuttle Radar Topography Mission (SRTM) Digital Elevation Model (DEM), processed using ArcGIS software.

3.5 Land Use and Land Cover (LULC)

The study area is divided into agricultural land, uncultivated land, water bodies, forests, and settlements according to the Land Use and Land Cover (LULC) map displayed in Fig. 7. The picture is categorized using the Maximum Likelihood Classifier, and the Supervised Classification technique was used with training samples totaling over 70 for each class. By considering more than 60 points for each class, the confusion matrix approach was used to assess the accuracy of LULC. With a Kappa coefficient of 0.84, the land use and cover map have an overall accuracy of 87%.

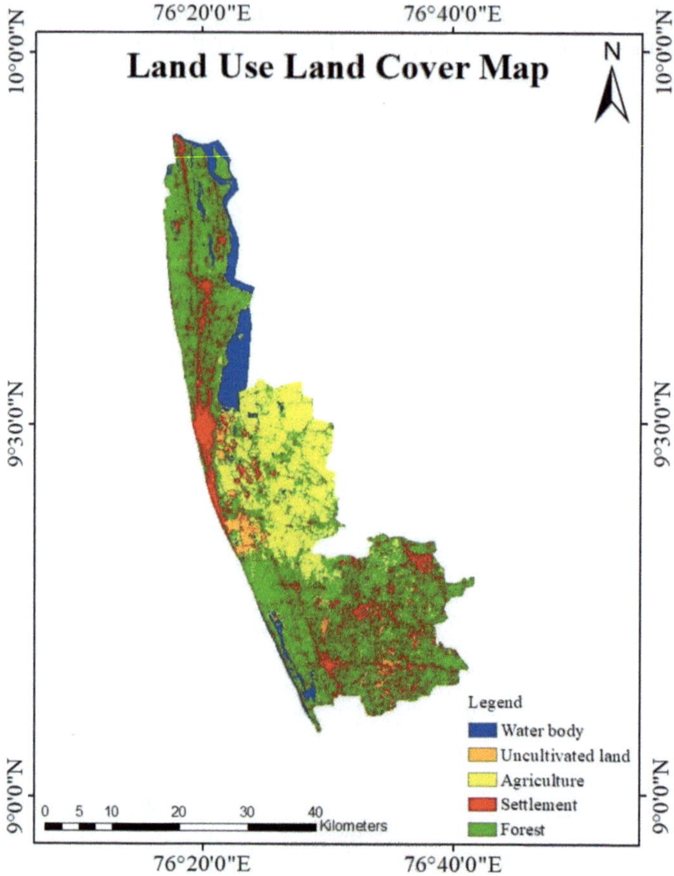

Fig. 7. Land use Land Cover map of Alappuzha district (July, 2018). Source: Constructed by authors, prepared using Landsat 8 imagery (downloaded from USGS Earth Explorer) and classified using supervised classification in ArcGIS.

3.6 Agricultural Area Map

The agricultural area map of Alappuzha district shown in Fig. 8 highlights the spatial distribution of agricultural land across the region. These areas are shown in yellow, clearly demarcating regions under cultivation. Based on land use and land cover classification, agricultural zones are predominantly concentrated in the central part which is the Kuttanad region due to a unique combination of geographical and hydrological factors such as fertile alluvial soil, regular flooding deposits nutrient-rich alluvium, creating highly fertile soil ideal for agriculture and availability of water, Kuttanad has many numbers of rivers, canals, and backwaters, offering plentiful irrigation. The map reveals that a substantial portion of land is dedicated to agriculture, particularly in the Kuttanad region, which is known for its extensive paddy fields and below-sea-level farming practices. The LULC map indicates that 59% of land area is forest, 16% of the area is agricultural land, 15% of land is settlement, 8% of land is covered by water body, and 2% is uncultivated land by overlaying this map with flood inundation or hydrological data, it becomes possible to evaluate the extent of agricultural loss and inform decision-making for disaster management, agricultural planning, and sustainable land use practices.

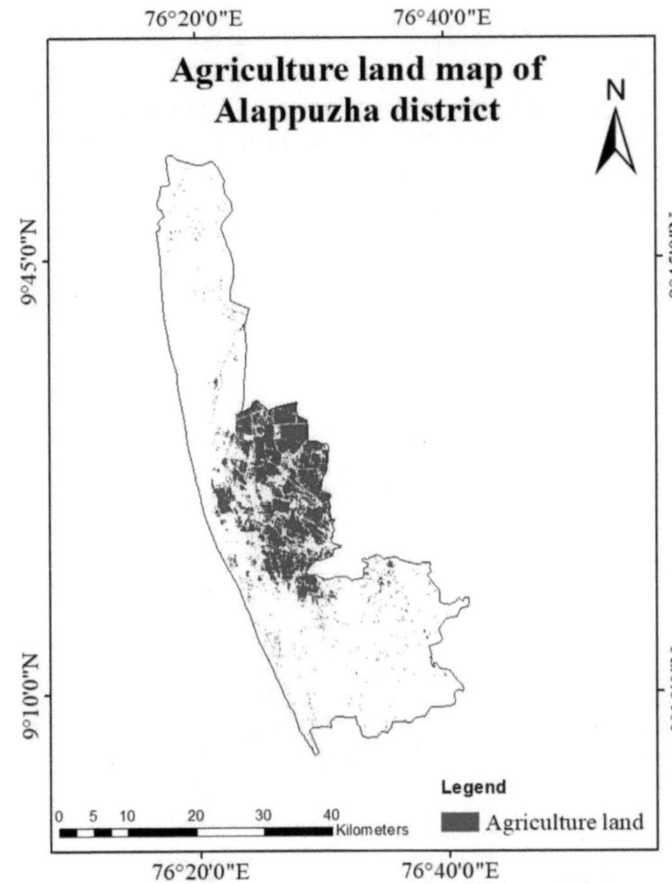

Fig. 8. Agricultural area map of Alappuzha district. Source: Constructed by authors, prepared using Landsat 8 imagery (downloaded from USGS Earth Explorer) and classified using supervised classification in ArcGIS.

3.7 Flood Inundation Map

The flood inundation map of Alappuzha district shown in Fig. 9 illustrates the spatial extent and intensity of floodaffected areas derived from a 30 m resolution SRTM DEM data. Inundation area of district is projected on DEM in the given map. 97.82% of the total agricultural land of the district is flooded. The inundation pattern reveals that the central and western regions of the district, particularly the low-lying Kuttanad area, experienced significant flooding. These zones are depicted in darker blue shades indicating higher flood levels. In contrast, the eastern highland areas show minimal inundation, represented in red and brown tones. The presence of existing water bodies is also mapped, helping to understand their contribution to the flood spread. The gradient from high to low inundation visually emphasizes the topographical control on water movement and accumulation. This map is crucial for flood risk assessment, emergency response planning, and evaluating the impacts on agricultural zones.

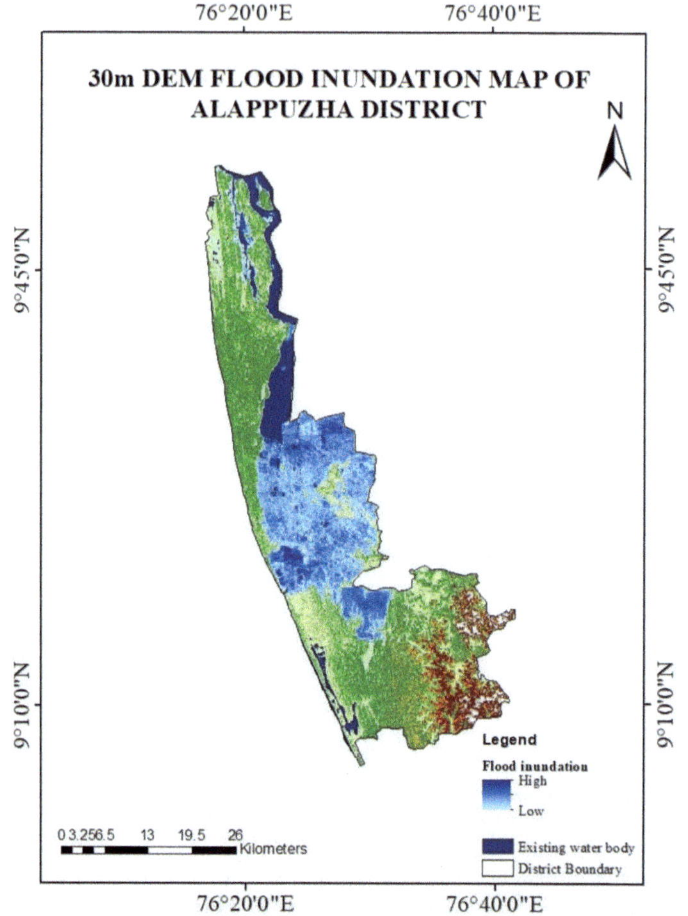

Fig. 9. Flood inundation map of Alappuzha district. Source: Constructed by authors, data derived using SRTM 30 m Digital Elevation Model (DEM) downloaded from USGS Earth Explorer in HEC-RAS, and map prepared in ArcGIS

3.8 Depth of Flood in Agricultural Area

The map shown in Fig. 10 depicting the depth of flood in agricultural areas of Alappuzha district provides a detailed spatial representation of floodwater penetration across farmland. The data indicates that flood depth varies significantly, with values ranging from 0 to 13.7 m. The highest flood depths, shown in darker blue shades, are concentrated in the central part of the district, primarily within the Kuttanad region, a well-known low-lying agricultural zone. This spatial distribution highlights the vulnerability of key agricultural areas to severe inundation. The northern and southern fringes show relatively lower flood depths. This map is essential for assessing floodinduced crop damage, as depth plays a crucial role in determining yield loss and crop survival. By overlaying flood depth data with agricultural land use, a more accurate evaluation of economic losses and planning for future flood mitigation strategies becomes possible. The results support targeted interventions for resilient agricultural management in flood-prone zones.

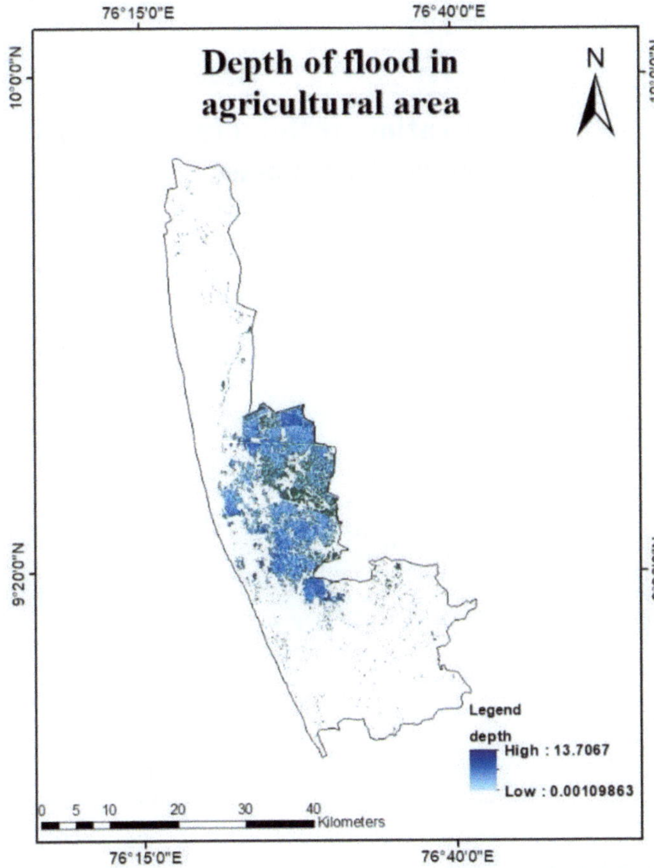

Fig. 10. Depth of flood in agricultural area in Alappuzha district. Source: Constructed by authors, data derived using SRTM 30 m Digital Elevation Model (DEM) downloaded from USGS Earth Explorer in HEC-RAS, and map prepared in ArcGIS

3.9 Duration of Flood in Agricultural Area

The flood duration map in Fig. 11 shows the agricultural areas in Alappuzha district illustrates the temporal extent of inundation across cultivated regions during the 2018 flood event. The duration of flooding ranges from short-term inundation of a single day to prolonged water stagnation of up to 13 days. Northern and central agricultural zones, particularly in the low-lying Kuttanad region, experienced extended flooding due to their below sea level elevation and poor drainage conditions. In contrast, southern and elevated regions saw shorter durations of inundation. Prolonged water stagnation poses significant threats to crop productivity by limiting oxygen supply, increasing plant stress, and fostering disease outbreaks. This spatial variation in flood duration is critical for accurately assessing agricultural loss and prioritizing areas for intervention. The map serves as a valuable resource for planning disaster response, agricultural resilience strategies, and infrastructure development aimed at mitigating future flood impacts on vital farming areas.

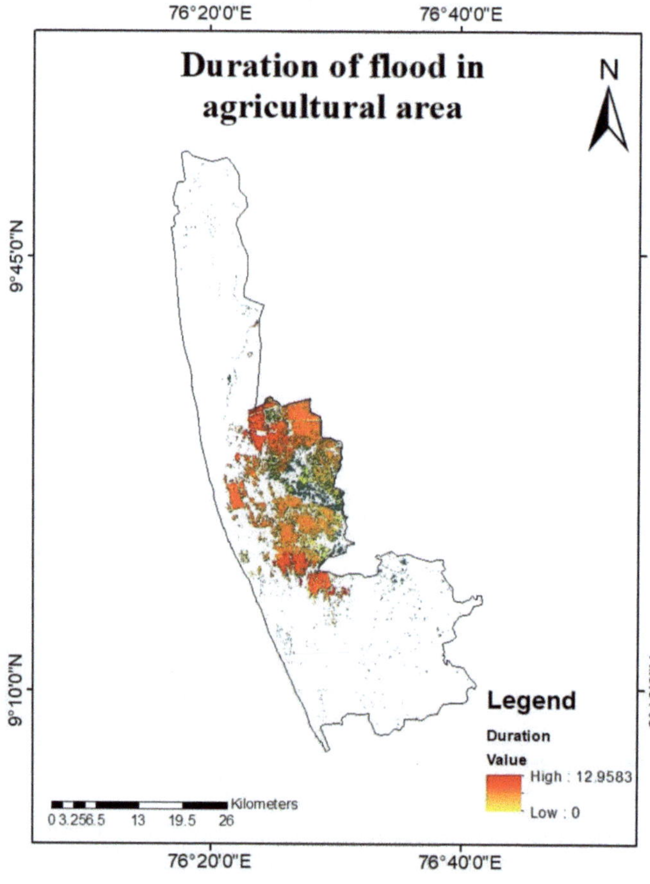

Fig. 11. Duration (days) of flood in agricultural area in Alapuzha district. Source: Constructed by authors, data derived using SRTM 30 m Digital Elevation Model (DEM) downloaded from USGS Earth Explorer in HEC-RAS, and map prepared in ArcGIS

4 Conclusion

In this study, hydrological modelling and geospatial analysis were used to examine how the 2018 flood event affected agricultural areas in Kerala's Alappuzha district. Flood extent, depth, and duration in relation to agricultural vulnerability were thoroughly evaluated through the integration of HEC-RAS, remote sensing, and GIS techniques.

- The central Kuttanad region, a low-lying area dominated by paddy cultivation, experienced the most intense and prolonged inundation, with water depths reaching up to 13.7 m and submergence lasting up to 13 days.
- The total agricultural land of Alappuzha is 16% out of which 97.82% of the agriculture land was affected by flooding.
- Flow direction, accumulation, and stream order maps provided insight into surface runoff and drainage behavior, aiding flood impact interpretation. The agricultural land in the north part of the district which is the Chertala Tehsil and south part of the district which is the Mavelikkara Tehsil was comparatively less effected during the 2018 flood.
- The central area of the district which is Ambalapuzha Tehsil and Kuttanad Tehsil was highly affected by flooding and this area comes under the Pampa and Manimala basin.
- Combining flood depth and duration data with the agricultural map highlighted specific farmlands highly vulnerable to crop damage and loss.

The spatial outputs and analytical methods developed in this study are valuable tools for disaster management authorities, agricultural planners, and policymakers. They support flood damage assessment, crop loss estimation, prioritization of relief, and development of long-term strategies for flood-resilient agriculture and climate-adaptive land use planning in vulnerable regions like Alappuzha.

References

1. Chung, S., Takeuchi, J., Fujihara, M., Oeurng, C.: Flood damage assessment on rice crop in the Stung Sen River Basin of Cambodia. Paddy Water Environ. **17**, 255–263 (2019)
2. Darem, A.A., Alhashmi, A.A., Almadani, A.M., Alanazi, A.K., Sutantra, G.A.: Development of a map for land use and land cover classification of the Northern Border Region using remote sensing and GIS. Egypt. J. Remote Sens. Space Sci. **26**(2), 341–350 (2023)
3. Desalegn, H., Mulu, A.: Mapping flood inundation areas using GIS and HEC-RAS model at Fetam River, Upper Abbay Basin, Ethiopia. Sci. Afr. **12**, e00834 (2021)
4. Hussain, M.A., Guha, P.: Flood threat on cereal crops production in irrigated and rainfed agriculture: a study of selected Indian states. Indian J. Agric. Res. **55**(6) (2021)
5. Kumar, K.H.Y., Daija, K.: Accuracy assessment of land cover classification and change detection using remote sensing and GIS. In: Int. Conf. on Emerging Trends in Engineering (ICETE-2021) (2021, November).
6. Ministry of Earth Sciences: Report on the exceptionally heavy rainfall over Kerala during 1st–19th August 2018. Earth System Science Organisation, Government of India. (2018)
7. Mohanty, M.P., Mudgil, S., Karmakar, S.: Flood management in India: a focussed review on the current status and future challenges. Int. J. Disaster Risk Reduct. **49**, 101660 (2020)

8. Pal, S.C., Chowdhuri, I., Das, B., Chakrabortty, R., Roy, P., Saha, A., Shit, M.: Threats of climate change and land use patterns enhance the susceptibility of future floods in India. J. Environ. Manag. **305**, 114317 (2022)
9. Patel, A.D., Patel, D.P., Prakash, I.: Flood modelling using HEC-RAS and geo-informatics technology in lower reaches of Shetrunji River, Gujarat, India. In: Proc. Natl. Conf. on Water Resources & Flood Management with Special Reference to Flood Modeling (2016).
10. Rutuja, B., Anant, P.: Flood inundation mapping of upper Krishna basin using hydrodynamic model (2023).
11. Shrestha, B.B., Sawano, H., Ohara, M., Nagumo, N.: Rice-crops flood damage assessment in the Pampanga River Basin of the Philippines. Adv. River Eng. **21**, 497–502 (2015)
12. Shrestha, B.B., Sawano, H., Ohara, M., Yamazaki, Y., Tokunaga, Y.: Methodology for agricultural flood damage assessment. In: Recent Advances in Flood Risk Management (2018).
13. Singh, A., Sreeparvathy, V., Debdut, S., Pregnolato, M., Wright, N.: A critical review of flood risk assessment in Kerala post-2018: methodological approaches, gaps, and future directions. J. Hydrol. Reg. Stud. **58**, 102262 (2025)
14. Sivan, L., Anitha, V.: Kerala Flood 2018 and the Socio-Economic Deprivation of Households in Alappuzha District (2024)
15. Sudheer, K.P., Thomas, J., Jainet, P.J., Nizar, S., Beegum, S., Sebastian, D.E., Pai, D.S.: From science to policy—toward an approach linking extreme rainfall events to climate resilience and policy development: a case study from Kerala. India. Dev. Environ. Sci. **14**, 143–172 (2023)
16. Suneela, S.S., Mathew, B., Sureshkumar, S.: The anomalous weather parameters that lead to the extreme rainfall of Kerala in August 2018. Meteorol. Atmos. Phys. **135**(4), 37 (2023)
17. Veerakumaran, G., Santhi, S.L.: Impact assessment of Kerala flood 2018 on agriculture of farmers in Edathua Panchayat, Kuttanad Taluk of Alappuzha District (2019).
18. Walia, A., Sharma, P., Nusrat, N.: Kerala Floods 2018: Impacts and lessons drawn. In: International Handbook of Disaster Research, pp. 1–17. Springer, Singapore (2022)
19. Win, S., Kawasaki, A., Zin, W.W.: Flood damage assessment for flood prone agricultural areas in the Bago River Basin. In: 8th Int. Conf. on Science and Engineering (2018).

Machine Learning and Space Technology Solutions for Climate Change Adaptation

Jagadeesha J. Chinagudi(✉)

Indian Space Research Organization, NRSC-RRSC-Bangalore, Bangalore 560076, India
cjagadeesha@gmail.com

Abstract. Climate change adaptation is critical to sustainability across social and geographical scales. Many actions that facilitate adaptation to climate change are executed to deal with current extreme events such as heat waves, floods,and cyclones. In industries, it is essential to adapt to sustainable business practices. Many times, planned adaptation initiatives are also not executed as stand-alone measures but are embedded within broader sectoral initiatives such as water resource planning, coastal defense, disaster management planning, and green engineering. As we move towards sustainable social networks across the world, geographically referenced location data becomes important. Micro-satellite constellations are providing continuous (real-time) and more economical inputs than large satellites for monitoring climate change adaptation. Even the ICT-with Industry 5.0 becomes a kingpin by adopting all that is advancing technically (both in AI and Quantum) and to put human in the loop.

The carbon nanotubes (CNT) and polymer-based design and fabrication of nanosensors, S-band interferometric synthetic aperture radar payloads, for earth observation, communication payloads and space electric propulsion systems that can adapt to electromagnetic interference while orbiting are the various space materials that can bring economically sustainable solutions to studying climate change and geohazards.

Spatial inputs from microsatellites around the world and Indian and global webportals assist in modeling solutions for natural resources and waste management using big-geodata analytics and spatial machine learning tools.

The international engineering curriculum design for the common problems facing society, such unequal distribution of energy, water, minerals, forests, and climate change, is also discussed.

Keywords: micro-satellites for geospatial infrastructure · machine learning on BigGeodata · Carbon nanotubes and international curriculum design

1 Introduction

They have a potential financial impact from climate change for almost every nation in the world. United States estimates its effect in trillions of dollars in the next decade. A necessary step would be to make a more granular climate insights generation to empower stakeholders to take a more data-driven approach to climate adaptation.

H. P. Dasari et al. (Eds.): NITK-CREST 2025, PTT, pp. 161–199, 2026.
https://doi.org/10.1007/978-981-95-3218-6_13

The food system is a massive employer in the world,employing about 40% of all global jobs. It's responsible for about 12% of global GDP, taking about 50% of all habitable land. This food system is responsible for about 25% to 35%of all global carbon emissions. If food is produced in a business-as-usual way, there is every likelihood that progress in reducing hunger, food affordability, and driving food insecurity gets slower. Sustainable food production helps reduce the number of undernourished people by 300 million. Continuing business-as-usual is not sustainable. System-level changes in food security are needed. Five possible system level solutions for sustainable food systems are (a) protecting, restoring, and improving natural capital including the land, the soil, the water, the vegetation, the wildlife, the ecosystem services to improve food production and food security; (b) accelerating innovation, technology and productivity improvements; (c) enhancing circularity in the food system by concentrating more on use of vapor shift technologies in rain fed farming; (d) guiding dietary shifts and consumer choices; (e) reducing emissions by following suitable land use crop mixes (regenerative farming).Climate adaptation pathway/conservation law can be achieved by (a) Climate-adaptive biodiversity conservation, must be followed with conservation of agricultural ecosystems and ecosystem services across the landscape.{b} biodiversity conservation and management must respond to ecological dynamics more adaptively and proactively (as per changed phenological time periods) [1].

An understanding of some of the key sources of uncertainty associated with global climate change is helpful for managers in semi-arid climatic region areas. A large area covering 10° to 30° latitude in both the hemispheres, between the tropical rain forests and deserts, is generally termed semi-arid tropical. It has an alternating wet and dry climate. This belt broadly covers 49 countries around the globe and is inhabited by about 600 million people. Some of the uncertainties discussed in the recent literature include: (a) The magnitude of changes in basic agro-climatic variables (degrees of aridity, stability of climate compared to the probable year, potential evapotranspiration, agronomy, ecology, soil moisture, etc. which are functions of temperature, precipitation, and biotic factors); (b) how much of this change is natural variability and how much is due to anthropogenic causes; (c) the strength of climate feedbacks (positive or negative);(d) the frequency and distribution of the effects of climate change; (e) the distribution and frequency of extreme weather events; (f) how natural systems will respond and how human societies' cultural aspects will respond in terms of mitigation or adaptation. As interdisciplinary research on climate change adaptation has expanded globally, there has been increasing recognition of the role of culture in shaping the processes and outcomes of climate change adaptation. Analysis of cultural dimensions was embedded in a study in semi-arid regions. This considered (a) multiple knowledge systems for farming in Botswana; (b) the dynamics of pastoralist values and livelihoods in Kenya; and (c) the interplay of caste and livelihood choices in India. In these works, which helps us understand barriers and enablers of adaptation, the researchers used a portfolio of social science research to analyze social differentiation and its relationship to adaptation. It was noticed that the notion of a static interlinkage of cultural and economic structures that shape the livelihood benefits of migration do not hold good anymore. Dynamics of power shifts bring changes in becoming enablers and barriers of climate change adaptation. Culture is inherently dynamic, functioning as enabling as well as constraining

[2]. The ICT in Industry 5.0 is bringing digital agriculture to make it sustainable in a circular economy using nature-based approaches. This acts as the cultural shift enabler for adaptation to climate change. These too need social processes to be understood, and they build sustainable social networks.

India needs to work on local resources for local livelihoods. Gross domestic product GDP cannot be relied upon, as natural and social capital are not pronounced in it. Instead, gross natural products should also drive its economy. One cannot afford to do development based solely on climate change scenarios. The strategy should be based on nationally determined contributions for GHG emissions, as well as co-benefits supplemented by nature-based solutions embedded into a circular economy.

2 Climate Change Adaptation: Policies, Strategies, Frameworks, Assessment, Measurement, for Any Country

The climate change adaptations are adjustments in response to climate stimuli. They are pursued with efforts globally (a country or watershed) to enhance adaptive capacity, strengthen resilience, and reduce vulnerability to climate change [3]. These adjustments may be planned or unplanned and may also take the form of institutional, behavioral, or technological changes. Adaptation decisions can be made by individuals, communities, corporations, governments, and international and transnational bodies.

The challenge with climate change and the impact of climate change is that these are occurring more often. They are happening over longer periods of time when there's a season—and they're having a wider-scale impact. In this situation, it's harder to manage them on an individual basis.

The process of adjustment to actual or expected climate change effects is called adaptation. In human systems, the tendency for adaptation is to seek moderation, avoid harm, or exploit beneficial opportunities. In some natural systems, human intervention may facilitate adjustments to expected climate and its effects.

Broadly, there are two types of adaptation. See Table 1 [4] which explains the transformative and incremental adaptations with the criteria specified. Table 1 and Fig. 1 together explains the response categories for these adaptation categories, depending on the magnitude of change they entail, so that we can design performance metrics.

Communities can implement effective climate change adaptation strategies through a variety of approaches: (a) Risk Assessments by (1) conducting thorough assessments to identify vulnerabilities to climate impacts (e.g., flooding, heatwaves), (2) by using spatial data analytics to prioritize areas and populations most at risk; (b) Community engagement by (1) involving local residents in the planning process to gather insights and build support; (2) climate risks and adaptation opportunities need to be made known through camps and education; (c) Infrastructure improvements by way of (1) upgrading infrastructure to withstand extreme weather events (e.g., enhancing drainage systems, reinforcing buildings),(2) investing in green infrastructure (e.g., parks, green roofs) to manage storm water and reduce heat; (d) Ecosystem-Based Approaches by(1) protecting and restoring natural ecosystems, such as wetlands and forests, which can buffer against climate impacts,(2) implement sustainable land-use practices that enhance resilience (e.g., agroforestry); (e) Policy and Planning by way of (1) developing and enforcing

Table 1. Criteria of Transformative and Incremental Adaption

N	Criterion	Transformative	Incremental
1	Innovative/Restructuring, Is this adaptation new to the region?	Adaptation is implemented or proposed for implementation in the region for the first time (implementation in other parts of the world is acceptable)	Adaptation is not new but is an important aspect of existing measures.
2	Multiscale/systemwide Does this adaptation involve large-scale changes?	Large territory, several regions, several countries	Small area, several experimental locations, limited community
3	Path shifting and restructuring, Does this adaptation have a long or short term of implementation?	Long term	Short term
4	Innovative, Does it require a big investment?	Yes	No
5	Persistent/ Systemwide, Does the implementation of this adaptation change (or will change in the future) attributes of a system?	Results of the implementation of this adaptation change or bring changes at different levels (local, country, region, global),system or part of the system (policy, economy, law, socio-environment etc.),condition of the system.	Results of the implementation maintain the essence and integrity of the existing technological, institutional, governance, and value systems.

land-use policies that consider climate risks, (2) incorporating adaptation strategies into local development plans and zoning laws; (f) Collaboration and partnerships through (1) other governments, NGOs, and businesses to share resources and best practices, (2) building social networks for knowledge sharing and collective action; (g) Financial mechanisms by (1) allocating funds specifically for adaptation projects,(2) exploring grants, partnerships, and insurance mechanisms to support initiatives; (h) Monitoring and evaluation by (1) regularly assessing with key performance indicators the effectiveness of adaptation strategies and adjust them based on feedback and changing conditions, (2)

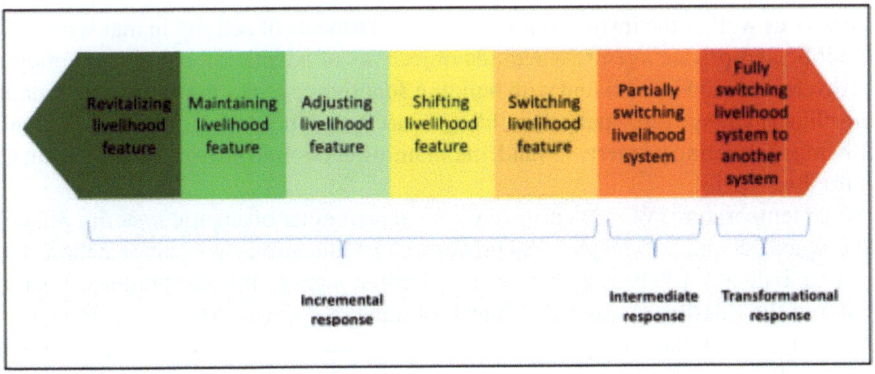

Fig. 1. Response categories organized according to the magnitude of change, they entail, from incremental (left) to transformational (right) while adapting to climate change [4].

establishing new indicators to track progress and community resilience. By combining all these strategies, communities can create a holistic approach to climate change adaptation, promoting resilience and sustainability for future generations.

The pathways toward adaptation, the technologies and the institutional forms of collective action are diverse. In order to ensure their sustainability and resilience in the context of uncertain futures (projected by climate change), represents a major challenge. Adaptation is observable in society. The adaptations are observed in markets. These markets are like insurance and reinsurance, coastal planning, health interventions, built environment, water resources, etc. These are generally not measured effectively to understand the lessening of uncertainty brought out by adjustments and adaptations within resource-based livelihoods.

The lessons learnt so far in climate change adaptation taken place around the world is [5]: "(a) the whole country should be covered for risk and vulnerability assessment of any climate change disaster and should essentially be backed by data; (b) National adaptation plans are more operational, helping institutions translate priorities into practical blueprints for action; (c) the adaptation networks should be locally led and locally owned as far as possible; (d) climate change mitigation and adaptation actions should be suitably combined; (e) overall investment in climate action should incentivize and create a enabling environment (as opposed to barrier) for private sector participation in national plans of action; (f) long term action plans to mitigate risks should be planned via national planning and budgeting processes, with a focus on agricultural sectors (water, soil, ecological services) at local areas; (g) Indicators and frameworks for monitoring and management of adaptation activities should be made available for researches throughout the nation" [5].

"Adapting to climate change involves cascading decisions across a landscape made up of agents from individuals, firms and civil society, to public bodies and governments at local, regional and national scales, and international agencies. As mentioned above, a broad distinction can be drawn between action that often involves creating policies or regulations to build adaptive capacity and action that implement operational adaptation decisions. The latter will often be constrained and influenced by a higher-level adaptation framework as well as the institutions that define all aspects of activity in that society. For both public and private agents, where the objectives of adaptation are explicit, they are often diverse. Actions associated with building adaptive capacity may include communicating climate change information, building awareness of potential impacts, maintaining well-being, protecting property or land, maintaining economic growth, or exploiting new opportunities" [5].

Water temperature (WT) is a critical physical parameter of aquatic systems, affecting the biological, chemical, and physical processes. In Hungarian environments these act as a sensitive indicator of climate change. The thermal regime of water bodies is resulting in shifts in the distribution and abundance of aquatic species, Also, sometimes in the potential changes of the timing of biological events. Reductions in summer precipitation have been found to lead to changes in the hydroperiod of lentic and lotic systems, further affecting their productivity and biodiversity. Therefore, continuous and detailed water temperature monitoring datasets are crucial for the effective management and conservation of these ecosystems in this Hungary region.

"Understanding the spatiotemporal dynamics of WT in inland waters (e.g., lakes and rivers) is essential for effective water resource management and ecosystem conservation. The significance of research aimed at reconstructing non-existing real (in situ) data pertaining to the thermal regime of inland waters is becoming increasingly common. The results concerning the reconstruction of water temperatures in both lakes and rivers), indicate a large potential for such a methodological approach, which is particularly applicable in long-term analyses. This knowledge can aid in predicting changes in water availability and quality, and their impact on local wildlife, thereby supporting efforts to preserve the unique ecological environment of the Pannonian ecoregion [Hungary] in Europe." [6].

But in India, Karnataka Region by UAS (University of agricultural sciences) Bangalore and EMPRI (Environmental Monitoring and Policy Research Institute under the Department of Forestry and Environment of the Government of Karnataka (GOK)) have shown that there is a shift in monsoon season rainfall amounts to the pre-monsoon category, up to 30% of the total rainfall in the region. The farm advisory is given accordingly for the past 8–10 years through KSNDMA [Karnataka State Natural Disasters Monitoring Agency-GOK] -Bangalore Master Control Canter for warning about flood, drought, cloudburst, lightning disasters, and the agro-climatic advisories (rainfall, cloudy conditions, soil moisture to a certain degree, etc.).

The following is observed in their (UAS and EMPRI) study [7]. "The State first half-century's (1901-1950) normal of 1204 mm has been reduced to 1140 mm during the second half of the century (1951-2000). Nevertheless, a few districts like Bengaluru, Kolar, and Tumkur are gaining in their mean annual rainfall, while some traditionally heavy rainfall-receiving districts like Kodagu, Chikmagalur, and South Canara are losing in their mean annual rainfall." "The eastern districts of the state are tending to be more dependent on North East monsoon than terminal rains of the South West monsoon. Consequently, individual crop growing area, growing period are changing. The normal sowing season rains are being delayed due to the shift of July rains to the August month and September peak rainfall is being shifted to October month. The maximum water available period for the grand growth period is shifting towards the end of September and beginning of October in many districts. Finger millet crop area (main food crop of southern Karnataka) in Chikmagalur district, Groundnut area in Chitradurga and Tumkur districts, Red gram in Bidar and Gulbarga districts is increasing. Whereas, Groundnut area in Belgaum and Gulbarga districts and Red gram area in Belguam and Tumkur is decreasing." Each district's geographical area in square kilometers as given in parentheses is: Bidar (5448); Kolar (8223); Gulbarga (10951); Chikmaglur (7201); Chitradurga (8440); Tumkur (10597); South Canara (8441); Kodagu (4102); Bangalore {2191}; Belgaum (132415).

In Cuttack District (3932 square km area), Orissa state, India, the studies on Climate change have revealed the following:

"CMIP5 (Coupled Model Inter comparison Project Phase 5) was employed using bilinear interpolation for downscaling to generate these climate projections with included weather parameters i.e temperature, rainfall and solar radiation for the study area. The model results revealed that the RCP projections of climate change showed an increase in seasonal maximum temperature, minimum temperature and solar radiation and a decrease in the rainfall condition in future for Cuttack district of Odisha for the years 2030, 2050, 2070 and 2090 and there will be an maximum increase in seasonal maximum temperature by 2.56°C and by 3.96°C in seasonal minimum temperature in the year 2090 under RCP 8.5 scenario".

"The district experiences tropical climate with mean annual rainfall of 1577 mm, mean maximum summer temperature of 39°C and mean minimum temperature of 11.5°C. The climate condition of the district is generally hot with high humidity during April and May and cold during December and January (District Survey Report, 2018). South West monsoon is primarily responsible for the rainfall. Rainfall is generally heavy during the monsoon, which occur during the months of July and August".

"Projection of futuristic climatic scenario: The changes in major weather parameters like temperature, rainfall and solar radiation in future years viz., 2030, 2050, 2070 and 2090 which were purposively selected to assess the climate change for every 20 years difference for Cuttack district of Odisha were projected under four possible climatic scenarios (RCP 2.6, 4.5, 6.0 and 8.5), which were used in Fifth Assessment Report of IPCC. We (Cuttack Team of researchers) used present day simulations (2001-2017) and future projections (2030, 2050, 2070 and 2090) of global climate at original GCM resolution (200 Km) from GFDL-ESM2G (NOAA, USA) GCM under CMIP5, and all RCPs namely, RCP 2.6, 4.5, 6.0 and 8.5. GCM data included monthly time series of maximum temperature, minimum temperature, solar radiation and rainfall. All GCM data were downloaded and later downscaled to 3 km resolution. The method of downscaling used is bilinear interpolation. Downscaling of the GCM data is often required for climate change impact studies. In general, the results of the present study with the increase in seasonal maximum and minimum temperatures by about 2.56 o C and 3.96 o C, respectively, by 2090 could be due to the increasing level of greenhouse gases in the atmosphere majorly because of burning of fossil fuels, deforestation and other man made activities. The Recent Sixth Assessment Report of the Intergovernmental Panel on Climate Change 2021, also reported that there is a chance of increase in the global warming level by 1.5 o C in the next two decades. The RCP projections of climate change showed an increase in seasonal maximum, minimum temperature and solar radiation in the future for Cuttack district. For the years 2030 and 2050, projected climatic condition under RCP 2.6, 4.5, 6.0 and 8.5 scenarios are likely to cause an increase in seasonal maximum temperature, minimum temperature, and solar radiation as compared to present conditions, but a non-significant trend of decrease in rainfall is shown under RCP 2.6, 6.0 and 8.5 scenarios for Cuttack district of East and South Eastern Coastal Plain agro-climatic zone of Odisha. However, RCP 4.5 scenario in the year 2030 is likely to cause increase in rainfall for Cuttack. For the year 2070, projected climatic condition under RCP 2.6, 4.5, 6.0 and 8.5 scenarios also likely to cause increase in seasonal maximum temperature, minimum temperature and solar radiation as compared to the present condition but a non-significant decrease in rainfall is showed for Cuttack district. For the year 2090, projected climatic condition under RCP 2.6, 4.5, 6.0 and 8.5 scenarios showed an increasing trend in seasonal maximum temperature, minimum temperature and solar radiation but, there is a non-significant trend of decrease in rainfall for Cuttack district." [8]

The above mention of three case studies one from Hungary, second from Karnataka state India and third from Cuttack district, Orissa state India reveals that climate change adaptation needs an integrated use of research studies carried out throughout the world. It should be based on agreed upon agro-ecological regions preferably [9]. All the countries in the world need to exchange their research studies along with their models if possible for enhancing the economic way of adaptations due to climate change. In order to put the results of the GCM-Global Climate Model for the use by agricultural, water, energy, infrastructure assets sector, monitoring and management committees /agencies need to put data in DIGITAL way in computer databases. As we are dealing with large geographical areas, Data Centers with Spatial databases and BigGeodata analytics using spatial statistics and various other mathematical tools and models is the need of the hour. This is discussed in the next section of this paper which deals with Machine Learning for Climate change studies and adaptation strategies made possible by digitization, *digitalization*, and digital transfer enabled by each country.

2.1 Satellite Data Studies from the Indian Space Research Organisation (ISRO) on Climate Change

The issues related to climate change can be well represented using space remote sensing and communication inputs. ISRO has done research on (a) perspectives of multidisciplinary observations and modeling of global climate change; (b) measuring spatial distribution of aerosols, clouds, and water vapor using indigenously developed lidar technologies; (c) spatio-temporal patterns of greenhouse gas (GHG) emissions over India and upscaling of methane from agricultural land use; (d) modeling spatio-temporal changes in the runoff of major river basins in India; (e) Himalayan glacial cover change; (f) Monitoring coralreefs, Arctic and Antarctic polar ice, Indian coasts, seawater intrusion in coastal aquifers, and sea level rise aspects in Gujarath and Andhrapradesh states,etc. [10].

A desktop analysis study using the data collected from Tungabhadra project and IPCC emission factors is carried out for the Tungabhadra sub-basin (Krishna Basin) using satellite based landuse and land cover categories showed that GHG emissions are within the limits [11]. It was observed that there are no net GHG emissions from the Tungabhadra sub-basin.

Indian Space Research Organisation, NRSC,under Geosphere Biosphere Programme, National Carbon Project (NCP) has measured soil vegetation atmosphere flux (SVAF) to determine soil carbon flux. The NCP addresses the estimation and modelling of the carbon pools in the soil, vegetation and SVAF of carbon. As part of NCP-SVAF a network of flux towers are being established in the different representative ecosystems. The carbon/water flux measurements by the flux tower are based on the eddy covariance (EC) method, a micro-metrological technique that provides a direct measure of net carbon and water fluxes between vegetated canopies and the atmosphere and vice-versa.With current technology, the EC method is able to measure mass and energy fluxes over short and long timescales (hour, days, seasons, and years) without disturbing the vegetation around it. Deployed as a network of measurements sites, the EC method when combined with remote sensing to upscale over larger region has the potential of quantifying how whole ecosystems respond in terms of carbon fluxes to various scenarios of climate regimes.

The EC flux data are an important input in estimating the net carbon balance of the forest ecosystem. This study which provides uniform continental-scale spectroscopic estimations of soil organic carbon is highly useful.This meets the increasing demand for information on the state of the soil that can be used in biogeochemical models and the monitoring of soil degradation [12]. State wise preliminary Desertification Status Maps (DSM) on 1:500,000 scale are prepared in Geographical Information System (GIS) environment. Geo-database was created in GIS using ArcGIS software package on a scale1:250 K. These maps have lambert conforml conical (LCC) projection and WGS 84 datum. Base layers of settlements, water bodies and administrative boundaries, are used as references from national resources database,NRDB datasets. The road and rail networks benck marks are used as reference bench marks from SAC NWIA project. Forest Survey of India (FSI) provided forest boundaries at 1:50000 scale that are used as a reference layer to delineate polygons, particularly within forest areas. Ground truth data and field checks were carried out in different crossections of the forest to finalize the maps. Quality Checking (QC) was carried out considering (a) accuracy of georeferencing (Image co-registration <2 pixels error), (b) uniformity in projection and datum (WGS 1984 L Conformal Conical), process & severity identification, (c) GIS database design and standards (MMU > 225 ha, topology checking, seamless mosaic, codification, cartographic elements, etc). Necessary corrections are incorporated. State-wise statistics are generated for different processes for two time frames.

Changes in desertification and land degradation classes were brought out for the timeframes 2011–13 and 2003–05. Insignificant land degradation observed in all Union Territories; hence, included in No Apparent Degradation (NAD) category. Further analysis shows that around 23.95% (2011–13) and 23.64% (2003–05) of desertification/land degradation with respect to total geographical area is contributed by Rajasthan, Maharashtra, Gujarat, Jammu & Kashmir, Karnataka, Jharkhand, Odisha, Madhya Pradesh and Telangana in descending order [13].

The Land use change probability surfaces for climate change impacts needs great deal of attention. The temperature and rainfall changes induced land use changes need to be monitored and managed with appropriate mathematical models using satellite data. The climate impacts should be seen in changes of agro-climatic zones/agro-ecological zones. The agro-climate zones mapping is carried out both by national agencies and international agencies (like ICRISAT). In India agricultural universities aid state government departments in preparing agroclimatic maps. India has been divided into 24 agro-climatic zone by Krishnan and Mukhtar Sing, in 1972 by using "Thornthwaite indices". The planning commission, as a result of mid-term appraisal of planning targets of VII plan (1985–90) divided the country into 15 broad agro - climatic zones based on physiography and climate.. Agriculture departments and concerned agencies/industries use this ACZ for agro-economic planning and management. The land use/ land cover change using satellite data should be ranked and weighted for its agro-climate variability in assessing land degradation/ change [14].

3 Machine Learning, Micro-Satellites Data, Geospatial Technology for Climate Adaptation

In order to have a dynamic solutions using a digital database for the world, one should possess Geospatial Knowledge Infrastructure (GKI) which aids in digital economies to survive sustainably. Data from space is fueling growth across industries, and its influence is set to grow. Industry growth depends on how government and industry collaborate. Miniaturization of satellites has led to exponential growth in private business, necessitating the development and evolution of relevant laws. Creating opportunities at the intersection of climate disruption and regulatory change is required. Climate risks, stakeholder demands, and the commitment of net-zero transitions are to be incorporated for business transformation.The geo-spatial data based foundational models are going to be used for all businesses.

GKI, geospatial knowledge infrastructure, integrates with and supports a geospatial dimension of the wider digital ecosystem. Satellites, drones, aerial remote sensing, and persistent remote sensing are providing the digital raster data that enables geolocation tagged to the data. Moreover, the Indian Drone Policy has enabled remote sensing using drones throughout India [15, 16]. Geolocation through an open positioning infrastructure, geospatial information, expertise, technologies, and analytics are important elements of the digital infrastructure. The latest developments using 5G and 6G improve localization, which must be designed with location, geospatial data, and geospatially enabled analytics at its core. Geospatial statistics, artificial intelligence, and machine learning play an important role in GKI data retrieval and analysis. Then, location becomes a powerful means to derive knowledge from across data and application sources, populated by various government departments,are sitting on the web. This loops right back to integrated digital policy. The six elements of a GKI are shown in Fig. 2

GKI is needed broadly for any nation for the following objectives to be achieved holistically: (a) To work toward the development and recognition of a new asset class for 'Resilient Infrastructure' in order to promote demand-side uptake and incentivize investments in resilient infrastructure; (b) Support nations affected by disasters in assessing damages to strategic and local infrastructure and in developing strategies for stronger,

Fig. 2. The Six Elements of a Geospatial Knowledge Infrastructure. Source: [17]

faster, and more inclusive recovery and reconstruction; (c) Develop 'Resilient Infrastructure Standards' in different risk contexts, sector-wise. These could be coupled with an independent certification mechanism to promote their worldwide adoption; (d) Develop metrics and methodologies to measure the resilience of infrastructure and produce a comparative report on the 'Status of Infrastructure Resilience' around the world.

ICT-enabled digital databases are poised to proliferate across all nations. For example, the work collaborated on "AI for Good" was demonstrated by the International Telecommunication Union recently. Google Research has reported recently, on Jan 29–2025, its research on foundational models used in a form that can be enabled by GKI for the generation of scenarios in economics for resilient infrastructure development for climate change adaptation [18]. There is a necessity to support the health and well-being of dynamic populations around the world. This requires researchers, governmental agencies, and industry enterprises to understand and reason over complex relationships between human behavior and local contexts. This support includes identifying populations at different levels of risk. Then realizing using grading methods as to where to target limited aid resources. Traditional approaches to these classes of problems often entail developing manually curated, task-specific features and models to represent natural and built environments, including human behavior. This can be challenging to adapt to new or even related tasks. Digital data in computers can change the innovation scenarios to invest for adaptations. This can be addressed by introducing the Population Dynamics Foundation Model (PDFM). This PDFM model aims to capture the relationships between diverse data modalities and is applicable to a broad range of geospatial tasks. For example: (a) Epidemic Spread and Health Facility Implementation; (b) Floods in real time in Urban Regions (as well as Rural Watersheds/Basins); (c) Ecosystem Resilience, Resistance, Adaptation Management; (d) Energy, Transportation, Mines Management.

It is reported in [18] “Models of population dynamics can be a powerful tool in understanding the impact of environmental, social, and economic factors on our well-being. “Greater population-level data could contribute to better planning and resource allocation, leading to improved outcomes for some of public health’s most vexing problems, including non-communicable diseases, behavioral health disorders and climate. related health impacts. A novel Population Dynamics Foundation Model (PDFM) that uses machine learning to synthesize rich, globally available geospatial data such as maps, busyness, and aggregated search trends, along with environmental signals such as weather and air quality. Model simplifies the modeling of a broad range of problems affecting populations around the world.This approach allows digital datasets to be at variable resolutions, consistencies, and qualities. For example, aggregated search trends might be insightful in heavily populated regions, while maps could be more useful in areas with lower internet penetration. We also show that incorporating embedding’s from PDFM can imbue existing models with geospatial *reasoning* capabilities. PDFM’s architecture leverages a graph neural network (GNN) because of its inherent ability to embed the network relationships across locations and handle missing data via message passing. We show that the embedding’s derived from the GNN are generally useful across a wide range of downstream tasks in spatial interpolation, extrapolation, and super-resolution problems. For these tasks, we show that a variety of relatively simple downstream models, including linear regression, simple neural networks, and gradient-boosted decision trees (GBDTs) all achieve similar performance when using PDFM-synthesized embeddings.

Also authors quote “PDFM-based embeddings were generated and used to improve an existing time series foundation model for geospatial forecasting.” To assess performance, the researchers conducted interpolation and extrapolation ***experiments*** at the postal code level, representing real-world applications with missing data. For the interpolation, we randomly chose 20% of counties and all their corresponding postal codes as the holdout set. In the interpolation experiment, the researchers randomly chose 20% of counties and all their corresponding postal codes as the holdout set. To simulate larger gaps in the data, the researchers performed an extrapolation experiment by selecting 20% of US states and all their associated postal codes as the holdout set in an extrapolation by selecting 20% of U.S. states and all their associated postal codes as the holdout set. For the super-resolution experiments, the researchers used only county-level labels for training and 20% of postal codes as the holdout test set.”

3.1 Micro-Satellites and Machine Learning

The super-resolution, extrapolation, and interpolation for monitoring water quality across the country’s regions is possible with Hyperspectral Remote Sensing, radar interferometry data, and drones’ high-resolution data. Then transforming water quality variables for super-resolution through artificial intelligence/machine learning methods and combining them with PDFM for downstream applications of health and water quality parameters. The geospatial data is available in about 50 countries. The space-faring nations, including India, have a number of geospatial portals possessing land resources, water resources, forests, atmospheric data, ocean data, cryosphere, etc., spatial data, and many more attributes of data such as population, administrative, political, and watershed/basins data. The spatial and non-spatial data aggregation is possible in geographic information system tools embedded in geospatial portals. The following is a list of some geoportals in India.

The following Indian web portals are useful for preparing Climate Change Adaptation Scenarios:

1. Government of India-CWC, https://indiawris.gov.in/wris/#/
2. Government of India, https://www.india.gov.in/
3. Government of India, Ministry of Jalshakti, http://jalshakti-dowr.gov.in/
4. Government of India, Ministry of Water Resources, http://nhp.mowr.gov.in/
5. Government of India, Central Water Commission, http://www.cwc.gov.in/
6. Government of India, Central Ground Water Board, http://cgwb.gov.in/
7. Government of India-ISRO-NRSC, https://www.nrsc.gov.in/
8. Government of India, http://ffs.india-water.gov.in/
9. Government of India-NRSC, https://bhuvan.nrsc.gov.in/home/index.php
10. Government of India, ISRO-SAC, Ahmedabad, https://www.mosdac.gov.in/
11. Government of India,ISRO-SAC, https://vedas.sac.gov.in/vcms/en/

In addition to the above, we have State Government based web portals and various research organization's web portals and World Wide Web portals for natural resources and anthropogenic resources management.

In an innovation of extracting from satellite data, Plastic Mulched Farmland Agriculture PMFA, people have tested three machine-learning algorithms.The studies have used random forest (RF), classification and regression tree (CART), and support vector machine (SVM) algorithms. In comparison of the PMF mapped results from these three machine learning methods, the CART(classification and regression trees algorithm was more suitable for long-term PMF monitoring. Since the PMF had some slight variations in different years (e.g., The thickness of plastic mulch, then, the CART algorithm was easy to interpret. Researchers have to slightly adjust the threshold to extract PMFA to improve the accuracy of long-term PMF mapping. See Fig. 3 for understanding and characterising using spectral signatures the plastic-mulched areas agriculture takes place with different background objects in the land.

The adaptation to climate change in various aspects/agencies of water sector in any country can broadly be summarized as shown in Table 2. Source: [20]. The spatial datasets obtained from satellite remote sensing are going to help provide information to GKI related to the water sector, both for water quality and quantity.

A summary of climate adaptation strategies for a few water use sector aspects is given here in Table 3 (Source: [21]).

In [22] IPCC it is reported as "GCMs depict the climate using a three-dimensional grid over the globe, typically having a horizontal resolution of between 250 and 600 km, 10 to 20 vertical layers in the atmosphere, and sometimes as many as 30 layers in the oceans. Their resolution is thus quite coarse relative to the scale of exposure units in most impact assessments, hence only partially fulfilling criterion 3. Moreover, many physical processes, such as those related to clouds, also occur at smaller scales and cannot be properly modeled. Instead, their known properties must be averaged over the larger scale in a technique known as parameterization. This is one source of uncertainty

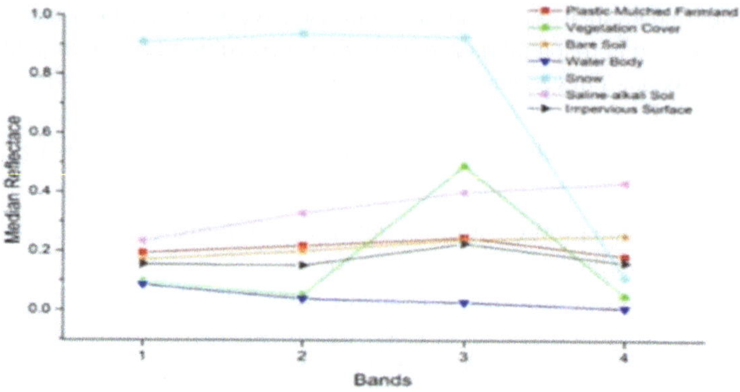

Fig. 3. Spectral separability of different ground objects including plastic mulched land (agriculture) using Sentinel-2 imagery MSI (Blue, Red, NIR and SWIR-2). Source: [19]

in GCM-based simulations of future climate. Others relate to the simulation of various feedback mechanisms in models concerning, for example, water vapor and warming, clouds and radiation, ocean circulation, and ice and snow albedo. For this reason, GCMs may simulate quite different responses to the same forcing, simply because of the way certain processes and feedbacks are modelled Global climate models GHG scenarios" [22].

The developed nations are using these GCM downscaled outputs on rainfall, temperature, radiation etc. for operating their water reservoirs. They may downscale with few algorithms embedded in the information through hydrologic models simulation and get information on water volumes available in reservoirs. This is evident from Table 3. In developing countries like India, one is not using these simulated outputs of GCMs for operating the water reservoirs. It is to be noted that all the reservoir operations of around 5000 large dams (out of 5250 total) are *not* digitally and automatically operated in India (Figs. 4 and 5).

The so-called "small satellites" vary widely in terms of mass, volume, orbital altitudes, station-keeping, and de-orbit capabilities. These satellites have been in use for the past decade for many different types of applications. For instance, their manufacturing and usage can range from (a) one-of-a-kind student experiments that are typically cube satellites (CubeSat's) to (b) the application of relatively larger 250- to 500-kg satellites deployed within sizeable constellations. In between these two extremes, there are also small satellites for (a) scientific experiment-carrying satellites (b) amateur radio relay, (c) business-to-business data relay satellites, (d) satellites used for capturing data relayed from Earth-based features / patterns. The launch of large-scale commercial constellations can piggyback small satellite deployments. Also small satellite launch vehicles SSLV are available in India both in ISRO and private enterprises. These SSLV utilize small spacecraft with a high degree of functionality in order to support applications such as Earth remote sensing, telecommunications, and atmospheric and scientific research that involve, by far, the greatest number of small satellite deployments.

Furthermore, many space programs are focused on the development of micro-, Nano-, and pico- satellites for educational purposes. SmallSats or small spacecraft have a mass

Table 2. Example Adaption Options for Key water sectors. Source [20]

Water Use Sector	Example Climate Change stimulus	Example Stake Holder	Example Adaptations		
			Accepting Losses	Preventing Effects	Changing Uses and or Locations
Municipal	Supply is reduced, and demand increases	Municipal officials	Drought contingency planning	Increase water intake pumping capacity	Promote water Conservation
Agriculture	Supply is reduced	Individual farmers	Supplement income by seeking off-farm employment.	Construct or enlarge farm ponds to store more water.	Shift to low-water-using sprinklers
Industry	Supply is reduced	Individual industries	Seek government subsidies for inceased operating costs	Increase water storage capacity	Recycle cooling water
Rural domestic	Ground water supply is reduced,and demand increases	Households	Allow lawns and garden s to deteriorate	Deepen wells	Conserve water in the home and yard
Commercial navigation	Lake levels decline	Shipping companies	Contingency plans for interruption of shipping	Adapt shipboard loading and unloading facilities for lower lake levels	Reschedule shipments to coincide with seasonal peak water level periods
Energy	Lake levels decline	Power generation utility	Substistute energy sources to replace lost generation capacity	Increase pumped storage capacity	Encourage energy conservation by consumers
Recreation and tourism	Lake levels decline	Marina owners	Temporally close Marinas	Install floating docks at Marinas	Relocate Marinas
Commercial fishery freswater	Streamflow reduction reducehabitat quality	Governement regulators	Close the fishery	Construct resrvoirs to increase flows	Reduce the number of licensed fishers

less than 180 kg. They are about the size of a large kitchen refrigerator. The Small Sats differentiation is based on a variety of size and mass. They are named as: (a) Minisatellite, 100–180 kgs, (b) Microsatellite, 10–100 kgs, (c) Nanosatellite, 1–10 kg, (d) Picosatellite, 0.01–1 kg, (e) Femtosatellite, 0.001–0.01 kg. CubeSat's are a class of nanosatellites that

Table 3. Summary of CC adaption strategies and the corresponding model. Source [21]

Category of Strategies	Reference	Study area	Adaptation strategies	Climate Model/GHG emission scenaario	Hydrologic, hydraulic, and water management models
Updating reservoir control curves and reservoir manageme nt	Sant Anna et al.(2022)	Lievre River Basin in Quebec, Canada	Updating reservoir operation rules & flood control	Ensemble of 828 GCMs/RCPs 2.6,4.5,6,8.5	Simulation model HecRes Sim optimisation and stochastic dynamic programming
	Sterle et al.(2020)	Trucker River Basin-US	Updating the reservoir operating curves for changes in snowmelt.	GCM(Getting er et al.2016)	Hydrologic-PRMS.Reservoir operations: River Ware
	Stagge and Moglen (2017	Washington, DC-Metropolitan Area - US	Adopting a supply system and reservoir operating rules	5GCM and stochastically generated streamflow and participation	Hydrologic Routing and Reservoir Operations -OASIS: Optimization Algorithm - SMSEMOA
	Steinschn eider & Brown (2012)	WestField river Basin,Springfield, MA-USA	Best-Guess strategy based on future CC-Dynamic Strategy	GCMs-Downscaled/ A2	Optimisation algorithm-Hydrologic-DHSVM
	Koch et al. (2010)	Part of the Elbe River Basin, Czech Republic	Reservoir water management	GCM ECHAM5 A1/B	Hydrologic SWMM manangement and WBalMo
	Medellin Azura et al.	California State US	Updating operation rules of a complex optimized water management system (ground,surface, and reallocated water)	GCM downscaled /A2	Hydrologic: no model optimization CALVIN
Adopting water allocation	Reinheim er et al (2016)	Upper Yuba River Basin, California	Redefining the WYT thresholds: AHD and AHDF techniques	4GCMs/RCP 8.5 from CMIP5	WEEP,Optimisation :MILP
	Yanget al.(2014)	Indus river basin Pakistan	Adapting Surface water allocation schemes:Econom ics and Water allocation economics combined	Historical data and Deta change from literature	Indus basin Model,Revised MultiYear-IBMR-MY-Hydro-agro-economic model
	Null and Viers (2013)	Californias Sacramento Valley &SanJaoquin Valley	Maintaining historical WYT thresholds with changing flow regimes* modifying WYT thresholds for changing flow regimes	GCM statistically downscaled/A 2 and B1	Hydrologic Variable Infiltartion Capacity(VIC)

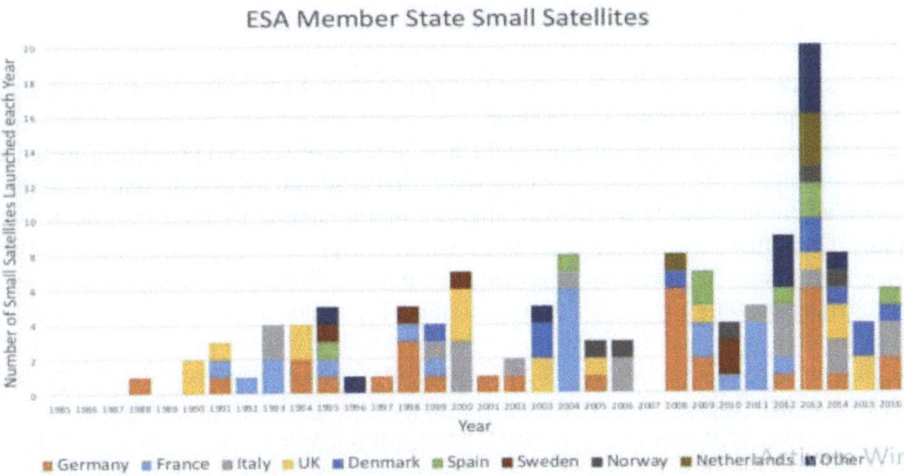

Fig. 4. Small satellites (<200 kg) launched by ESA Member States from 1985–2016. Source: [23]

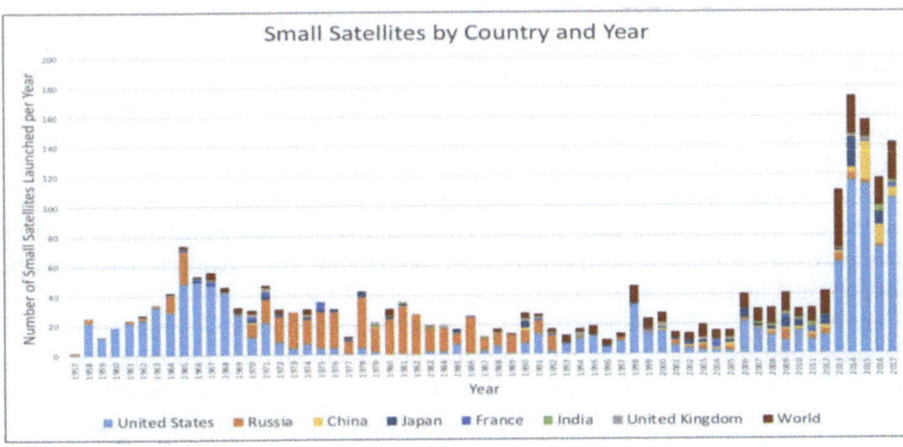

Fig. 5. Small satellites (<200 kg) Launches per year from countries around world. Source: [23].

use a standard size and form factor. The standard CubeSat size uses a "one unit" or "1 U" measuring 10x10x10 cm and is extendable to larger sizes up to 27 U or may be higher in the near future. The development of CubeSats is now taking place in a collaborative mode among government, industry, and academia to attain ever-increasing capabilities. CubeSats now provide a cost-effective platform for science investigations, new technology demonstrations, and advanced mission concepts using constellations, swarms, and disaggregated systems [24].

Small satellites of say, up to 100 to 200 kgs are less costly. Large satellite constellations is ten times costly than small satellites when number of temporal revisit of satellites on earth regions is considered. New space activities bring in innovations of academia, industry and startups.

The space commercialization effect took place with the transfer of space activities from the public to the private sector. Space activities were initially government-financed (based on tax money).They were the main driving force of national prestige. The major space companies either directly or indirectly started to search for private space activities as soon as public funding (especially after the moon landing by the USA) decreased rapidly. Once these activities proved to be viable morphosis of investment and governance took place. The primary governance of public-private partnerships started getting changed to fully privately financed entities. Examples are Intelsat and Inmarsat, which eventually gained access to private equity funding without public sector participation eventually. This process created a rapid shift from governmental to more private commercial activities.

The number of startups has increased in the past few years in the space sector. They are guided by IN-SPACe and ISRO. The space reforms aim to make India a leading space economy globally through increased private sector participation.

IN-SPACe aims to get small satellites manufactured in India through startups with innovative application ideas. The Chairman IN-SPACe has given a visionary statement in the Business Today Special Issue of February 2024 that: "1. ISRO continues to be a beacon of innovation. But the game changer in the future will be through the participation of the Indian private sector. 2.Dedicated manufacturing sectors, supported by the states, with plug and play facilities for small satellites, launch vehicle subsystems, ground station equipment, payloads, and subsystems catering to global market requirements. 3. An Indian Earth Observation Constellation, being managed and operated by private players.4. Downstream applications are developed by NGEs (non-government entities) with the use of space data and images. These can be to monitor changing climate, extreme weather, use of natural resources, address challenges in agriculture, clean energy, drinking water and many more" [25].

The National Geospatial policy is a very good initiative by the Government of India [26], as getting the spatial data was very tedious, costly and the procedure was very lengthy. So that has been reduced. Even if somebody has to pay some amount and get it downloaded, that would also be okay, but the data should be available. So in this policy, it has been made sure that the data is available to the general public. There are several types of Big Data: (a) structured data, (b) unstructured data, (c) Semi-structured data, (d) geospatial data, (e) machine or operational logging data, (f) open-source data. These big geospatial data is going to give impetus to following technical activities (i) effective and efficient management of data in data centers. (ii) seamless workflows from ingestion of geospatial data to actionable analytics on-the-fly. (iii) need for advanced extract load transform (ELT) tools along with on-demand analytics for concurrent processing. The blue economy needs a push as most of the resources in marine region (authorized national marine areas) are exploited more than 80%. Coastal Marine Spatial Planning (CMSP) in which India needs to understand coastal areas as to where the country needs to put up industries and of which type. CMSP is a process of analyzing and allocating the spatial and temporal distribution of human activities in marine areas to achieve economic, ecological, and social objectives that have been specified through a political process.

Based on space data, we need to make an informed decision as to where we need to plant mangroves and where are coral reefs. The constellation of small satellites is needed. Our ships and even other countries' ships are being monitored through AIS methods in large satellites. New space satellites are going to take over soon as they are cheaper and temporarily have higher revisit capability. The Indian Space Policy 2023 marked a transformative phase, welcoming Non-Government Entities (NGEs) into core space activities.

Once space commercialization has proved to be viable, the initial public-private partnerships are more and more transferred to fully privately financed entities. Examples are Intelsat and Inmarsat, that got access to private equity funding without public sector participation eventually. This process created a rapid shift from governmental to more commercial activities. The business information provided by private industries in India and even Satsearch newly formed in 2023 is very low. They do not include the help needed for technology roadmaps and service innovations necessary for the success of small satellite constellations.

SatSearch announces in its website like this: "1. In today's global, diverse, and fast-moving space industry, it can take hundreds of man-hours to find the best products and suppliers for any new mission; 2. You can also see very similar products and services (producing near-identical results) vary in price between 1X to 10X depending on the supplier; 3. Transparency on the market lets you get a clear view of price-to-performance and can save significant time and money in space missions; 4. We can help you short-cut discovery processes dramatically and we have supported over 400 missions to date to do just that; 5. Helping thousands of engineers from more than 150 countries. Satsearch is a specialist supply chain discovery platform for the space industry offering multiple benefits for space engineers – ALL 100% FREE." ALSO; it says, "Request – easily ask for quotes, further information, and proposals in seconds". Once you've identified a relevant product or service that meets your needs, *satsearch* makes it quick and easy to submit requests to the supplier. With just a few clicks you can ask for pricing, documents, lead times, technical specifications, or anything else you might need for your engineering and mission plans. Only share what information you're happy to give out, though the more details we have, the more relevant your results will be. This service is also 100% FREE and available 24/7."

Small satellite pioneers like Surrey Technologies, Planet Labs and Star link are the pioneers in the field of small satellites and knowledge management systems are yet to be developed both for technology and business transfers. Space X has launched over 6,000 *Starlink satellite*s into low Earth orbit (LEO). The first 1,440 satellites are distributed across 72 orbital planes, with 20 satellites in each plane. Starlink is a satellite internet constellation operated by SpaceX, providing global satellite Internet access. It aims to provide high-speed Internet to underserved and remote areas. Satellites use phased array antennas for communication and Hall-Effect Thrusters to maintain orbits and facilitate deorbiting. Recent Launch was on May 18, 2024, adding to the operational constellation. Planet Labs, based in California, operates a constellation of Earth-imaging satellites. These satellites capture high-resolution images of the Earth's surface. These images after processing are used for monitoring crop health, yield prediction, and precision farming; in environmental monitoring for tracking deforestation, urban development, and natural disasters; in climate research for studying climate change impacts. Planet Labs has nearly 150 working satellites in its constellation. Planet Labs provides daily satellite imagery and Earth data analytics. In India too such efforts are on by newly formed IN-SPACe which is promoting NGEs-Non Government Entities like private industries, startups, universities innovations to thrive by taking appropriate support from ISRO in availing its infrastructure and knowledge. If one visits the Planet Labs website, it can be explored for their daily satellite imagery and insights.

The *new space* assets started coming up recently.IN-SPACe is the nodal agency in DoS, responsible for promoting, enabling, authorizing as well as supervising space activities of the Non-Government Entities (NGEs). IN-SPACe has already launched a *seed fund scheme* for startups using space technology. The scheme started in December 2023.The scheme has the provision of providing financial assistance of up to Rs. 1 crore in addition to mentorship support, training and networking opportunities. There are various educational institutions in India offering specialized courses in satellite technology. Graduation courses in Aerospace are available in few dozens of institutes in India, both government as well as private universities. Few small satellites weighing less than a few kg(say 2–10 kg)are already launched by private institutes like PESIT,Anna university,SRM university and RVCE.Indian-Global startups/enterprises like Skyroot Aerospace,Agnikul Cosmos, Bellatrix Aerospace, Pixxel, Brahmos Aerospace, Tata Advanced System, Dhruva Space, etc. are already involved in space/aerospace assets, launch vehicles, and small satellites.

3.2 Geospatial Knowledge Infrastructure(GKI) Readiness Index for India

GKI Index for India now (2025) stands at 27th out of 50 countries analysed. See Table 4 for understanding India's position. Source [27].GKI Readiness Index is used as a tool to understand how much of the digital infrastructure assets and utility are contributing to the economic development of a country. India, with its ranking as 27th out of 50 countries, stands in a challenger position with 56.46 points out of 100.The interpretation of points and weights given to the Pillars of GKI index suggests that foundation data pillar is too low with less utility. The sub-pillars of the Foundation Data Pillar also need attention to improve India scores. The sub pillars of the Foundation Data pillar are: (a) fundamental data themes like the scale and frequency of updates of available thematic layers-forests, minerals, agriculture, soil moisture, soil nutrients, geomorphology, energy grid, social grids, hydrogeology, etc.; (b) positioning infrastructure like satellite-based positioning systems, satellite-based augmentation systems, and ground-based augmentation (RTK); (c) open and linked data like open and free data and integrated data collection flow; (d) earth observation infrastructure like satellite platforms/sensors, data reception, and data resolution. See Table 4 for understanding pillars and sub-pillars of GKI index [27]

This pillar, if improved, can be integrated to work with PDFM as discussed earlier and downstream assets enhancement for sustainable use can happen. The geospatial maps in Geoportals of India and algorithmic support software and relevant hardware are provided in geoportals of India (Section). The proof of concept on models chosen from geoportals of India and test and analysis for effective use of geoportals with appropriate plugins and sub models, needs to be propagated possibly through academics to the GKI industry. The land use/land cover, as determined by algorithms developed at NITK on

Table 4. The Pillars of Geospatial Knowledge Infrastructure Readiness Index

Pillars	Sub-Pillars	Factors	Weighting (%)
Pillar 1: Integrated Policy Framework	Core Geospatial Policies	National Geospatial Policy; Space Policy; Remote Sensing/Earth Observation Policy; UAV/Drone Policy; GNSS Policy	10%
	Enabling Policies	Open Data Policy; Science & Technology and Innovation Policy; Digital Strategy; AI Strategy; Telecommunications/ 5G Strategy; Data privacy/Protection Policy; PPP Strategy	
	Geospatial Mandate in Sectoral Policies	Mention of Geospatial in sectoral policies (land, infrastructure, cities, etc.)	
	UNIGIF Adoption	Country level action plan	
Pillar 2: Foundation Data	Fundamental Data Themes	Scale and frequency of update of available thematic layers	40%
	Positioning Infrastructure	Satellite-based positioning systems; satellite-based augmentation systems; ground-based augmentation systems (RTK)	
	Open & Linked Data	Open and free data; integrated data collection flow	
	Earth Observation Infrastructure	Earth Observation infrastructure and data resolutions	
Pillar 3: Partnerships & Collaborations	International Collaborations	Member of GEO, W3C, OGC; Participation in UNGGIM Regional Measures; EO Satellite/PNT Collaboration	5%
	National Geospatial Collaborations	Stakeholders involved; type of collaboration	
Pillar 4: Industry Leadership	Industry Capacity	Existence of industries in different geospatial technology segments; market capitalization; growth of innovative companies; companies embracing disruptive ideas; ICT services exports[27]	10%
	Innovation Promotion Programs	Incubation and accelerator programs	
	Industry Networks	Industry networks; knowledge networks	
Pillar 5: Applications, Analytics and Modeling	Stages of Geospatial Technology Architecture	Platforms and portals; delivery mechanism	15%
	User Adoption[28,29,30]	Mapping or service level; asset management/business process modeling level; analytics and workflow level; system integration level; enterprise level	
	4IR Adoption and Integration[31]	Adoption of emerging technologies; investment in emerging technologies; government online services; government promotion of investments in emerging technologies; legal framework's adaptability to emerging technology	
Pillar 6: Geospatial Dimension to Digital Infrastructure	Spatial Data Access	Presence, maturity level and effectiveness of SDI	20%
	Standards	Type of standards	
	Institutional Capacity	Foundation and knowledge creation courses; skill set of graduates; ease of finding skilled employees; updates in education curricula and investment in the skills needed for jobs and "markets of tomorrow"[32]	
	Capacity Building	Extent of staff training	
	Digital Readiness	E-participation; digital skills among active population[33]	

hyperspectral data, needs to be tested and validated for use in downstream applications of PDFM like soil moisture regimes and temperature heat maps which are needed to be continuously updated for determining agro-climatic zones because of climate change [28, 29].

A continuum computing approach may become a necessity soon to monitor climate change. ISRO has put in its geoportals data on several reservoir levels and validated a few for field level gauges. This needs to be extrapolated for all 5000 dams by using Interferometric Synthetic Aperture Radar (InSAR) data. SARAL Altika waveform data, Ka-band (35.75 GHz) radar altimeter, is put in SAC web portal MOSDAC on inland water levels. The water level product for inland water bodies (presently for the Brahmaputra River and Ukai Reservoir) has been generated using the SARAL-ALTIKA waveform data [30]. These continuous real-time transmissions of data and computing may appropriately solve water reservoir operations for climate change adaptation.

There exists a natural continuum across the evolving science ecosystem (example, climate change adaptation monitoring and management) spanning large-scale instruments, experimental facilities, observatories, and sensor networks, all streaming data; high-speed networks and network services; and a range of computing capabilities along the continuum, from edge, to in-network, to large-scale data centres. One example application that is already proved is (a) continuum computing workflow is to use streaming data from an actual oilfield along with subsurface flow simulations and ML-based optimizers to manage oil production, reduce environmental impacts, and (b) urgent computing for air quality variations because of wildfires or temperature inversions in a region (like Delhi or California) [31].

GKI Readiness Index and GKI assets carry the potential to implement nature-based solutions for climate change adaptation.

Potential impacts of nature-based solutions for climate change adaptations. Nature-based solutions can be implemented at different scales, depending on local conditions. They can help address some of our biggest challenges, including adapting to risks associated with hazards such as fires, floods, and heat waves, as well as emissions mitigation, resource reallocation through LLMs and also for possible carbon sequestration. See Table 5 for nature-based solutions. Source [32].

Table 5. Twenty three Nature-Based Solutions. Source: [32]

Regulating Services		Provisioning Services	Cultural Services
Protective Services	Other Regulating Services		
Disaster Risk Reduction	Benefits obtained from regulation	Products obtained from ecosystems	Nonmaterial benefits obtained from ecosystems
▪ Flood protection ▪ Landslide protection ▪ Avalanche protection ▪ Soil erosion protection ▪ Heat wave mitigation	▪ Carbon capture and sequestration ▪ Pollution mitigation ▪ Heat mitigation ▪ Water quality improvement ▪ Water regulation ▪ Disease regulation ▪ Pollination	▪ Provision of food and free water ▪ Livestock fodder ▪ Fuelwood ▪ Genetic resources ▪ Biochemicals ▪ Other harvestable resources	▪ Opportunities for recreation and ecotourism ▪ Spiritual and religious benefits ▪ Aesthetic benefits ▪ Cultural heritage ▪ Educational benefits

Nature-Based Solutions		Hazard Risk Reduction							Water Outcomes		Atmospheric Outcomes		
		Fluvial Floods	Riverine Floods	Coastal Floods	Coastal Erosion	Landslides and Erosion	Wildfire	Extreme Heat	Water Quality	Water Quantity	Carbon Sequestration	Air Quality	Aggregate Benefit
Watersheds	Slope stabilization	1	1	1	2	4	0	0	3	1	1	0	14
	Forests	3	3	1	1	4	4	4	4	4	4	4	32
	Agroforestry/ silvopasture	3	3	0	1	3	3	2	3	3	3	3	24
	Grasslands and other vegetation	3	3	1	1	3	3	1	3	2	3	1	22
	Farmland best practices	3	3	0	0	3	2	1	4	2	2	3	20
	Riverbanks, riparian areas	3	3	1	1	3	2	1	3	3	2	1	22
	Inland wetlands	2	3	0	0	2	2	1	4	3	4	2	23
	Floodplains and bypasses	3	4	2	1	1	2	1	3	4	2	1	23
	Setback levees (riverbeds and riparian areas)	1	4	2	1	2	0	1	3	3	2	1	19
Coastal	Mangroves	1	1	4	4	0	0	2	3	0	4	1	20
	Marshes and other coastal wetlands	1	2	4	4	0	2	1	3	0	4	1	22
	Living shorelines	0	1	2	4	0	0	1	2	0	3	1	12
	Coral reefs	0	0	3	4	0	0	0	1	0	1	0	9
	Oyster Reefs	0	0	2	4	0	0	1	3	0	1	0	10
	Sediment transport management	1	2	1	4	0	0	0	3	0	2	0	13
	Seagrasses and submerged aquatic vegetation	1	1	3	3	0	0	1	3	0	3	0	14
	Sandy beaches and dunes	1	0	3	4	0	0	1	1	0	0	0	10
Urban	Urban canopy	3	2	1	1	1	1	4	3	3	3	4	24
	Urban green spaces (parks, water plazas)	3	2	1	1	1	1	4	3	3	3	4	25
	Bioretention areas/rain gardens/bioswales	4	2	1	1	1	1	2	4	4	2	3	24
	Constructed and urban wetlands	4	3	1	1	1	1	3	4	4	3	2	25
	Green roofs, facades, walls	3	2	0	0	0	0	3	2	3	3	2	16
	Permeable pavement, urban water harvesting	4	2	1	0	1	0	1	2	3	0	0	14

NBS may be implemented through a variety of land management categories, ranging from private lands to protected areas.

Effect:
- 0 No effect
- 1 Low effect
- 2 Intermediate effect
- 3 Strong effect
- 4 Very strong effect

Maximum potential positive impact of a nature-based solution under ideal conditions (e.g., enough space, maturity)

3.3 New Materials Study

Micro-satellite constellations are capable of doing what all large satellites used to do earlier. Use of nanomaterials in space technology needs to be explored well. The prime necessity of lightweight materials and smart nanomaterials like CNTs is to alleviate the interference caused by space weather and enable small satellite constellations to do their best for Earth resource monitoring and management. It is especially true for LEO satellites.

Also, adding CNTs to aerospace materials could protect them against lightning strikes. In a recent research study, MIT experts have strengthened these advanced aerospace materials with what they call "nanostitching." Rows of CNTs perpendicular to the carbon microfibers fill the spaces between them, reinforcing the fibre layers without piercing them. The researchers showed that these materials are not only 10 times stronger than those that do not contain nanotubes, but they are also more than 1 million times more electrically conductive, which suggests that they might protect aircraft from lightning strikes [33]. The use of Cubesats constellations with InSAR-type sensors is going to be enormous if one wishes to monitor and manage all infrastructure projects undertaken by the Government of India in PM Gatishakti or even others. Delivering green infrastructure (GI) can provide multiple benefits that enhance the performance, efficiency, and resilience of linear infrastructure assets and can reduce whole-life costs along transport, energy, and water networks. The SAR interferometry technology has been developed for monitoring river currents (along track interferometry), roads, and bridges (using across track interferometry) and persistent scatterer mode remote sensing. There is a need to develop architecture for PNT(navigation) and SAR interferometry techniques to fulfill the objectives of logistics from anywhere to anywhere in an optimal way during the operational stage of multimodal connectivity. Already, some fifty multimodal digital infrastructure platforms are planned across India in India's PM Gatishakti project to help villagers move optimally with their agricultural goods. The PNT architecture is going to be national in scope and includes the home ministry, the intelligence community, as well as civil, commercial, and international users and systems using both Indian and global systems. This should happen even before India adopts digitization/ digitalization/digital transfer of all its businesses/ infrastructure assets at a 50% level. In order to make encrypted GNSS signals for most civil applications, one also has to go for QKDs apart from making use of CNT technologies. CNTs help microsatellites for communication constellations become cheaper and more reliable because of LEO orbit precision.

The chemistry and mechanism of material degradation due to AO, UV, VUV (atomic oxygen, ultraviolet, and vacuum ultraviolet) attack are complex and not fully understood. In general, the degradation process can be described as the removal of carbon, hydrogen, or other atoms in the material through processes involving abstraction, UV, or replacement of atoms. However, the anticipated uses of nanocomposites for space applications highlight the critical need for short-and long-term testing of materials in the space environment. The process to space-qualify a material is long, expensive, and usually application-specific. Until then, the effects of the space environment on CNTs and composites can be estimated from ground-based experiments and the effects on carbon materials similar in structure, such as pyrolytic graphite and other carbon composites [34].

Nanosensor development for environmental contaminants is growing rapidly (water quality monitoring, air quality monitoring (microplastics) etc).Monitoring water quality uses magneto-plasmonic SERS (surface enhanced Raman spectroscopy) substrates. SERS substrates are roughened silver/copper/gold surfaces. SERS substrates can also be obtained using renewable source luminescent POM (polyoxometalate) based sensors also. Functional features of POMs in sensors could be (a) electrochemical sensors, where they enhance selectivity and sensitivity; (b) Optical Sensors uses them for their luminescence properties, making them suitable for optical sensing; and (c) piezoelectric sensors, where they can be integrated into piezoelectric devices for sensing applications.

The prime necessity of lightweight materials and smart nanomaterials like CNTs is to alleviate the interference caused by space weather and enable small satellite constellations to do their best for Earth resource monitoring and management. It is more so for LEO satellites. The NanoFEEP thrusters are designed to face the constraints of limited volume, weight, and power that can be made available on a 1 U CubeSat. Then, the main objective of the thruster design becomes having good power efficiency, the optimum use of space, and the aim to achieve a lightweight propulsion system. Highly miniaturized Field Emission Electric Propulsion (FEEP) possesses a CNT neutralizer chip. The uncontrolled divergent spacecraft charging due to the inclusion of NanoFEEP thrusters has to be controlled. Then, one must develop a cold, highly efficient electron source that will emit the same number of electrons as the number of ions emitted by the thrusters. This electron source is called a neutralizer. Technologists and scientists have decided to develop a new kind of electron source to address the power, weight, and volume limitations of CubeSats. The hollow cathodes and thermionic cathodes, if used, become oversized and consume more power. Analogous to the working principle of the NanoFEEP thruster, a high electric potential is applied between an emitter (silicon chip) and an extractor electrode (grid), but with an opposite polarity. The neutralizer uses a silicon chip with deposed Carbon Nano Tubes (CNTs). This is in place of using metallic needles wetted with propellant, as in FEEP thrusters. One neutralizer chip could compensate for the total emitting current of one thruster over the whole thrust range. The targeted power consumption of the neutralizer for the full thrust range is set below 400 mW. The CNTs act as the field-enhancing component. As CNTs have high aspect ratio and small diameter, the starting voltage of the neutralizer (several hundred volts) is much lower than that of FEEP thrusters (kilovolts) [35] In one of the Chinese satellites, Taiji-1, launched in August 2019, the RF ion micro-thruster system (μ RIT) achieved the neutralization control requirements of ion micro-thrusters for the past 100 test missions in space. The combined use of carbon nanotube field emission neutralizer (CNTN) and μRIT has no performance degradation, and the neutralization control strategy is effective. Even the micro-propulsion system (μRIPS) used for compensating non-gravitational forces on the satellite was successful [36]. In another research study in improving the accuracy requirements of propulsion required for high-precision exploration missions (like gravitational wave detection, Earth gravity field detection), the guidance for neutralizer configuration and structure design in the propulsion system are obtained. Results have shown that (a) aging treatment is helpful in improving the working stability of the neutralizer, (b) the background electric field of the emitter has a strong correlation with the instability of the emission current, (c) the gate aperture has an influence on both emission and extraction performance, and (d) the 1 mm gate aperture has the best performance under experimental conditions [37]

4 Climate Change Science and Engineering Research Studies-Conflicting Reports

There are around 200 cases in the anamoly of scientific results on climate changes observed as per IPCC Intergovernemntal Panel on Climate Change) Vs those observed by the NIPCC (Non-governmental International Panel on Climate Change) and other sources. I am mentioning only two or a few of them here.

ONE: Reference [38] quotes as "Human-induced warming reached approximately 1°C (likely between 0.8°C and 1.2°C) above pre-industrial levels in 2017, increasing at 0.2°C (likely between 0.1°C and 0.3°C) per decade (high confidence). Global warming is defined in this report as an increase in combined surface air and sea surface temperatures averaged over the globe and over a 30-year period."

There is an 80% likelihood that the annual average global temperature will temporarily exceed 1.5 °C above pre-industrial levels for at least one of the next five years, according to a new report from the World Meteorological Organization (WMO). This is a stark warning that we are getting ever closer to the goals set in the Paris Agreement on climate change, which refers to long-term temperature increases over decades, not over one to five years [39].

Reference [40] quotes that "It has been reported that anthropogenic contributions in rising the global temperature is only 0.01degree centigrade. This is supported by thermodynamic calculations of the ocean as sinks"[40].

The anthropogenic effects are discussed in [40] in detail. The adiabatic theory allows one to evaluate quantitatively the influence of anthropogenic emission of carbon dioxide on the Earth's climate. The increase in the surface temperature at sea level caused by the doubling of human-induced CO_2 emissions will be less than 0.01 °C, which is negligible compared to the natural temporal fluctuations of global temperature. It has been deduced that even a considerable increase in the anthropogenic emission of CO_2 and other green house gases practically does not change the global atmospheric temperature.

Evaluating the climatic consequences of anthropogenic CO_2 emissions, one has to take into consideration that the emitted. CO_2 dissolves mostly in oceanic water (according to Henry law) and then is fixed into carbonates. Together with anthropogenic CO_2, part of oxygen is also removed from the atmosphere (about 2.3 g per 1 g of carbon). If the ocean and plants consume all this additional CO_2, after the year 2000 this would lead to a reduction in atmospheric pressure by approximately 0.34 mbar and cooling of the climate by, -8.2×10^{-2} °C $\approx$ 0.01 °C.

In reality, however, the metabolism of plants should almost completely compensate for the disruption of equilibrium by mankind and restore the climate balance [40].

TWO: Mean Global Sea Levels: Quotes from reference [41] "Even more revealing was the globally distributed sea-level time series study of Jevrejeva et al. (2006),who analyzed information contained in the Permanent Service for Mean Sea Level database using a method based on Monte Carlo Singular Spectrum Analysis and removed 2-to 30-year quasi-periodic oscillations to derive nonlinear long-term trends for 12 large ocean regions, which they combined to produce the mean global sea level (gsl) and gsl rate-of-rise (gsl rate)curves depicted in Fig. 6.

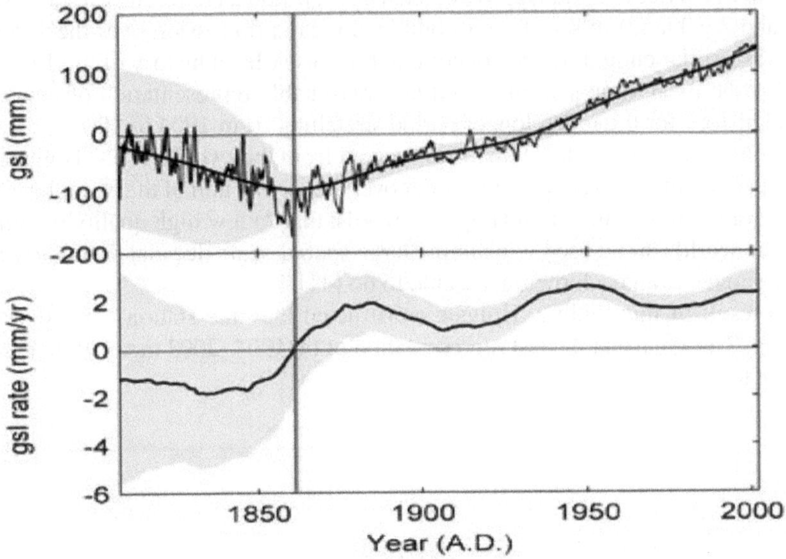

Fig. 6. Mean global sea level (top),with shaded 95% confidence interval,and mean gsl rate-of-rise(bottom),with shaded standard error interval, adapted from Jevrejeva et al.(2006) [41].

"The figure clearly shows no acceleration of sea-level rise since the end of the Little Ice Age. Jevrejeva et al. say "global sea-level rise is irregular and varies greatly over time," but "it is apparent that rates in the1920-1945 period are likely to be as large as today's." In addition, they report that their "global sea-level trend estimate of 2.4±1.0 mm/year for the period from 1993 to 2000 matches the 2.6±0.7 mm/year sea-level rise found from TOPEX/Poseidon altimeter data." With respect to what the four researchers describe as "the discussion on whether sea-level rise is accelerating," their results pretty much answer the question in the negative."

"The observations described above make us wonder why late twentieth-century global warming-(if it were as extreme as the IPCC claims it has been)-cannot be detected in global sea-level data. The effects of the warming that led to the demise of the Little Ice Age-which the IPCC contends should have been considerably less dramatic than the warming of the late twentieth century-are readily apparent to the right of the vertical red line in the Fig. 6. Likewise, although the atmospheric CO_2 concentration experienced a dramatic increase in its rate-of-rise just after 1950 (shifting from a 1900–1950 mean rate-of-rise of 0.33 ppm/year to a 1950–2000 mean rate-of-rise of 1.17 ppm/year),the mean global sea-level rate-of-rise did not trend upwards after 1950, nor has it subsequently exceeded its 1950 rate-of-rise."

"In concluding our examination of the peer-reviewed sea-level science, we report the findings of the most recent study of Holgate (2007). In a previous paper, Holgate and Woodworth (2004) derived a mean global sea-level history from 177 coastal tide gauge records that spanned the period 1955-1998.In an attempt to extend that record back in time another half-century, Holgate chose nine much longer high-quality records from around the world (New York, Key West, San Diego, Balboa, Honolulu, Cascais, Newlyn, Trieste,and Auckland) to see if their combined mean progression over the 1955–1998 period was similar enough to the concomitant meansea-level history of the 177 stations to employ the mean nine-station record as a reasonable representation of mean global sea-level history for the much longer period stretching from 1904 to 2003."

"In comparing the sea-level histories derived from the two data sets, Holgate found their mean rates-of-rise were indeed similar over the second half of the twentieth century; this observation thus implied, in Holgate's words, that "a few high quality records from around the world can be used to examine large spatial-scale decadal variability as well as many gauges from each region are able to do [41]."

As a result of this finding, Holgate constructed the nine-station-derived wavering line in Fig. 7 as a reasonable best representation of the1904–2003 mean global sea-level history of the world."

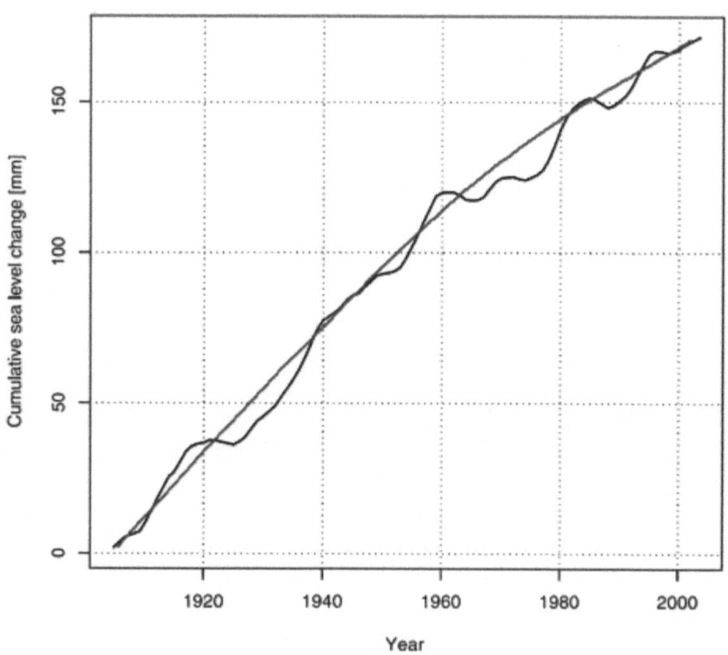

Fig. 7. Cumulative increase in mean global sea level (1904–2003) derived from nine high-quality tide gauge records from around the world. Adapted from Holgate (2007) [41].

"Whichever way one looks at the findings of Holgate—either as two successive linear trends (representative of the mean rates of-rise of the first and last halves of the twentieth century) or as one longer continuous curve (such as we have drawn)-the nine selected tide gauge records indicate that the mean rate of global sea-level rise has *not* accelerated over the recent past, and has probably fallen. Additional information on this topic, including reviews on sea level not discussed here, can be found at http://www.co2 science.org/subject/s/subject_s.php under the heading Sea Level."

Hence if we can rely on the fact that the changing climate has existed for millions of years. We have to adapt to it. Hence,the topic chosen in this paper is Climate Change adaptation based on the science and engineering of the actual climate/ weather change that is happening. The societal response needs to be streamlined toward actionable adaptation possibilities based mostly on nature-based solutions and circular economy concepts of sustainability.

THREE: In [41] the quote is "Green hydrogen and ammonia cannot make the planet green. There are lots of myths, mis-concepts, self-serving assumptions and intentional or unintentional conspiratorial ideas being propagated around the globe by erroneously attributing accelerated climate warming to increased CO_2 emission. The so-called Green Hydrogen and Ammonia would only accelerate climate warming by emitting stronger Green House Gases (GHG) and consuming more energy than they can generate. The main and stronger GHG is water vapor and not CO_2 gas. The water vapor in the ambient air is about 4000 times more than CO_2 hence attributing climate warming to CO_2 and proposing so called Green Hydrogen and Ammonia as substitutes for Fossil fuels is seriously erroneous. Besides, CO_2 is the key nutrient to the green biosphere and biogenesis; hence any attempt to eliminate it will lead to massive misery for mankind" [41].

In another paper [42] Kumargoud Mulemane quotes "It is well documented fact that the Solid Particulate Matters (SPM) in the troposphere have been gradually reducing in the last four decades and humidity levels have been significantly increasing during the same period. SPM of the right composition and size acts as cloud condensation nuclei (CCN) and adequate quantities of SPM are essential to facilitate a proper precipitation and the formation of raindrops of adequate size ensuring the natural Hydrologic cycle.This facilitates dissipation of atmospheric heat to the space. The presence of fine hydrophobic aerosol particulates from biomass and forest fire generate CNN of smaller cloud droplets, necessitating super-saturation of water vapor and increasing the size and residence time of clouds in the troposphere, which enhances the greenhouse effect with positive feedback, impeding the ambient heat dissipation process (malfunctioning of the natural hydrological cycle). The planet is therefore experiencing extreme weather, unsteady precipitation, maldistribution of water recirculation, cloudbursts, floods, droughts, heat waves, and snow downpours. It is my hypothesis that the deficit of the right SPM (hygroscopic natural salts and sulfates) and excess concentration aerosol (hydrophobic fine generated from biomass and forest fire), in the troposphere for the past five decades is the double jeopardy cause of accelerated climate warming and not due to gradual increase of CO_2 concentration. The immediate scientific solution lies in restoring hygroscopic (hydrophilic) SPM level in the troposphere to pre-1975 level through experimentally controlled effective and efficient injection of non-toxic fine particles of natural salts and sulphates without causing air pollution and without disturbing the solar system" [42].

It is to be noted here that GHG water vapor is 4000 times more than CO_2 in the atmosphere. Other than Kumargoud Mulemane, author of [41 and 42] it was also observed by KREC Suratkal Hydrolgist and water resources engineer Prof. Laxman Nandagiri, Department of Water Resources and Ocean Engineering, KREC (NITK) Suratkal, in 2021 during my personal communication with him. One best thing is the availability of satellites (both large as well as micro-satellites) to systematically measure the integrated water vapor in the atmosphere at different layers. Research studies on this water vapor contribution to climate change need to be understood. Also, it is to be noted that Prof. U.R. Rao, former Chairman ISRO, conducting his astrophysics and physics-based studies had opined that there is no influence of variation of solar flares on the temperature increase in the Earth's systems. Professor Rao encouraged NIPCC report to be spread in Kannada in Karnatka Legislative Assembly. A brief report in Kannada was circulated among political members in the Legislative assembly.

In yet another research study aimed at climate change from Dr. HSM Prakash the quotes" "Youtube video reference CREACTIVE Group of Space Scientists [43]

Ministry of Water Resources, Government of India, in one of its publication states as "Studies have been carried out by NIH to analyze the trends of variation in temperature over India/Indian sub-continent and the results have been compared with global trend [43]. The quotings from [43] is "An analysis of temperature data of 125 stations distributed all over India shows an increase of 0.420 °C, 0.920 °C and 0.090 °C in annual mean temperature, mean maximum temperature, and mean minimum temperature, respectively, over the last 100 years." "However, the trends are varying on a regional basis. It has been observed that the changes in temperature in India/Indian-Subcontinent over last century are broadly consistent with global trend of increase in temperature. Scenarios of future climate change are usually developed using the Global Circulation Models with different scenarios of GHG emissions. GCMs are complex 3-dimensional models of the land, atmosphere and oceans. GCMs are invaluable tools for identifying climatic sensitivities and changes in global climate characteristics; the major problem of the current generation of GCMs is the limitation of their spatial resolution. A single grid of GCM may encompass hundreds of square kilometers and include mountainous and desert terrain, oceans and land areas. Usually, the output of GCMs is given for a scale much larger than that of even a large watershed. There are more than 200 GCMs available, which have been developed by different agencies. The variation in outputs from them varies anywhere from 20 to 30 percent. Input data requirement for the these GCMs are generally same but the output results vary and sometimes with slight variation in input parameters (which may be due to different data collection agencies) the results are contradictory giving confusing future climate scenarios. Despite recent improvements in the modeling of the climate dynamics with complex and large-scale models, the use of GCMs is still limited in evaluating regional details of climatic changes. For generating future climate scenarios on regional basis there are downscaling models called RCMs which use output of GCMs. However, RCMs do not give basin-level scenarios" [43].

The Geological Survey of India (GSI) Report [44] reports that we are in a geological cooling age, and the rise in temperatures is the serrated variation locally in the earth system. Also GSI DDG DrHSM Prakash reports that "Based on Geometeorological studies till date it is observed that the activities of some of the major volcanoes around

India are stagnant. Hence not much of vapour is coming into the atmosphere around India." In Fact DrHSM Prakash gives prediction of monsoon using Volcano activity surrounding India region. This says that one has to study water vapour content in the atmosphere to see its effect on climate change. Satellite images are capable of measuring water vapour in the atmosphere [45].

5 NEP 2020 and Climate Change Education Pedagogy

In the Industry 5.0 Scenario a single objective like Climate Change adaptation alone cannot provide sustainable scenarios. The world is moving towards Intelligent Automation and Artificial Intelligence. The digital transfer for becoming a digital country is yet to be achieved in a State-wide sense. The human interface of Industry 5.0 is capable of providing sustainable solutions to sustainable social networks. Real-time continuous data, a continuum of computing for all homogeneous (sustainable forest ecosystems) and heterogeneous networks (energy grids of electric, hydroelectric, thermal, bioenergy, hydrogen, solar energy, etc.) to get quick decisions for human civilization can happen only at the end of Industry 5.0.All the related science and technologies need to be imparted as education. A glimpse of Industry 5.0 framework can be had from observing the following Fig. 8 [46].

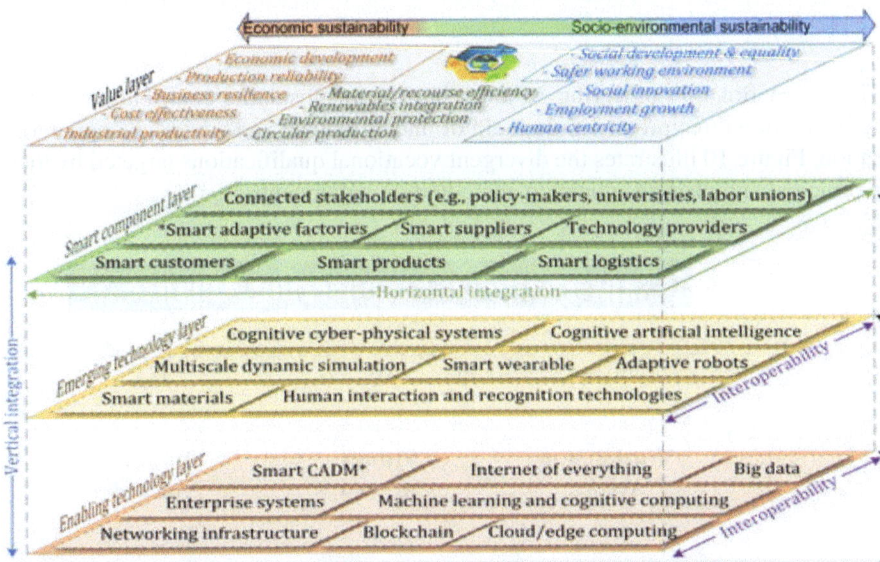

Fig. 8. The Reference Model of Industry 5.0. Source: [46]

The New Pedagogy should encompass not only pyramid of Bloom's taxonomy (Cognitive as in Washington accord adopted in India) but also Affective and Psychomotor Pyramids. See Fig. 9

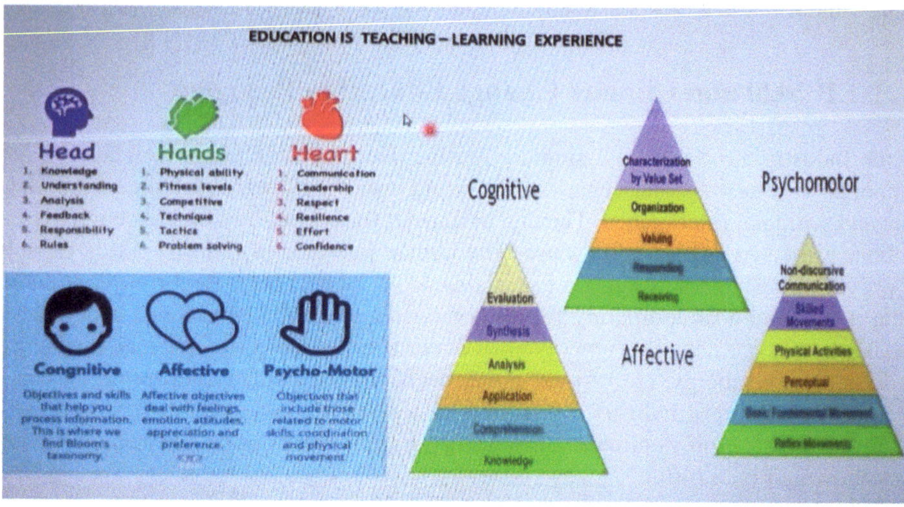

Fig. 9. Cognitive,Affective, Psychomotor Pyramids for Understanding Learning Outcomes. Source [47].

The International Engineering Alliance has made substantial contributions to fostering international reciprocity in the field of undergraduate and advanced engineering education. Figure 10 illustrates the divergent vocational qualifications targeted by these three principal engineering education accreditation agreements.

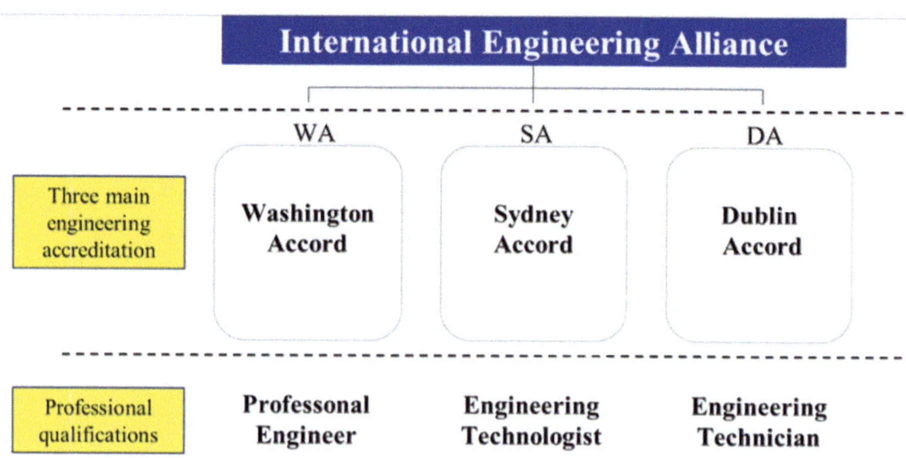

Fig. 10. Introduction to the international engineering education accreditation system. Source: [48]

In [48], it is quoted as "The accreditation requirements have shaped the education on sustainability in the engineering discipline. To be more specific, the international accreditation of engineering education under the Washington Agreement system and the establishment of the marine engineering discipline on this basis hold great practical significance for the sustainable career development of graduates. Adhering to the standards of the engineering education accreditation system is not only a crucial measure of graduates 'capabilities but also a key foundation for achieving cross-border recognition of their academic qualifications. Marine engineering, as an integral part of the maritime field, requires graduates to possess expertise in marine engineering, ship science, and technology, as well as the ability to think independently and respond effectively to emergencies".

The attainment of these accreditations enables graduates to acquire internationally recognized qualifications, thereby enhancing their competitiveness in employment and providing them with global career development opportunities. International accreditation holds immense significance for individual graduates while also establishing a foundation and guarantee for sustainability in engineering education.

The objective of this study is to compare the curriculum structures of these case studies, analyze the similarities and differences in undergraduate engineering education between China and the UK, and explore their respective characteristics and advantages in engineering talent development. Additionally, using the "Marine Engineering English" module as an example, this study provides a comprehensive explanation of how an independent course can meet the accreditation requirements and align with the overall curriculum framework. This study serves as an important practical case in the transnational accreditation of engineering education within the framework of the Washington Accord" [48].

In order to achieve this international agreement on curriculum for sustainability, especially on sustainability and climate change adaptation systems design, monitoring and implementation, it is good to go to basics of curriculum design with learning outcomes generated from industry workplaces also. It is given in detail in Figs. 11, 12, 13 and 14.

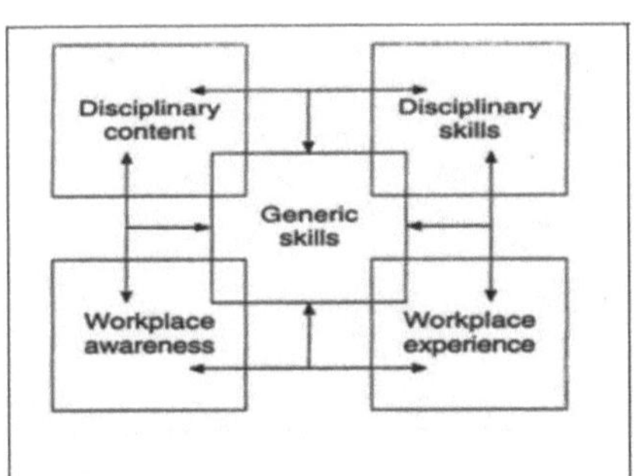

Fig. 11. Model of Course Provision

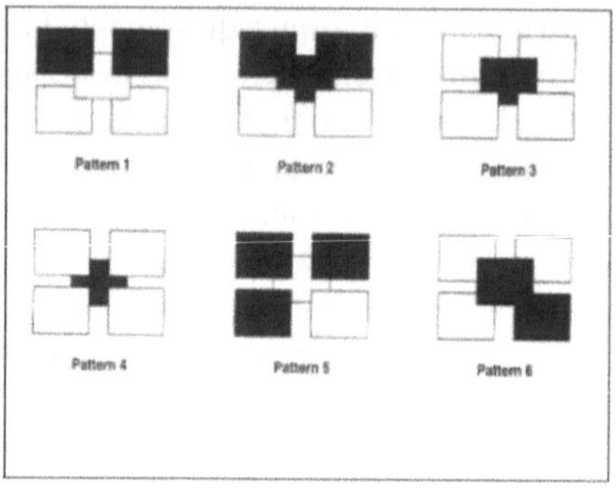

Fig. 12. Patterns of Course Provisions

MANAGEMENT OF SELF	MANAGEMENT OF INFORMATION
• Manage time effectively • Set objectives, priorities and standards • Take responsibility for own learning • Listen actively and with purpose • Use a range of academic skills (analysis, synthesis, argument, and so on) • Develop and adapt learning strategies • Show intellectual flexibility • Use learning in new or different situations • Plan/work towards long-term aims and goals • Purposefully reflect on own learning • Clarify with criticism constructively • Cope with stress	• Use appropriate sources of information (library, retrieval systems, people, and so on) • Use appropriate technology, including IT • Use appropriate media • Handle large amounts of information/data effectively • Use appropriate language and form in a range of activities • Interpret a variety of information forms • Present information/ideas competently (orally, in written form, visually) • Respond to different purposes/contexts/audiences • Use information critically • Use information in innovative and creative ways
MANAGEMENT OF OTHERS	MANAGEMENT OF TASK
• Carry out agreed tasks • Respect the views and values of others • Work productively in a cooperative context • Adapt to the needs of the group • Defend/justify views or actions • Take initiative and lead others • Delegate and stand back • Negotiate • Offer constructive criticism • Take the role of chairperson • Learn in a collaborative context • Assist/support others in learning	• Identify key features • Conceptualize issues • Set and maintain priorities • Identify strategic options • Plan/implement a course of action • Organize subtasks • Use and develop appropriate strategies • Assess outcomes

Fig. 13. Generic Skills Framework –Sustainable Curriculum Element

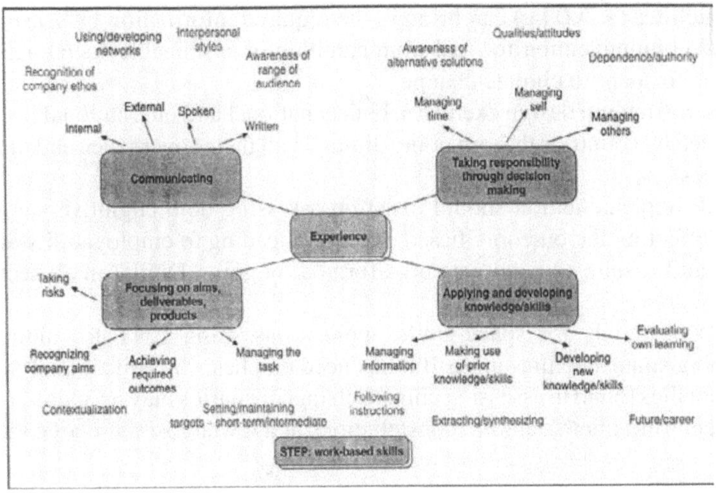

Fig. 14. Work Experience Skills Framework

AICTE in India is of the opinion that: "To lead in fields from AI to quantum computing, more Indians need to opt for M.Tech and PhDs" It is one part of the world opinion.

The conflict here is whether we should proceed from fundamentals or adopt the syllabus of M.Tech and PhD research problems for a selected set of undergraduates. These undergraduates should be industry-specific as well as possess a degree of generative digital skills. These are yet to be addressed.

Under NEP-Curriculum Development for a New Generation of Digital Science/Engineers(ing) in sustainable networked societies, key competencies should be clearly defined in relevant policy documents and/or school curricula, including a wide range of clear, concrete, and detailed learning outcomes linked to teaching practice, taking into account both cognitive and non-cognitive components. There is a need to transform teachers' attitudes and practices in the classroom in order to 'unpack' the competencies to be taught and assessed. Learning Management Systems should be remodeled to include emotional intelligence and problem-solving through design philosophy (affective pyramid instead of only cognitive pyramid elements).UNESCO and other world bodies, including India, are geared towards emotional intelligence assessment for educational purposes. Service Learning and Learning through service LTS models should be used in arriving at a pedagogy of teaching and learning.

6 Conclusions

Broadly, the following can be concluded from the research carried out so far using geospatial and attribute data.

1. Geospatial Knowledge Infrastructure readiness index levels determine the levels of climate change adaptation possible in any country.

2. Microsatellites (<200 kg) can bring in the required information by sensing, navigation, and communication for real-time handling of sustainable social networks that are trying to adapt to climate change.
3. The new curriculum design exercises, both at national and international levels, should be initiated by countries that are going digital in all their governance and management businesses.
4. The NEP implementation should insist on assessing both cognitive and emotional metrics affecting the outcome-based education leading to employability and beyond This should sustain societal networks formed by going Digital in all sectors of the economy.
5. The new materials for space-grade applications using CNT for aiding startups developing small satellites up to 100 kgs need emphasis by educational institutes.
6. The anomalies found in assessing climate change scenarios may be addressed properly using data from micro-satellite constellations at a lower cost than large satellites.

Acknowledgements. I hereby acknowledge my 1979ers classmates group for having encouraged me by reading and suggesting improvements where ever possible to send it for possible publication in the NITK-CREST Proceedings to be brought out by Springer Nature Journal.

References

1. Deloitte: Turning point: Feeding the world sustainably- The costs and opportunities of long-term food system transformation, November 2024, https://www.deloitte.com/content/dam/assets-shared/docs/about/2024/deloitte-turning-point-feeding-the-world.pdf, (2024)
2. Few, R., Spear, D., Singh, C., Tebboth, M.G.L., Davies, J.E., Thompson-Hall, M.C.: Culture as a mediator of climate change adaptation: Neither static nor unidirectional. WIREs Clim. Change. **12**, e687 (2021). https://doi.org/10.1002/wcc.687
3. McKinsey, Water Resilience: Closing the Funding Gaps for Utilities, p 10, March 2025, www.McKinsey.com (2025)
4. Zant, M., Schlingmann, A., Reyes-García, V., García-del-Amo, D.: Incremental and transformational adaptation to climate change among Indigenous Peoples and local communities: a global review. Mitig. Adapt. Strateg. Glob. Change. **28**, 57 (2023). https://doi.org/10.1007/s11027-023-10095-0
5. UNDP, 7 things we have learned about adapting to climate change-And how they inform our Climate Promise-published by UNDP Climate; In Climate Change Adaptation, January 17th, (2021)
6. Li, H., Wuf, S., Sun, J., Zhu, S., Zhoub, Q., Sojka, M., Viktor, R.: 150-year daily data (1870–2021) in lakes and rivers reveals intensifying surface water warming and heatwaves in the Pannonian Ecoregion (Hungary). Toth, J. Hydrol. Region. Stud. **56**, 101985 (2024). https://doi.org/10.1016/j.ejrh.2024.101985. Available online from April (2024)
7. Rajegowda, M.B., Babu, B.T.R., Janardhanagowda, N.A., Muralidhar, K.S.: Impact of climate change on agriculture in Karnataka. J. Agrometeorol. **11**(2), 125–131 (2009). https://doi.org/10.54386/jam.v11i2.1237
8. Srilaxmi, B., Rath, B.S., Pattnaik, S., Paikaray, R., Khadke, L.: Projection of futuristic climatic scenario of Cuttack District of Odisha Using CMIP5. Int. J. Environ. Clim. Change. **12**(7), 10–17 (2022). https://doi.org/10.9734/ijecc/2022/v12i730698

9. Thenkabail, P. S., Teluguntla, P. G. et al.: Global cropland-extent product at 30-m resolution (GCEP-30) derived from Landsat satellite time series data for the year 2015 using multiple machine-learning algorithms on Google Earth Engine Cloud, U.S. Geological Survey Professional Paper, 63 p, https://doi.org/10.3133/pp1868. (2023)
10. ISRO-NNRMS, ISRO-DOS-GOI, India: Earth Observation Applications in Climate Change Studies. Bulletin of the National Natural Resources Management System, NNRMS. **2020**(8), 35 (2020)
11. Jagadeesha, C.J., Manavalan, P., Ravi, N., et al.: Climate change impact scenarios on irrigated agriculture systems: management using space inputs. In: Proceedings of the International Conference on Emerging Scenarios in Space Technology and Applications (ESSTA 2008), pp. 583–595. Sathyabama University, Chennai, India (2008)
12. Stevens, A., Nocita, M., Toth, G., Montanarella, L., van Wesemael, B.: Prediction of soil organic carbon at the European scale by visible and near infrared reflectance spectroscopy. PLOS ONE. **8**(6), e66409 (2013)
13. Indian Space Research Organisation, Space Applications Centre, Ahmedabad, Department of Space, Government of India, Desertification and Land Degradation Atlas of India (Based on IRS AWiFS data of 2011–13 and 2003–05), and Ministry of Environment, Forest and Climate Change, Government of India, India, June (2016)
14. Jagadeesha, C. J.: The efficacy assessment of land degradation because of climate change using spatial data, Environmental Management Policy Research Institute (EMPRI), Government of Karnataka, India National Symposium on Climate Change NSCC (2019)
15. Ministry of Civil Aviation, New Delhi, India, The Drone Rules, 2021, Notification, CG-DL-E-26082021-229221, PART II—Section 3—Sub-section (i),G.S.R. 589 (E),Gazette of India, 25th August, (2021)
16. Ministry of Civil Aviation, New Delhi, India, The Drone (Amendment Rules), 2022, Notification, CG-DL-E-26082021-229221, PART II—Section 3—Sub-section (i), G.S.R. 108 (E),Gazette of India, (2022)
17. GWF, UNSD: The Power of Where- Jointly organised by Geosptaial World Forum and United Nations Statistics Division 2021, https://www.geospatialworld.net/consulting/gki-campaign.html; https://interoperable-europe.ec.europa.eu/sites/default/files/news/2021-06/GKI-White-Paper.pdf (2021)
18. Agarwal, M., Sun, M., Kamath, C., Muslim, A., Sarker, P., Paul, J. et al.: General geospatial inference with a population dynamics foundation model, © 2025Google Research, arXiv:2411.07207v4 [cs.LG] 30 Jan (2025)
19. Xiong, Y., Zhang, Q., Chen, X., Bao, A., Zhang, J., Wang, Y.: Large scale agricultural plastic mulch detecting and monitoring with multi-source remote sensing data: a case study in Xinjiang, China. Remote Sens. **11**, 2088 (2019). https://doi.org/10.3390/rs11182088. www.mdpi.com/journal/remotesensing
20. de Loe, R., Kreutzwiser, R., Moraru, L.: Adaptation options for the near term: climate change and the Canadian water sector. Glob. Environ. Change. **11**, 231–245 (2001)
21. Mohamed Elgendy, S.M.; Sonia Hassini, A.M.; and Paulin Coulibaly, Review of Climate Change Adaptation Strategies in Water Management, ASCE, J. Hydrol. Eng., 2024, 29(1): 03123001,(2024)
22. IPCC Data Distribution Center-GCMs https://youtu.be/8EOnP8QqkCg (2010)
23. Lal, et al.: Global Trends in Small Satellites. Institute of Defense Analysis, USA (2017)
24. NASA Report: NASA/TP-2024-0001462, State-of-the-Art, Small Spacecraft Technology, http://www.nasa.gov/smallsat-institute/sst-soa, February 2024 pp 1–3 (2024)
25. Goenka, P.: The Sun is Rising over India's Space Sector. Business Today, 106–108 (2024)
26. Ministry of Science and Technology, New Delhi, India, National Geospatial Policy of India, 2022, Notification, CG-DL-E-28122022-241463, Extraordinary, PART II—Section 3-Sub-section (ii), S.O. 6095(E), Gazette of India, 28 December, (2022)

27. Geospatial World - Geospatial Knowledge Infrastructure Readiness Index and Value Proposition in World Economy, Society, and Environment, jointly produced by Geospatial World-Advanced Knowledge for Sustainability and United Nations Statistics Division, © Geospatial Media and Communications, (2022)
28. Suresh, S., Lal, S.: A metaheuristic framework-based automated spatial-spectral graph for land cover classification from multispectral and hyperspectral satellite images. Infra. Phys. Technol. **105**, 103172 (2020). https://doi.org/10.1016/j.infrared.2019.103172
29. Goswami, P., Rakesh, V., Patra, G.K., Prakash, V.S.: Real-time quantitative rainfall forecasts at Hobli level over Karnataka: evaluation for the winter monsoon 2010. Curr. Sci. **102**(10), 25 (2012)
30. ISRO-SAC: https://www.mosdac.gov.in/inland-water-height, Meteorological and Oceanographic Satellite Data Archival Centre, Space Applications Centre, ISRO, India.
31. Parashar, M.: Everywhere & nowhere: Envisioning a computing continuum for science, The International Conference for High Performance Computing, Networking Storage and Analysis, Denver, CO, USA, (2023). https://sc23.supercomputing. org/ (2023)
32. IDA-Nature based solutions: Evidence For Hazard Risk Reduction And Ecosystem Services-Scaling Up Implementation and Accelerating Learning, IDA-Institute for Defense Analysis, Science and Technology Policy Institute, USA. This work is made available under a Creative Commons Attribution (CCBY) 4.0 License: http://creativecommons.org/licenses/by/4.0.(2024)
33. Belyayev, Pronenko, V., Ivchenko, N., Zakharchuk, Y., Korepanov, V.: Miniature sensors for CubeSats' scientific missions at the 8th IAA Symposium on Small Satellites for Earth Observation. Small Satellite Technologies, Sensors and Instrumentation, Berlin, Germany (2011)
34. Watson, K.A., Connell, J.W.: Polymer and carbon nanotube composites for space applications. In: Dai, L. (ed.) Carbon Nanotechnology, pp. 677–698. Elsevier BV, Amsterdam (2006). https://doi.org/10.1016/B978-044451855-2/50022-X
35. Bock, D., Tajmar, M.: Highly miniaturized FEEP propulsion system (Nano FEEP) for attitude and orbit control of CubeSats. J. IAA Acta Astron. **144**, 422–428 (2018). https://doi.org/10.1016/j.actaastro.2018.01.012
36. He, J., Liu, P., Gao, R., Xue, C., et al.: Research on the neutralization control of the RF ion micro propulsion system for the 'Taiji- 1' satellite mission. Plasma Sci. Technol. **22**, 094002 (2020). https://doi.org/10.1088/2058-6272/ab7dd5
37. Wang, Y.X., Wu, Z.W., Huang, T.K., et al.: Operating characteristics of carbon nanotube neutralizer for space exploration missions. J. Deep Space Explor. China. **11**(2), 151–158 (2024). https://doi.org/10.15982/j.issn.2096-9287.2024.20240004
38. Allen, M.R., Dube, O.P., Solecki, W., Aragón-Durand, F., Cramer, W., Humphreys, S., Kainuma, M., Kala, J., Mahowald, N., Mulugetta, Y., Perez, R., Wairiu, M., Zickfeld, K.: Framing and Context. In: Masson-Delmotte, V., Zhai, P., Pörtner, H.-O., Roberts, D., Skea, J., Shukla, P.R., Pirani, A., Moufouma-Okia, W., Péan, C., Pidcock, R., Connors, S., Matthews, J.B.R., Chen, Y., Zhou, X., Gomis, M.I., Lonnoy, E., Maycock, T., Tignor, M., Waterfield, T. (eds.) Global Warming of 1.5°C. An IPCC Special Report on the impacts of global warming of 1.5°C above pre-industrial levels and related global greenhouse gas emission pathways, pp. 49–92. Cambridge University Press, Cambridge, UK and New York, NY, USA (2018). https://doi.org/10.1017/9781009157940.003. https://wmo.int/media/news/global-temperature-likely-exceed-15degc-above-pre-industrial-level-temporarily-next-5-years
39. Sorokhtin, O.G., Chilingar, G.V., Khilyuk, L.F.: Global Warming & Global Cooling: Evolution of Climate on Earth. Elsevier Publishers (2007)
40. Idso, C., Singer, S. F., Bast, J. L., Bast, D. C.: Climate Change Reconsidered, 2009 Report of the Nongovernmental International Panel on Climate Change (NIPCC) (2009)

41. Mulemane, K. G.: Can Green Hydrogen & Ammonia make the planet Green? No, but Solar Energy and Biomass can!, SNAMES (Society of Naval Architects and Marine Engineers Singapore), 42nd Annual Journal 2022/2023, Singapore (2023)
42. Mulemane, K.G.: Scientific solution to decelerate climate warming. Int. J. Eng. Res. Technol. (IJERT). **10**(12) (2021) http://www.ijert.org, ISSN: 2278-0181
43. Government of India, Ministry of Water Resources: National Water Mission under the National Action Plan on Climate Change (Draft), Report of Sub-Committee on Policy and Institutional Framework, New Delhi, (2008)
44. Mathur, U.B.: Climate Change: Past, Present and Future. Published by Geological Society of India GSI, Bangalore (2010)
45. Prakash, H. S. M.: Geological truth about climate change, March 16, 2025, Karnataka Physics Teachers Association, Youtube link: https://www.youtube.com/live/uyHedlQEyso?si=okiVILMLYzeF1Qro
46. Ghobakhloo, M., Iranmanesh, M., Mubarak, M.F., Mubarik, M., Rejeb, A., Nilashi, M.: Identifying Industry 5.0 contributions to sustainable development: A strategy roadmap for delivering sustainability values roadmap for delivering sustainability values. Sustain. Prod. Consum. **33**, 716–737 (2022). https://doi.org/10.1016/j.spc.2022.08.003
47. Shivaprasad, S.M.: NEP-2020 India towards becoming the Viswa Guru invited talk by KSCST-GOK. Karnataka State Higher Education Academy and JNCASR Bangalore (2021)
48. Zhang, J., Yuan, H., Da Zhang, Y.L., Mei, N.: International engineering education accreditation for sustainable career development: a comparative study of ship engineering curricula between China and UK. Sustainability. **15**, 11954 (2023). https://doi.org/10.3390/su151511954

Experimental and Numerical Investigation on PCM-Integrated Concrete Blocks for Reducing Global Carbon Footprint

Nitin Gotiya and Somashekhar S. Hiremath(✉)

Department of Mechanical Engineering, Indian Institute of Technology Madras, Chennai, Tamil Nadu 600036, India
somashekhar@iitm.ac.in

Abstract. Global energy consumption increases drastically and poses significant environmental challenges. The building sector alone is responsible for almost 30% of global CO_2 emissions. One of the major contributors to this emission is the air conditioning system, which provides thermal comfort to occupants in the building. Currently, there are more than 2 billion air conditioning units in the world and this could be tripled by 2050. According to the International Energy Agency (IEA), space cooling is responsible for around 1 billion tons of CO_2 emissions through electricity use.

The incorporation of Phase Change Materials (PCMs) into building blocks is an innovative solution to overcome this problem. PCMs are highly effective in controlling temperature and maintaining interior thermal conditions by absorbing and releasing heat during phase transitions. However, integrating PCMs into building materials contain significant challenges that avoid leaks or dips. One alternative approach to overcome the challenge is to create hollow concrete blocks and fill them with PCMs to provide structural integrity and improve thermal performance.

This study evaluates the performance of PCM-integrated concrete through experiments and numerical analysis. The result indicates the potential of advanced materials to build a sustainable, energy-efficient building and mitigate CO_2 emissions that align with global warming and climate change.

Keywords: Carbon emission · Concrete · Energy saving · Phase change material · Thermal comfort

1 Introduction

The growing global demand for energy, due to rapid urbanization and improving living standards, has increased the worldwide consumption of energy, which has a significant environmental impact, such as greenhouse gas emissions, Air pollution, and Refrigerant emissions. 30% of worldwide CO_2 comes from the construction sector as shown in (see Fig. 1), which is a significant contributor. This is mostly because Heating, ventilation, and air conditioning (HVAC) system. Use a lot of energy. Space cooling is a rapidly expanding energy demand, and this alone is predicted to produce 1 billion tons of CO_2 emissions yearly, as shown (see Fig. 1) with over 2 billion air conditioners in use now and predictions that this number will nearly triple by 2050 [1]

H. P. Dasari et al. (Eds.): NITK-CREST 2025, PTT, pp. 200–212, 2026.
https://doi.org/10.1007/978-981-95-3218-6_14

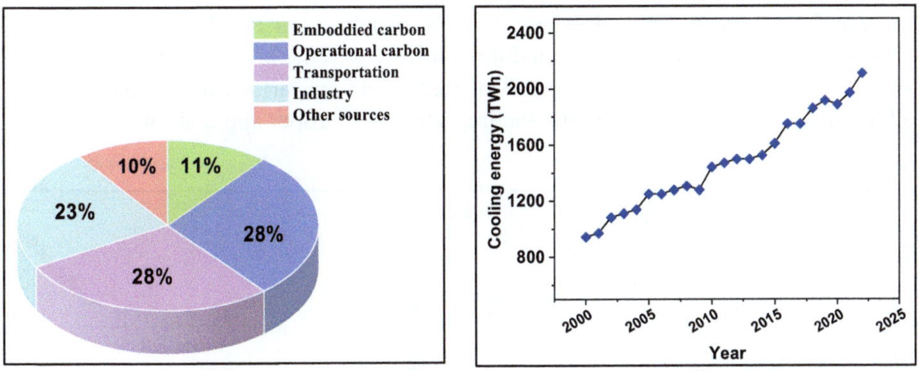

(a) (b)

Fig. 1. Global carbon emissions and energy demand (a) Annual global carbon emissions on different sector (b) global electricity demand for air conditioning (adapted from https://ourworldindata.org/air-conditioning-causes-around-greenhouse-gas-emissions-will-change-future)

The building envelope plays a vital role in managing a structure's energy performance by moderating heating and cooling demands between indoor and outdoor spaces to meet thermal comfort needs. Acting as a protective barrier, the envelope significantly influences the thermal regulation of interior environments. It is an essential aspect of energy-efficient design, accounting for nearly half of a building's heating and cooling requirements. On average, individuals spend over 90% of their time indoors—whether at home, in workplaces, or in commercial spaces making thermal comfort crucial for both well-being and productivity. Thermal comfort refers to a person's psychological state of satisfaction with the indoor thermal conditions. This comfort level is subjective and cannot be fully captured by a specific temperature range. There are several factors affecting thermal comfort extend beyond just air temperature; they include humidity levels, mean radiant temperature, air movement, clothing insulation, and physical activity. These are the essential parts of living; we cannot compromise on these characteristics, so the only thing we can modify is the permanent solution by changing the design and method of habitat. Innovative approaches are necessary for increasing building energy efficiency without compromising occupant comfort in order to lessen these effects.

Integrating phase change materials (PCMs) into wall structures is an emerging technique that has the potential to increase thermal mass and boost the overall thermal efficiency of building envelopes. PCMs have become one of the most promising options for passive thermal energy storage among the several methods. PCMs absorb and release heat during phase transitions (see Fig. 2), thereby reducing the load on active HVAC systems [2, 3]. Building temperature variations may be considerably reduced using PCMs, which might result in a 10–30% decrease in heating and cooling consumption of energy [4]. The integration of PCMs into building materials presents several challenges, including inadequate dispersion, deterioration of mechanical properties, and leakage during phase transitions [5]. Research has shown that incorporating phase change materials (PCMs) into bricks can lower surface temperatures by as much as 6 °C and cut down heat transfer by up to 30%. However, to fully benefit from PCMs without introducing drawbacks, they must be integrated thoughtfully into the brick structure. Numerous studies have explored the use of PCM-infused bricks in various hot climate regions, highlighting both their potential and the importance of proper application.

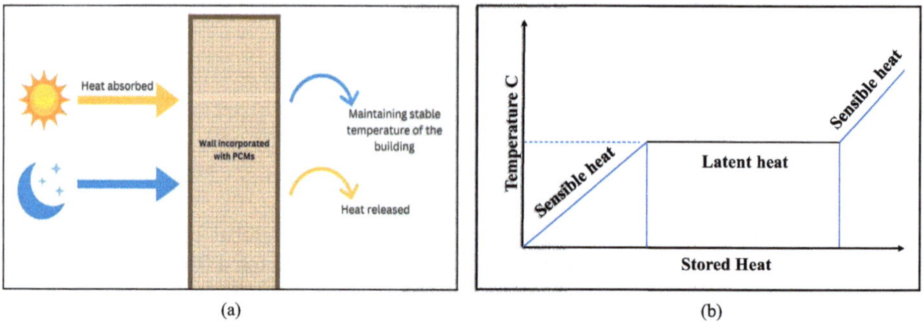

Fig. 2. Heat transition region (a) Schematic diagram of PCM integrated concrete wall (b) Shows the change of thermal energy storage with temperature (adapted from Yang et al., 2016)

Generally, three encapsulation methods have been found to address the limitation Microencapsulation involves enclosing PCM particles in a polymer shell, by enables uniform distribution inside concrete, but it may also raise prices and have an impact on workability [6]. Microencapsulation—such as the use of PCM-filled tubes or pouches—offers simpler integration and a lower chance of leakage; it may also result in thermal bridging and more complicated design [7]. The embedding of encapsulated PCMs into hollow concrete blocks has been investigated, this technique offers the benefits of increased thermal performance and structural strength PCM-incorporated concrete bricks, particularly those that use macroencapsulated PCMs, have reduced heat transfer and lowered the interior temperature by 5 °C and has better mechanical stability and enhance the thermal performance [8]. Positioning PCMs closer to the interior surface was found to maximize thermal performance [16, 17]. Table 1 shows the main characteristics of PCM used in different studies. PCM's thermophysical characteristics are often examined to make sure they are appropriate for the construction application being researched. The most widely used technique for testing the PCM is the differential scanning calorimeter (DSC). A common method for analyzing the thermo-physical characteristics of PCMs is DSC. The PCM's melting and solidification temperatures, enthalpies, heat storage capacity, and specific heat may all be determined using this approach [17].

This study explores how phase change materials (PCMs), when sealed in capsules, perform once they're built into concrete bricks. The focus is on how well they manage heat, and by testing these materials in real conditions, the research aims to shed light on whether PCM-enhanced bricks are practical for use in buildings and how to best include them without causing thermal issues.

Table 1. Thermophysical characteristics of phase change material

PCM type	Melting temperature [°C]	Heat of fusion [kJ/kg]	Thermal conductivity [W/(m-K)] (Liquid/ Solid)	Density [kg/m^3] (Liquid/ Solid)	Specific heat [kJ/(kg-K)] (Liquid/ Solid)	Reference
Paraffin	27–29	245	0.2	770/880	2	[9]
BioPCM	28.85	219	0.2	860	1.97	[10]
Paraffin wax	44	174.2	0.13	783/830	2.53/2.44	[11]
Eicosane	36-38	202	0.15/0.39	780/815	2.46/1.92	[12]
PureTemp 23	22.23–24.17	170.71	0.15/0.25	830/910	2.06/1.56	[13]
OM 35	35	160	0.16/0.2	870/900	2.71/2.31	[12]
OM 32	31.85	200	0.145/0.2198	870/928	2.3/1.95	[10]
HS29	26-29	190	0.55/1.05	1530/1681	2.62	[14]

2 Materials and Methods

2.1 Materials

In this study, the primary ingredient is Ordinary Portland cement (OPC) 53 Grade, which complies with IS12269:2013 standard and fine sand was used to make a concrete mixture. Paraffin wax petroleum jelly (white) was used as a phase change material due to easy availability, high latent heat storage capacity, thermal stability, and compatibility with the ambient temperature range corresponding to summer weather. PCMs are materials that change phases by either absorbing or releasing heat at a constant temperature The most popular phase transformation processes are solid-liquid-solid (Melting and solidification) and liquid-vapor-liquid (evaporation and condensation cycle). There are several types of material used for thermal energy storage (see Fig. 3). The more common PCMs are classified as organic and inorganic materials, such as paraffin, fatty acids, and eutectic mixtures [4] PCMs have two primary benefits when it comes to energy storage and release: (i) a lot of heat storage with little temperature variation (ii) variation in temperature flattening due to isothermal phase change behaviour

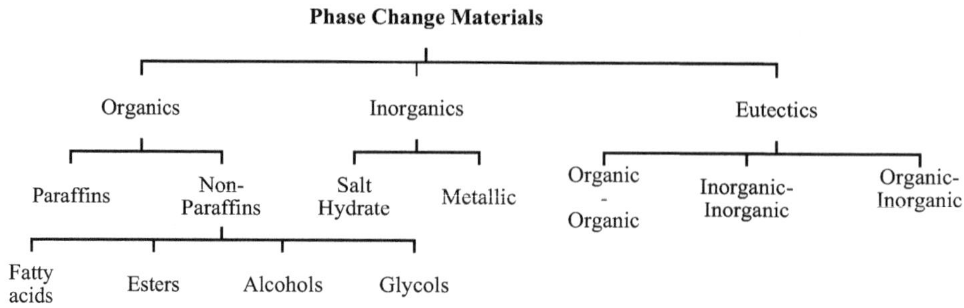

Fig. 3. Classification of phase change materials (adapted from Kurdi et al., 2021)

2.2 Concrete Brick Preparation

A common brick type used for wall construction around the world is concrete brick. The PCM-integrated concrete bricks were prepared using a methodical procedure that included conventional concrete casting techniques, thermal processing, and additive manufacturing. In this study, the concrete brick made of cement, sand, and water mixed with a ratio of 1:1.5:3 according to the Indian standard with standard block dimension of 190 × 90 × 90 mm (Length × width × thickness) Three types of brick were prepared concrete brick with no PCM added (Block C), other two PCM –15 Longitude(Block A), PCM-15 Latitude(Block B), containing cavities filled with Petroleum jelly equivalent to 15% of the total block.

Using 3D printing technology, custom molds were created to guarantee repeatability and precise geometry. Two mold sets were created. PCM mold 5 × 70 × 70 mm (Length × width × thickness) and the concrete mold. The molds were created using a fused deposition modeling (FDM) 3D printer using polylactic acid (PLA) filament. After this PCM was gently heated in a controlled environment until fully liquefied. After pouring the molten PCM into the 3D-printed PCM molds, it was allowed to cool to ambient temperature. After the PCM blocks solidified, they were carefully demolded and placed aside for concrete embedding. Sand, water, and regular Portland cement were used to make the concrete mixture (see Fig. 4). To prevent leakage of PCM during its phase transition, the solidified PCM blocks were embedded between the concrete mixture inside the larger mold. Care was taken to encapsulate the PCM block within the concrete matrix completely. The blocks spent seven days being cured in water. In order to guarantee moisture uniformity and stability for precise thermal assessments, the specimens were oven-dried at 40 °C for an hour before testing after the water curing procedure.

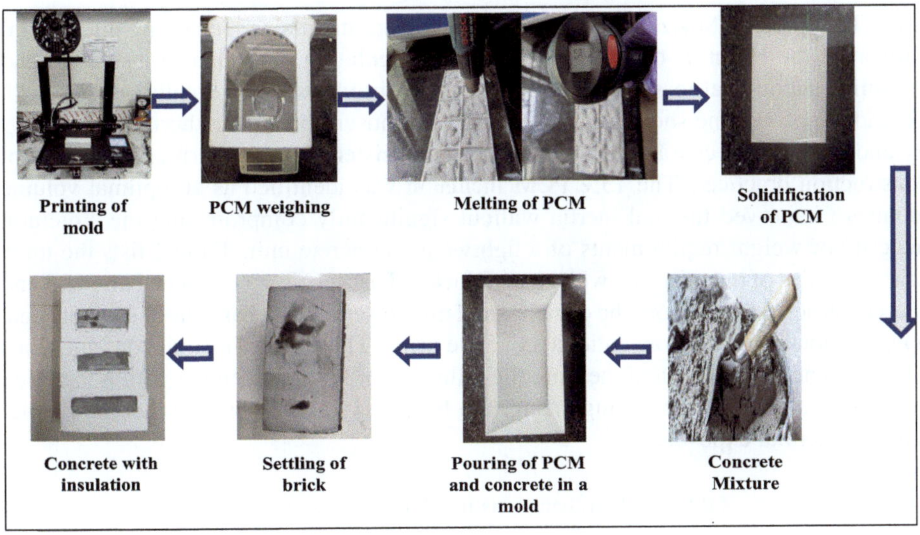

Fig. 4. Fabrication of phase change material integrated concrete block

2.3 Experimental Group

This study involved the development and evaluation of three types of concrete bricks, each designed to perform under high ambient temperatures typical of Chennai's summer climate. The aim was to assess how the presence and orientation of phase change material (PCM) influence the thermal response of the brick structure. All bricks shared the same outer dimensions of 190 mm × 90 mm × 90 mm and were produced using OPC 53 grade cement, along with river sand and water in standard proportions. The phase change material used in this investigation was white petroleum jelly, selected for its suitable melting point (between 38 °C and 56 °C) and latent heat storage properties. The three bricks prepared were as follows (see Fig. 5).

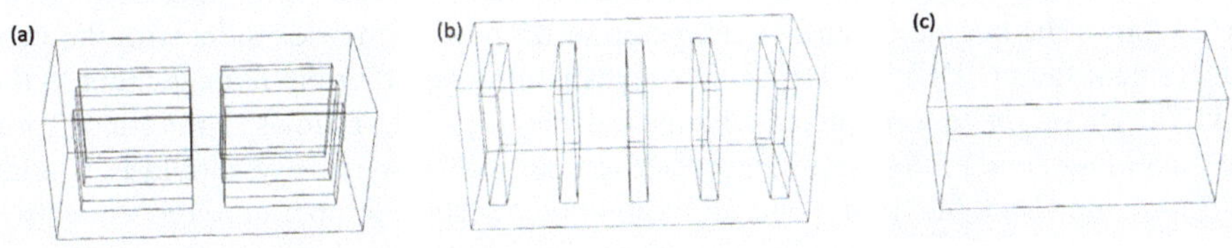

Fig. 5. Schematic view of the tested concrete block

Block A (PCM-15 Longitude) In this configuration, a cavity was embedded along the length of the brick (aligned with the 190 mm axis), into which a solidified block of petroleum jelly was inserted. The PCM insert measured 70 mm × 70 mm × 5 mm, accounting for roughly 15% of the total brick volume. This design allowed for the evaluation of longitudinal heat transfer performance. Block B (PCM-15 Latitude) Structurally similar to Block A, this brick featured a PCM cavity along the width (90 mm axis). The same PCM size and volume fraction were maintained, enabling direct comparison of the effect of orientation on thermal regulation. To enhance insulation and minimize lateral heat loss during testing, all three bricks were externally covered with expanded polystyrene sheets. This material was chosen for its low thermal conductivity and lightweight characteristics, which are consistent with modern energy-efficient construction practices. The 15% PCM inclusion was identified as an optimal volume to ensure improved thermal inertia without significantly compromising the structural integrity or weight requirements of a lightweight concrete unit, Table 2 lists the main characteristics of paraffin wax white soft. Block C This specimen contained no PCM and was constructed solely from the cement-sand mix. It served as a baseline to evaluate heat transmission through a conventional concrete brick. These experimental configurations were designed to isolate and measure the influence of PCM placement within the brick body, with a focus on observing changes in heat delay and surface temperature under uniform thermal exposure

Table 2. Main characteristics of paraffin wax white soft

Properties	Value/ property
Form	Whitish jelly
Transition temperature (°C)	38–56
Thermal conductivity (W/mK)	0.2
Density (kg/m^3)	900
Latent heat of fusion (kJ/kg)	190
Specific heat (kJ/kgK)	2.1

2.4 Thermal Performance Testing

The advantage of PCM were calculated as the difference between the inner and outer surface temperature of the brick at each cycle The temperature differential between T_i and T_o shows the degree of surface temperature drop on each brick side, with the help of decrement factor (DF) low DF value suggests improved thermal management in the brick. The DF might be computed mathematically by Eq. (1) as follows: $T_i$max and $T_i$min Brick maximum and minimum inner surface temperature respectively. Similarly, $T_o$max and $T_o$min are Brick maximum and minimum outer surface temperature respectively.

$$\mathrm{DF} = \frac{Ti\max - Timin}{To\max - To\min} \tag{1}$$

2.5 Numerical Model

Numerical simulations for the present study were carried out using EnergyPlus version 25.1.0, a widely recognized software for evaluating building energy performance. A three-dimensional model with dimensions of 8.4 m × 5.1 m × 3 m was developed using SketchUp 2025 to replicate a practical setup of the Passive Energy Innovation Laboratory

(PEIL) at MSB, IIT Madras (see Fig. 7). The building envelope was designed to include bricks constructed with and without phase change material (PCM) mortar to assess the influence of latent heat storage on thermal performance and overall energy efficiency.

The climate file for Chennai, India, was used to obtain weather data for the simulation from Climate.OneBuilding.Org. This data offers location-specific hourly meteorological information, such as wind speed, solar radiation, relative humidity, and outside air temperature, that is essential for precise building energy modeling. Realistic boundary conditions for evaluating the building's thermal behavior and HVAC energy demands throughout the year are ensured by using this site-specific meteorological data.

The phase change behavior of PCM was modeled using the Conduction Finite Difference algorithm embedded in EnergyPlus. This one-dimensional numerical method employs an implicit finite-difference scheme to solve the transient heat conduction equation, accurately simulating enthalpy-temperature relationships during phase transitions. The simulation model represents a single thermal zone with an internal volume of approximately 126 m^3. It features six identically sized, north-facing windows, ensuring uniform daylight exposure and external thermal gains. Two wall configurations were considered: one integrated with PCM and the other without, to allow comparative performance analysis. For HVAC modeling, the Ideal Loads Air System in EnergyPlus was utilized. This system provides a simplified, yet effective representation of air conditioning by supplying or removing heat and moisture at 100% efficiency, blending outdoor air with recirculated exhaust air to maintain thermal comfort.

Occupancy was modeled with six individuals operating within the space from 8:00 AM to 11:00 PM daily. The thermal comfort settings were defined with a cooling setpoint of 28 °C and a heating setpoint of 16 °C, thus maintaining indoor air temperatures within this range. The HVAC system was constrained by a maximum supply air heating temperature of 30 °C and a minimum cooling temperature of 15 °C. Additionally, the model incorporated a ventilation rate of 1 air change per hour (ACH) to simulate fresh air intake following indoor air quality standards.

3 Results and Discussion

3.1 Surface Temperature Variation

The experimental analysis aimed to investigate the thermal behavior of three concrete brick configurations subjected to one-sided heating using a heat gun set to 70 °C (see Fig. 6). The bricks tested included Block A, Block B and Block C, All samples were insulated on four faces using expanded polystyrene sheets to prevent lateral heat loss and focus heat transfer in a single direction. Temperature readings from the opposite face were collected at 10-min intervals for a total duration of 60 min using a non-contact infrared thermometer. From the recorded data, all bricks demonstrated a progressive increase in surface temperature over time; however, the magnitude and rate of temperature change varied among the three configurations. Block C, the control, exhibited the highest rate of thermal conduction and the greatest final surface temperature, reaching approximately 34.9 °C at the end of 60 min. This was expected, as the absence of PCM meant that heat energy was transferred directly through the concrete matrix without thermal storage or buffering (see Fig. 8).

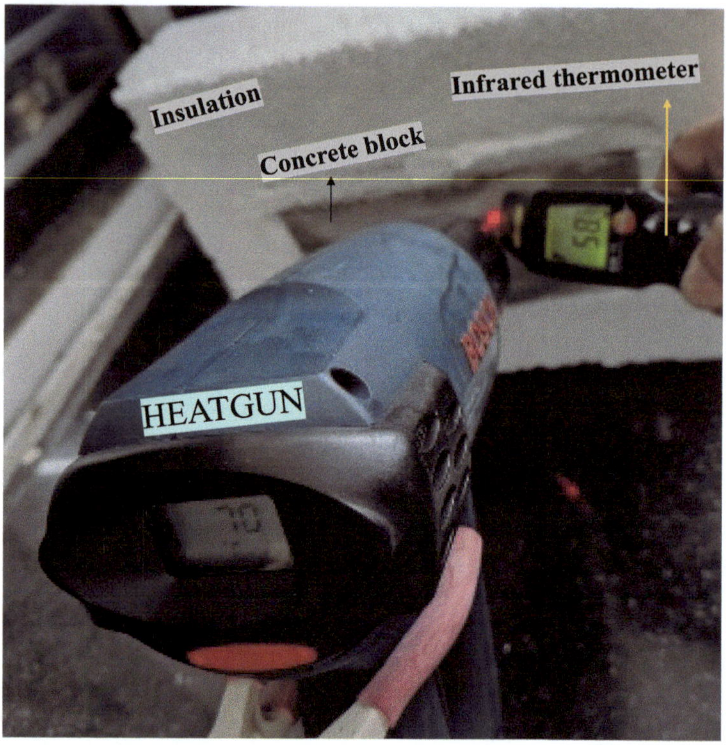

Fig. 6. Temperature rise at the insulated face of concrete bricks over a 60-min heating period using a 70 °C heat gun

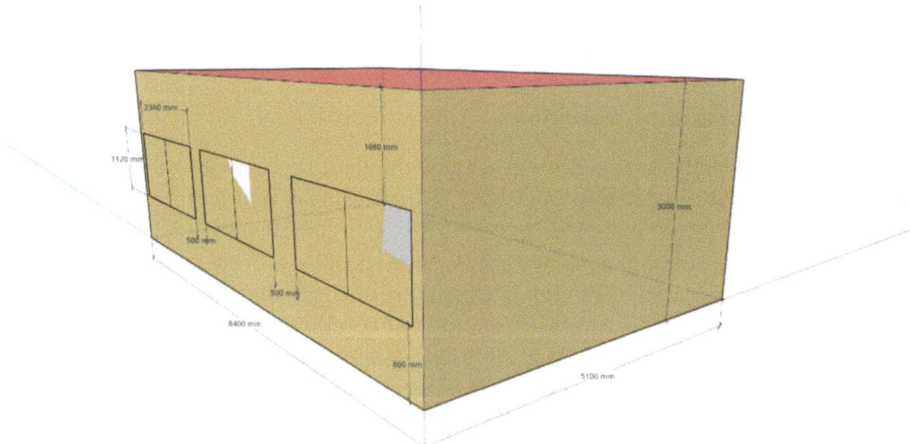

Fig. 7. Design of simulation model

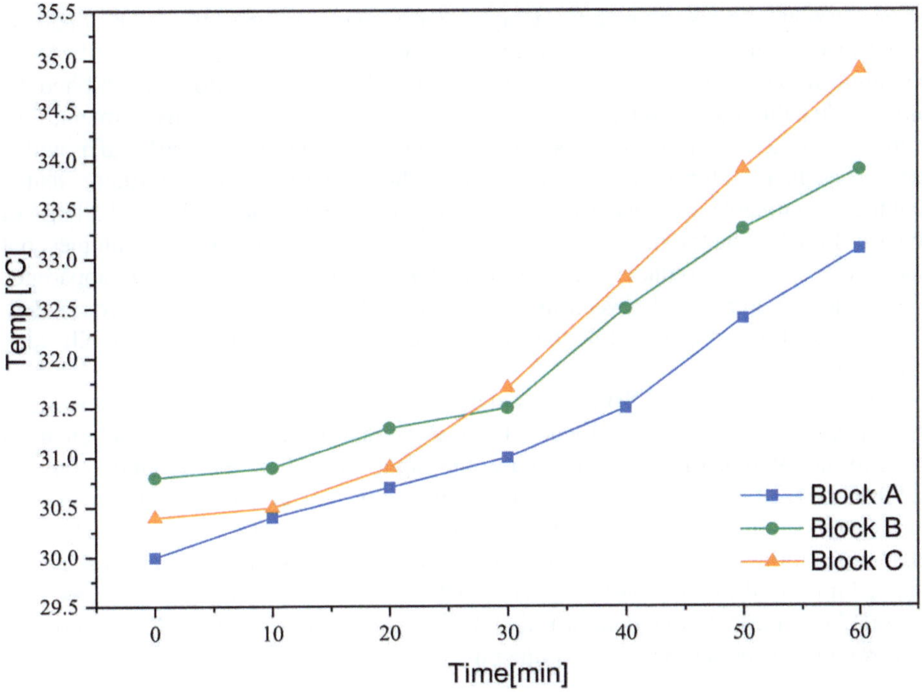

Fig. 8. Temperature variation of the concrete block with all samples

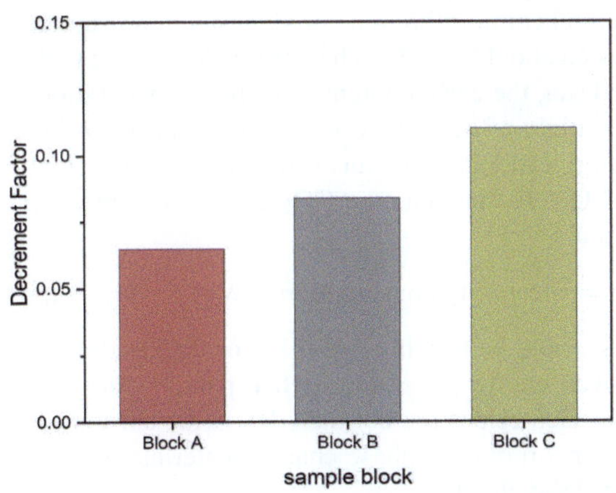

Fig. 9. Decrement factor of different test samples

Block B, showed improved thermal resistance compared to the control, with its back surface temperature rising more gradually and peaking at 34.0 °C. The slower heat transfer is attributed to the latent heat capacity of the petroleum jelly, which absorbs energy during its phase transition, delaying the temperature increase on the opposite side. However, the performance of Block B was exceeded by Block A, in which the PCM was aligned longitudinally, in the same direction as the heat flow. Block A exhibited the most gradual thermal response, reaching only 33.0 °C by the end of the experiment. This configuration demonstrated the most effective reduction in temperature rise and provided the greatest thermal lag among the three bricks. The temperature curves further revealed that the most notable differences between the blocks emerged during the final 30 min of the test. Initially, the temperature increase was relatively similar across all samples, but once the PCM began to undergo phase transition, its influence became more apparent

The temperature gap between the control and Block A widened to approximately 1.9 °C by the 60-min mark, while the gap between the control and Block B stabilized at roughly 0.9 °C

The decrement factor (DF) was calculated for each of the three bricks. DF measures the capacity of a material to attenuate thermal oscillations from the external environment to the internal surface. In this case, the input fluctuation was assumed constant at 70 °C on the heated face, and the output fluctuation was derived from the measured surface temperatures (see Fig. 9). Block C exhibited the highest DF, approximately 0.11, indicating minimal resistance to heat flow. Block B, containing PCM in a lateral layout, showed improved thermal buffering with a DF of 0.084. Block A, which featured longitudinal PCM orientation, demonstrated the best performance with a DF of just 0.065, confirming its superior ability to dampen incoming heat.

3.2 Quantitative Analysis of HVAC Loads

Using Chennai meteorological data, a year-long EnergyPlus simulation was run to create a baseline for building energy consumption. The findings showed that, in order to maintain indoor thermal comfort through HVAC systems, a structure made of traditional concrete blocks needed about 667.83 kWh per year. According to the Indian Power Sector's CO_2 Baseline Data, the carbon intensity of power generation in various locations and the energy mix might affect the CO_2 emissions linked to electricity consumption in India. India's energy generation has an average carbon intensity of 0.82 $kgCO_2/kWh$ (kilowatt-hour) of CO_2. To calculate the CO_2 emission from electricity consumption, we can use this formula,

$$CO_2 \text{ emission (kg)} = \text{Electricity consumption (kWh)} \times \text{Carbon intensity} \left(\text{kgCO}_2/\text{kWh}\right)$$

The CO_2 emissions would be roughly 547.6 kg, or 0.55 metric tons, each year which emphasizes how much energy is required when passive heat regulation is not there. Although PCM-enhanced bricks are not included in the current simulation, the testing results indicate that incorporating phase change materials into wall assemblies could significantly reduce HVAC loads.

4 Limitations and Insights for Future Work

A key limitation of this study is the **selection and integration of PCM,** where **leakage during phase change** remains a concern. While petroleum jelly was chosen for its suitable thermal range, ensuring reliable encapsulation without compromising the brick's integrity is essential. The **design and placement of PCM** within the brick also require optimization to balance thermal performance and structural strength. For future work, the thermal efficiency of PCM bricks could be enhanced by **incorporating nanoparticles** to improve conductivity and heat storage capacity. Additionally, **3D concrete printing** offers potential for precise PCM placement and customized internal geometries, enabling advanced thermal regulation and scalable fabrication.

5 Conclusion

Phase change materials show great promise for improving construction energy efficiency in hot areas like Chennai when used with hollow-core concrete bricks. According to experimental findings, using petroleum jelly as a PCM successfully slowed down heat transmission and lowered surface temperatures. Aligning the PCM with the direction of heat flow produced the best results.

A typical concrete building in Chennai of needs about 667.83 kWh of cooling per year, according to EnergyPlus models. This huge demand emphasizes how urgently passive thermal solutions are needed. Using coal-based thermal power to supply this demand would produce roughly 547.6 kg, or 0.55 metric tons of CO_2 annually, according to carbon emission estimates.

By adding Phase Change Materials (PCMs), indoor temperatures were lowered by 2 °C, According to previous experimental and literature data, the integration of PCM may lower cooling loads by 3% to 6% energy saving of per °C, which reduced the annual cooling energy used by 10%. This results in a yearly energy savings of about 66.78 kWh. Its initial cooling use was 667.83 kWh/year, or approximately 547.6 kg (or 0.55 metric tons) of CO_2 emissions per year. The PCM integration contributes to a more sustainable and climate-responsive building design by increasing energy efficiency and reducing the carbon footprint proportionately. This results in a significant decrease in overall energy use and emissions across all energy sources. This passive cooling approach is affordable, scalable, and consistent with India's objectives to reduce carbon emissions in the built environment and increase building efficiency.

References

1. Zhang, Y., Zhou, G., Lin, K., Zhang, Q., Di, H.: Application of latent heat thermal energy storage in buildings: state-of-the-art and outlook. Build. Environ. **42**(6), 2197–2209 (2007)
2. Baetens, R., Jelle, B.P., Gustavsen, A.: Phase change materials for building applications: a state-of-the-art review. Energ. Buildings. **42**(9), 1361–1368 (2010)
3. Kurdi, A., Almoatham, N., Mirza, M., Ballweg, T., Alkahlan, B.: Potential phase change materials in building wall construction—a review. Materials. **14**(18), 5328 (2021)
4. Kenisarin, M., Mahkamov, K.: Solar energy storage using phase change materials. Renew. Sust. Energ. Rev. **11**(9), 1913–1965 (2007)
5. Hunger, M., et al.: The behavior of self-compacting concrete containing micro-encapsulated phase change materials. Cem. Concr. Compos. **31**(10), 731–743 (2009)
6. Zhang, Y., Sun, X., Medina, M. A.: Thermal behavior of concrete masonry unit walls outfitted with macro-encapsulated PCMS: an experimental study. SSRN 4220110 (2022).

7. Al-Yasiri, Q., Szabo, M.: Effect of encapsulation area on the thermal performance of PCM incorporated concrete bricks: a case study under Iraq summer conditions. Case Stud. Construct. Mater. **15**, e00686 (2021)
8. Sun, X., Jovanovic, J., Zhang, Y., Fan, S., Chu, Y., Mo, Y., Liao, S.: Use of encapsulated phase change materials in lightweight building walls for annual thermal regulation. Energy. **180**, 858–872 (2019)
9. Saikia, P., Azad, A.S., Rakshit, D.: Thermal performance evaluation of building roofs embedded PCM for multi-climatic zones. In: Sustainable Energy Technology and Policies: A Transformational Journey, vol. 1, pp. 401–423. Springer, Singapore (2017)
10. Hasan, M.I., Basher, H.O., Shdhan, A.O.: Experimental investigation of phase change materials for insulation of residential buildings. Sustain. Cities Soc. **36**, 42–58 (2018)
11. Saxena, R., Rakshit, D., Kaushik, S.C.: Experimental assessment of Phase Change Material (PCM) embedded bricks for passive conditioning in buildings. Renew. Energy. **149**, 587–599 (2020)
12. Navarro, L., Solé, A., Martín, M., Barreneche, C., Olivieri, L., Tenorio, J.A., Cabeza, L.F.: Benchmarking of useful phase change materials for a building application. Energy Build. **182**, 45–50 (2019)
13. Kumar, S., Arun Prakash, S., Pandiyarajan, V., Geetha, N.B., Antony Aroul Raj, V., Velraj, R.: Effect of phase change material integration in clay hollow brick composite in building envelope for thermal management of energy efficient buildings. J. Build. Phys. **43**(4), 351–364 (2020)
14. Chihab, Y., Garoum, M., Laaroussi, N.: Dynamic thermal performance of multilayer hollow clay walls filled with insulation materials: toward energy saving in hot climates. Ener. Built Environ. **5**(1), 70–80 (2024)
15. Ritchie, H.: Air conditioning causes around 3% of greenhouse gas emissions. How will this change in the future? https://ourworldindata.org/air-conditioning-causes-around-greenhouse-gas-emissions-will-change-future (2024), last accessed 2025/11/28.
16. Ramakrishnan, S., Muthukrishnan, S., Sanjayan, J.: An innovative method of integrating phase change materials in buildings for thermal energy storage via additive manufacturing. Energy Proceedings. **17**, 202. ISSN 2004-2965, Applied Energy Symposium, USA (2021)
17. Wang, H., Lu, W., Wu, Z., Zhang, G.: Parametric analysis of applying PCM wallboards for energy saving in high-rise lightweight buildings in Shanghai. Renew. Energy. **145**, 52–64 (2020)

Stability Analysis of Metro Tunnel Against Settlement and Deformation Characteristics for Various Loading Conditions

M. S. Abhish(✉), M. H. Prashanth, and Sandi Kumar Reddy

National Institute of Technology, Mangalore, Karnataka, India
abhish.ms@gmail.com, {mh_prashanth,skreddy}@nitk.edu.in

Abstract. Metro tunnels are strategic elements in transportation and are increasingly being seen as an environmentally preferable means of providing infrastructure to densely populated urban areas thus posturing several utility networks in any metropolitan city. This rapid urbanization and increasing reliance on metro tunnel systems for public transportation have necessitated a thorough understanding of the stability of metro tunnels, particularly against stability and deformation. This study focuses on the stability analysis of metro tunnels by examining the settlement and deformation characteristics under various geological and loading conditions. The methodology consists of a combination of numerical modeling by finite element method (PLAXIS 2D) and analytical approaches to evaluate the impact of an interaction effect, tunnel depth, and external loads on tunnel behaviour. For this work operational metro tunnel systems were considered for a particular stretch. Key parameters such as ground displacement (horizontal and vertical displacements), internal forces (axial force, shear force, and bending moment), and lining deformation are analysed for a critical stretch to assess the potential risks of structural failure for a specific case study. This study highlights the critical factors influencing tunnel stability and provides insights into effective mitigation and environmental sustainability strategies to minimize lining failures.

Keywords: Stability analysis · Metro tunnel · FEM · Deformation · Settlement

1 Introduction

Metro tunnels are important components of transportation and are becoming more and more recognised as an environmentally friendly way to supply infrastructure to crowded urban regions, hence posing several utility networks in any metropolitan city Goel (2001). Considering that there is little space for modification above ground, tunnels have grown in demand as a viable substitute that makes better use of the underground space (Fig. 1). To make use of the limited space in crowded urban areas, multiple structures have been designed and built in close vicinity to existing tunnels in recent years Nasri (2017). During tunneling, the stress and strain states of the soil and the lining continuously vary, which has an essential effect on how the soil and lining perform (Hashash et al. 2001). However, the majority of tunnels are built in cities, where they run the risk of being

H. P. Dasari et al. (Eds.): NITK-CREST 2025, PTT, pp. 213–226, 2026.
https://doi.org/10.1007/978-981-95-3218-6_15

overcrowded. Three primary approaches are experimental, numerical, and analytical used to track these settlements. The use of numerical techniques to develop numerical solutions for predicting settlement as a result of tunnelling has been growing in popularity throughout the past ten years. Soil deformation during tunnelling can be calculated using numerical methods are the environmental sustainability strategies. These techniques combined with the field findings will enable further research into surface settlement, influencing causes, and settlement forecasts (Strokova, 2010).

Fig. 1. Inner view of Bangalore metro tunnel

By looking at the stabiltity and deformation characteristics of the soil mass and lining components, this study attempted to assess the stability of a critical metro tunnel for stretch (Bangalore metro East West and North South corridor). Also being studied are looked at factors that could affect the metro tunnel's behaviour, such as the interaction effect, tunnel depth, and external loads (Fig. 2).

Researchers study ground movement caused by tunnelling in sands using centrifuge testing, and evaluate how tunnel dimensions and volume loss affect soil deformation in greenfield conditions (Marshall & Haji, 2015). The ground deformation during tunnelling is predicted using numerical techniques like finite element analysis. The elasto-plastic behaviour, the complexities of the building processes, and the nonlinear interaction between the tunnel and the surrounding soils can all be emphasised by the numerical simulation method (Zhiguo et al., 2011).

Additionally, the framework for executing the individual effects of the settlement parameters is prepared by the Finite Element approach (Hashash and Whittle, 1996). Numerical methods are commonly used not only foresee soil movement but also to predict how tunnels will interact with underground or ground-based structures (Chen et al., 2011). By computing construction and ground conditions such geometry, initial

Fig. 2. Bangalore metro currently operational after Phase 1

stresses, ground behaviour, excavation stages, etc., these methods aim to compute and monitor ground movement at every location of the tunnel and surrounding ground (Leca, 2007). In additional to starting and boundary conditions that are identical to actual field conditions and time effects, the finite element approach accounts for heterogeneous ground layers with more complicated models (Leca, 2007). Although being a useful tool for forecasting ground movement caused by tunnelling, the Finite Element approach has limited applicability when it comes to forecasting ground displacement in soft soil (Rowe et al., 1982). The Finite Element approach can be utilised to analyse the stability of the tunnel direction, and numerical analysis is used as well for soil movement (Vermeer et al., 2011).

2 Methodology

Numerical methods such as finite element analysis, is used to predict the ground movement during the tunneling. Finite element software which is commercially available has been used to create the 2D planar strain model. The input parameters like soil properties and structural properties are used from Bangalore metro. The PLAXIS program then allocates these parameters to every soil element and structural element layer. The first PLAXIS 2D deformation and stress analysis software for Windows was made available in 1998 (Shinde et al. 2019).

2.1 Numerical Model

The models are created with PLAXIS 2D software considering 2D plane strain model with 15 node element type. The soil mass is represented by linearly elastoplastic model using Mohr Coulomb criteria. The dimensions of the soil model are considered as 60 m width and 40 m height. The range of depth of tunnel found in the Bangalore Metro Phase 1 project is 15 m to 18.5 m. Hence, the tunnel has been placed at three different depths such that the crown of the tunnel is at a depth of 15 m, 17 m, 19 m, from the ground (Fig. 3).

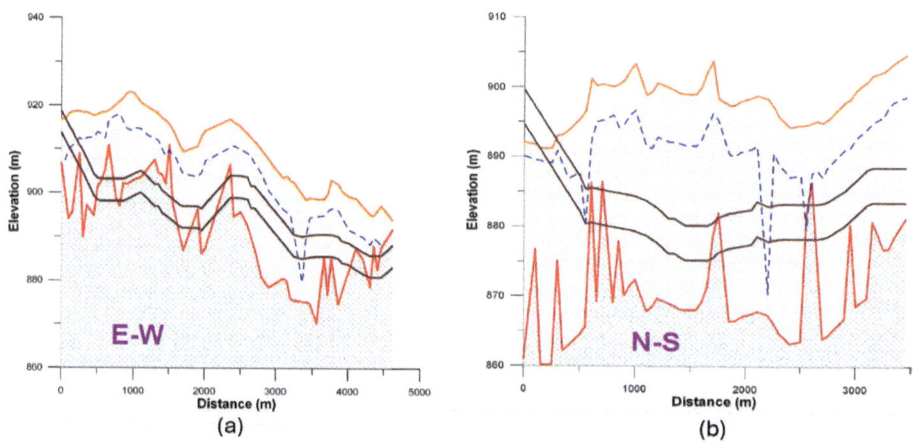

Fig. 3. Alignment of Bangalore metro corridor (Sekhar and Mohan Kumar 2009)

The excavated diameter of the tunnel is kept at 6.44 m. To avoid the boundary effect in numerical analysis, the outer boundary was placed at a distance of 3D times away from the centre of the tunnel.

2.2 Input Parameters

In finite element analysis, the selection of material qualities is crucial to accurately representing the system. In this work, the Mohr Coulomb Model is used to replicate the behavior of soil. PLAXIS requires the following soil characteristics in order to do calculations based on this model (Tables 1 and 2): the soil's Young's modulus (E), Poisson's ratio (v), internal friction angle (φ), cohesiveness (c'), and dilatancy angle (ψ) and structural element parameters (Huang et al. 2020) (Poudel and Kumar Yadav 2022).

These parametric values are chosen doing intensive literature-based review and the given soil condition from previously published research papers and articles from Bangalore metro work. One of the work From (Sitharam and Anbazhagan 2008) work subsurface investigation report has been used which were executed in one of the main projects at Bangalore Metro utilizing the traditional standard penetration test (SPT) at five points and the Multichannel Analysis of Surface Wave (MASW) survey at the same locations. In phase 1 purple line between Cubbon Park and Magadi Road is underground

Table 1. Input of soil properties for Bangalore metro (Raghavendra and Mohan 2014)

Layers	Depth (D) [m]	Density (γ) [kg/m^3]	Young's modulus (E) [N/m^2]	Cohesion (C) [kPa]	Poisson Ratio (v)	Friction angle (φ) [°]
Filled up soil	0–3.2	2000	3.25x10^{11}	5	0.3	35
Clayey sand	3.2–8	1900	1.11x10^{11}	20	0.3	30
Greyish silty sand	8–28.5	2000	4.08x10^{11}	15	0.3	35
Weathered Rock	>28.5	2000	5.23x10^{11}	400	0.2	42

Table 2. Input for structural element parameters (Raghavendra and Mohan 2014)

Specifications	Thickness (t) [m]	Poison ratio (v)	Unit Weight (W) [kN/m/m]	Axial stiffness (EA) [kN/m]	Flexural rigidity (EI) [kNm^2/m]
Segmental lining	0.280	0.15	8.2	81.23X 10^7	73.88X 10^3

for 4.8 km in section UG1 (on the east-west route). Between Sampige Road and National College, the green line's 4.0 km underground segment UG2 (on the north-south route) is located. The boring diameter of 6.44 m and tunnel diameter of 5.6 m with the thickness of 0.28 m. The bottom of the tunnel was a depth of 15-18 m with centre to centre distance of 15 m approximately (Raghavendra and Mohan 2014).

2.3 Initial Condition

Assuming the earth pressure at rest, $K_0 = 1$, the initial stresses have been determined using Jacky's formula, where $K_0 = 1\text{-}\sin\varphi$. The bottom and lateral borders are positioned adequately close to our site of interest (40 m × 60 m) (Fig. 4), where the impact of tunnel excavation would be negligible. The boundary condition is such that the top surface is allowed to move easily while the displacement boundary is fixed at the base and restrained from drifting horizontally on the lateral sides.

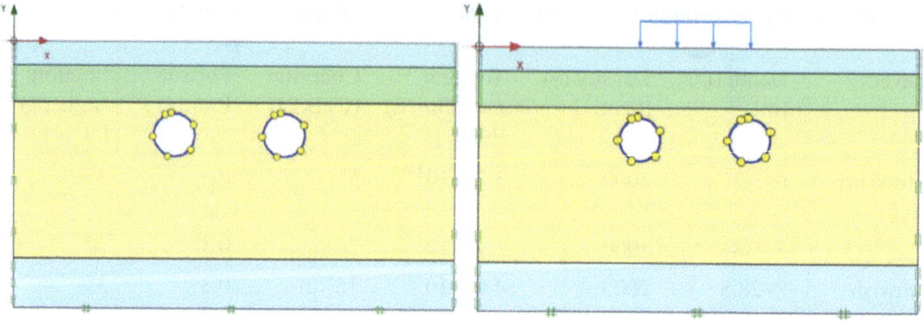

Fig. 4. Numerical model of Bangalore metro tunnel with soil profile (a) without surcharge (b) with surcharge

2.4 Mesh Generation

The model is built using the plane strain model, while taking into account an identical cross section and no deformation in the z-direction. Fine element distribution and enhanced mesh refinement are utilized to construct the mesh (Poudel and Kumar Yadav 2022).

2.5 Boundary Conditions

It is expected that stress concentrations occurs around the tunnel, therefore the mesh is refined in those areas. The default global coarseness parameter (medium) is adopted. The generated mesh is displayed in the output program. The model was provided with fixities boundary conditions in Plaxis 2D with boundaries at the two vertical edges that allow displacement in vertical directions but disable it in horizontal ones, and a base which possesses zero displacement in both directions.

3 Validation

Prior to assessing the influence of governing parameters on the ground settlements induced by tunnelling, the surface settlement trough obtained by numerical modelling has been compared with the surface settlement through from analytical methods proposed in literature.

A closed-form solution had been proposed by Peck and Schmidt to calculate the surface settlement in soil due to an underground opening (Schmidt 1969). Further, (Chow, 1994) had postulated an empirical relationship between surface settlement, tunnel diameter, and soil properties

$$S = \frac{\gamma D^2 Z_0^2}{4G\left(y^2 + Z_0^2\right)}$$

where S designated the vertical surface settlement (m), γ is the soil density (kN/m³), G is the shear modulus in Pascal (Pa), and D is the tunnel diameter in (m).

$$S_v = \frac{V_s}{\sqrt{2\pi}Kz_0}e^{\frac{-y^2}{2K^2z_0^2}}$$

where Vs is the volume of the settlement trough per meter of tunnel advance (m^3/m), defined as a percentage volume loss of the unit volume V of the tunnel, and was taken as 0.30% for low plastic silty soil (ML) in this study, K is trough width parameter and was taken as 0.5 for silty soil, y is the lateral distance from the tunnel centreline (m), and Z_0 is depth of the neutral axis from the surface

The results show that the values that are considered for validation is less than 15% which is within the acceptable limits (Table 3).

Table 3. Validation of analytical methods to numerical method (PLAXIS 2D)

Methods	Analytical value	Numerical value plaxis 2D)	ERROR %
Total displacement (Chow, 1994)	1301	1136	14.52
Vertical displacement (Peck & Schmidt formula, 1969)	17.64	20.35	13.32

4 Results and Discussions

In this stability analysis study results comparison was made with three conditions of tunnel depth 15 m, 17 m, 19 m with different loading conditions of (1) without surcharge (2) with surcharge of 6 kN/m (Fig. 5) for the overall numerical model as well as structural components like tunnel lining and study the results. This surcharge value is taken from the bangalore project DPR report. Below are the typical model screenshots taken at the time of analysis for displacement of soil model and lining elements (Fig. 6).

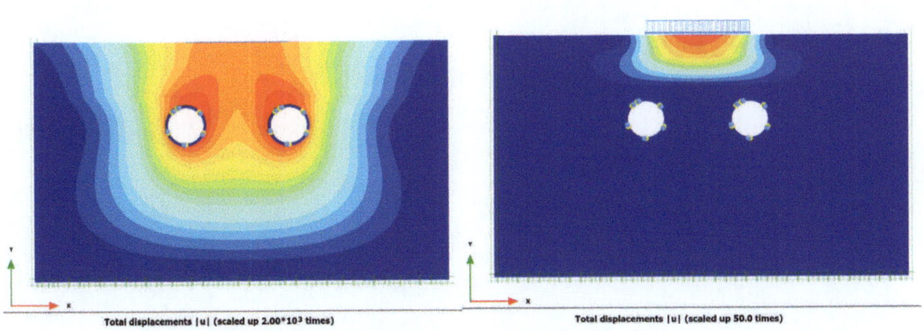

Fig. 5. Typical total displacement of without and with surcharge numerical model

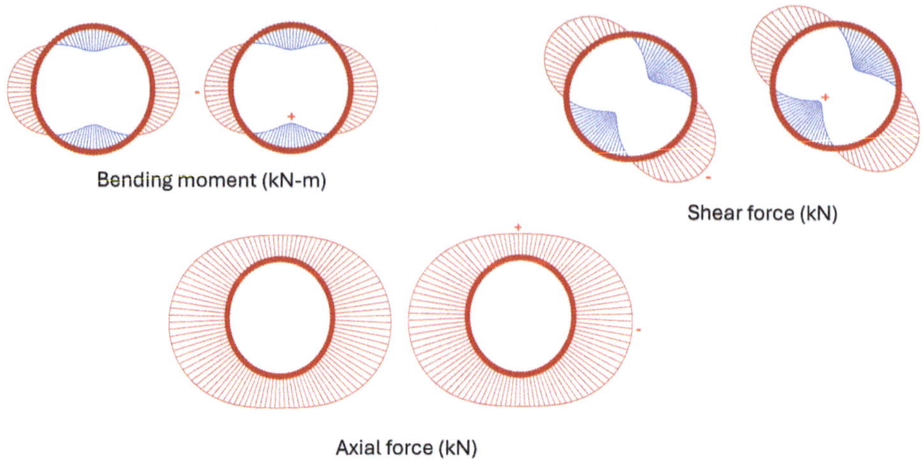

Fig. 6. Typical internal forces of numerical model

For better understanding the results where plot with depth along X axis and displacements on Y axis. The graphs are grouped into total displacement (settlement), horizontal displacement and vertical displacement for both soil model as well as lining element.

Overall Soil Model

When taking into consideration the entire soil model, the total displacement without surcharge is less affected at all depths, but the displacement increases with surcharge; however, the values for wet and dry are nearly equal (1052 mm) at 19 m depth. There aren't many differences in horizontal displacement, but in vertical displacement with surcharge, the values are highest for the dry state (406.9 mm) compared to the wet state, which is zero (Figs. 7, 8 and 9).

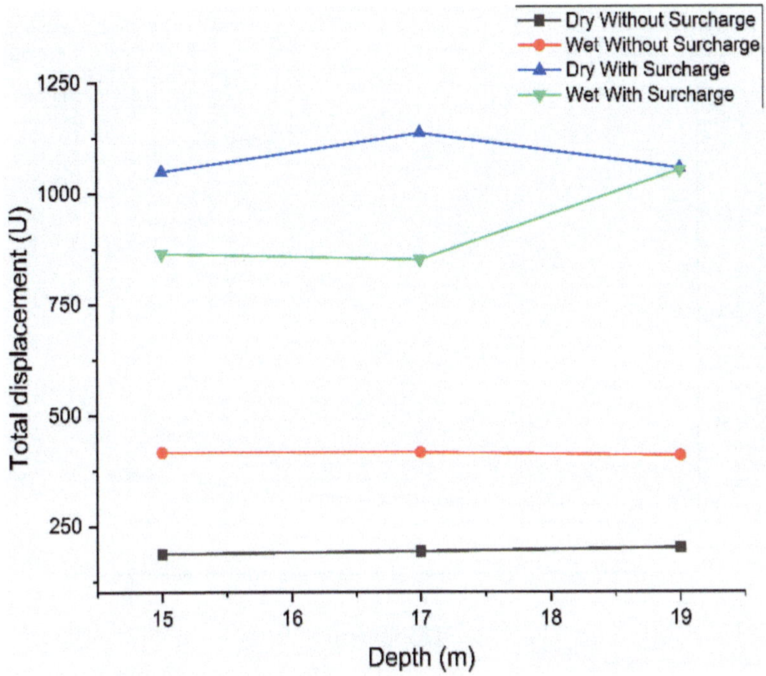

Fig. 7. Total displacement Vs Depth for soil model

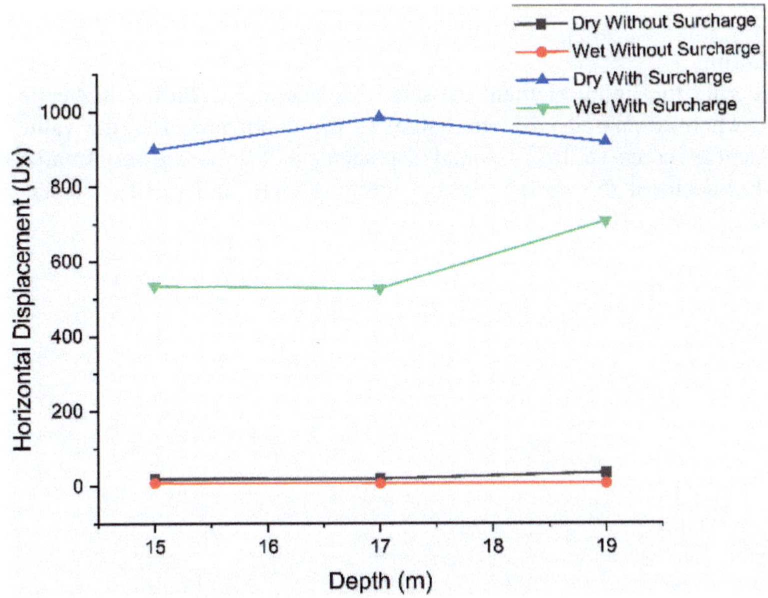

Fig. 8. Horizontal displacement Vs Depth for soil model

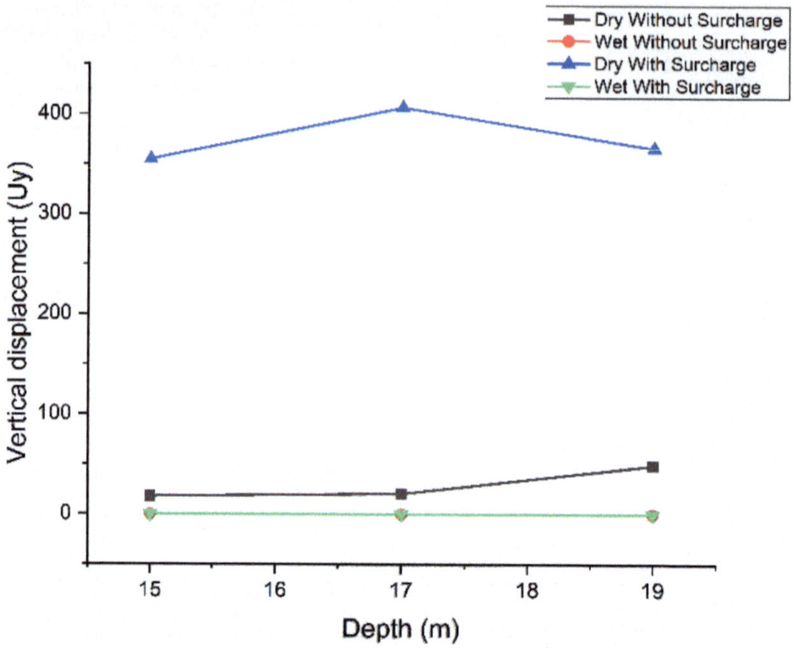

Fig. 9. Vertical displacement Vs Depth for soil model

Tunnel Lining

By considering the lining element the total displacement values is higher in without surcharge when compared with surcharge, as the depth increases the value of total displacement as increased. In horizontal displacement of lining are very small compared to total displacement (55 mm) of tunnel lining (6 mm) and vertical displacement is higher(40 mm) (Fig. 10).

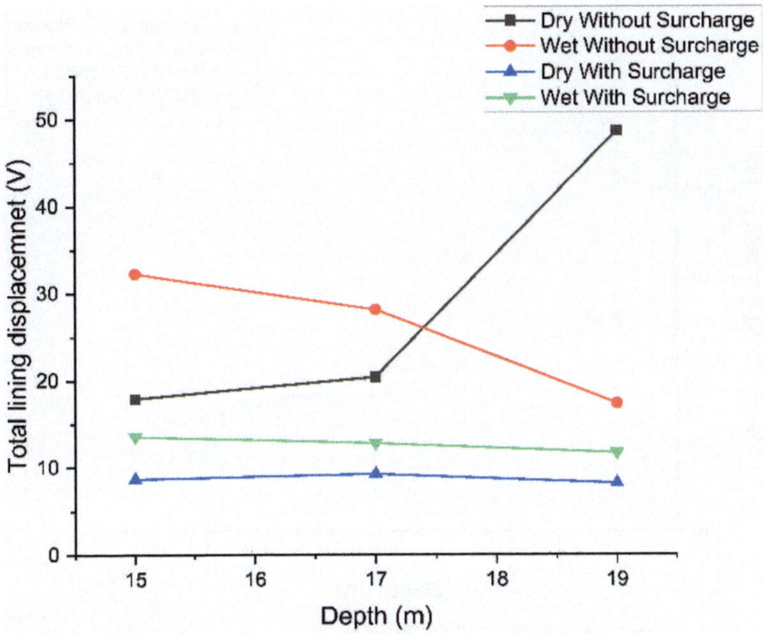

Fig. 10. Total displacement Vs Depth for lining element

Internal Forces

By considering the internal forces developed in the tunnel lining bending moment and axial forces are the major internal forces which are responsible for tunnel settlement and deformations as compared to shear force unless an horizontal loading as to be considered (Figs. 11, 12 and 13).

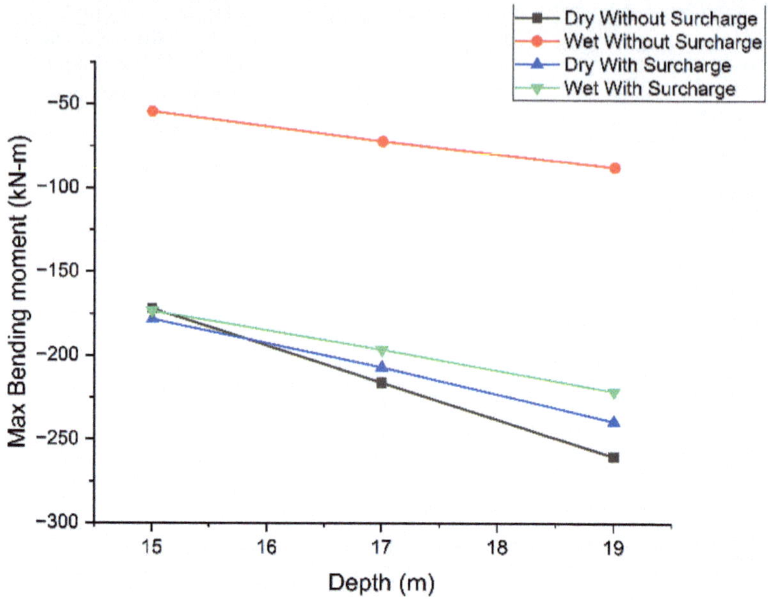

Fig. 11. Bending moment Vs Depth for lining element

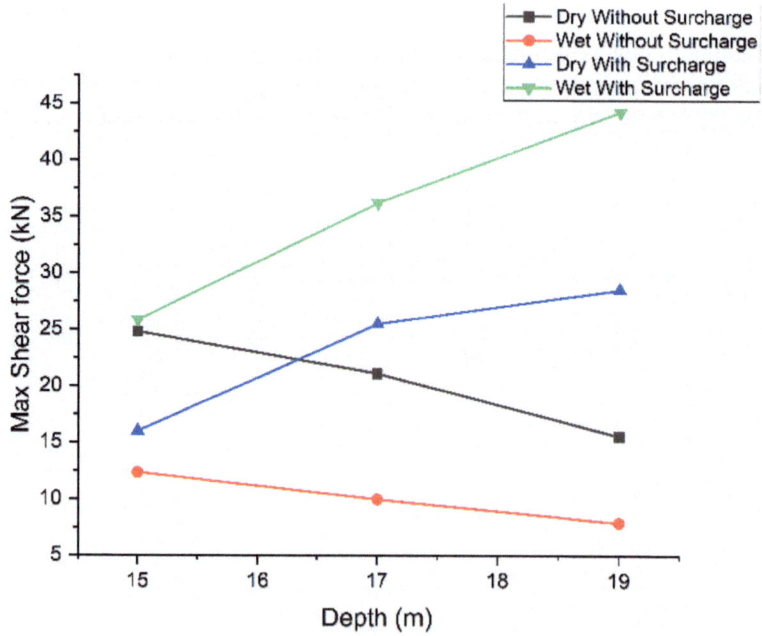

Fig. 12. Shear force Vs Depth for lining element

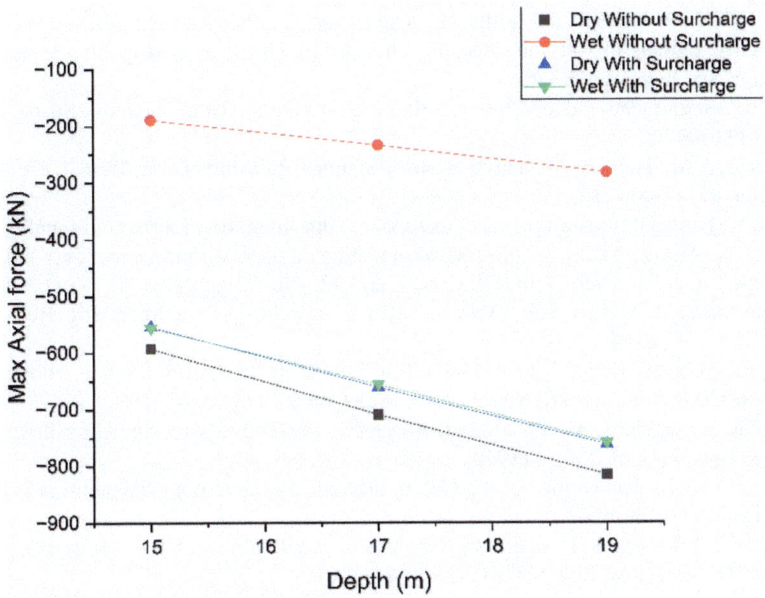

Fig. 13. Axial force Vs Depth for lining element

5 Conclusion

Prior to conducting the parametric study, the results of numerical model were compared to the existing empirical approaches for validation. PLAXIS 2D can provide a reasonable prediction of transverse surface settlements in layered soils of varying depth. With surcharge the total displacement (settlement) are the major parameter to be considered for stability of metro tunnel in urban areas. The displacement produced in lining element is independent of surcharge loading conditions. The major influence will be the bending moment and axial force on the internal forces of tunnel lining. Validation with respect to experimental/monitoring data has to be considered for stability analysis of metro tunnel.

References

1. Chen, R., Zhu, J., Liu, W., Tang, X.: Ground movement induced by parallel EPB tunnels in silty soils. Tunnel. Undergr. Space Technol., 163–171 (2011)
2. Chow, L.: Prediction of Surface Settlement Due to Tunneling in Soft Ground. M.Sc. thesis, University of Oxford (1994)
3. Goel, R.K.: Status of tunneling and underground construction activities and technologies in India. Tunn. Undergr. Space Technol. **16**(2), 63–75 (2001)
4. Hashash, M.A.Y., Whittle, A.J.: Ground movement prediction for deep excavations in soft clay. Geotech. Eng. (1996)
5. Hashash, Y.M.A., Hook, J.J., Schmidt, B., Chiang Yao, J.: Seismic design and analysis of underground structures. Tunn. Undergr. Sp. Technol. **16**(4), 247–293 (2001)

6. Huang, Z.K., Pitilakis, K., Tsinidis, G., Argyroudis, S., Zhang, D.M.: Seismic vulnerability of circular tunnels in soft soil deposits: The case of Shanghai metropolitan system. Tunn. Undergr. Sp. Technol., 98 (2020)
7. Leca, E.: Settlements induced by tunneling in soft ground. Tunn. Undergr. Space Technol., 119–149 (2007)
8. Marshall, A.M., Haji, T.: An analytical study of tunnel-pile interaction. Tunn. Undergr. Space Technol. **45**, 43–51 (2015)
9. Nasri, V.: Tunnel lining design and construction. 10th Annu. Breakthroughs tunneling (2017).
10. Poudel, A., Kumar Yadav, S.: Numerical modeling of shallow tunnel in soft ground: surface settlement analysis. Proc. 12th IOE Grad. Conf. **12**, 930–936 (2022)
11. Raghavendra, V., Mohan, S.R.: With and without support systems. Int. J. Eng. Res. Technol. **3**(5), 270–275 (2014)
12. Rowe, R., Lo, K., Kack, G.J.: A Method of Estimating Surface Settlement above Tunnels Constructed in Soft Ground. National Research Council of Canada (1982)
13. Schmidt, B.: Settlements and ground movements associated with tunneling in soil. Ph.D. Thesis, University of Illinois (1969)
14. Sekhar, M., & Mohan Kumar, M. S.: Geo-hydrological studies along the metro rail alignment in Bangalore. (May) (2009).
15. Shinde, N.T., Kuwar, S.R., Khairnar, A.P., Khairnar, Y.J., Shivde, S.R.: Tunnel analysis using Plaxis 2D. Int. J. Res. Eng. Sci. Manag. **2**, 4–5 (2019)
16. Sitharam, G.T., Anbazhagan, P.: Evaluation of low strain dynamic properties using geophysical method: A case study. Consult. Ahead. (2008)
17. Strokova, L.: Methods of estimating surface settlement during driving of urban tunnels. Soil Mech. Found. Eng. **47**(3) (2010)
18. Teachavorasinskun, S., Chub-uppakarn, T.: Influence of segmental joints on tunnel lining. Tunn. Undergr. Space Technol. **25**(4), 490–494 (2010)
19. Vermeer, P., Moller, S., Ruse, N.: On the Application of Numerical Analysis in Tunneling. Institute of Geotechnical Engineering, Stuttgart, Germany (2011)
20. Zhiguo, Z., Huang, M., Zhang, M.: Theoretical prediction of ground movements induced by tunneling in multi-layered soils. Tunn. Undergr. Space Technol., 345–355 (2011)

Water

Urban Flood Modelling: A Case Study in Mangalore Subcatchment

Dhanush Devadiga(✉), Samanyu P. Rai, and Pruthviraj Umesh

Department of Water Resource and Ocean Engineering, National Institute of Technology Karnataka Surathkal, Mangaluru, Karnataka 575025, India
dhanushkarkala2001@gmail.com

Abstract. Floods have been occurring more frequently and with greater severity in cities around the world, raising concerns about urban flooding. This also applies to Mangalore, an Indian coastal city in the state of Karnataka. This study intends to create an urban flood model for the Mangalore sub-catchment by applying PCSWMM (Personal Computer Storm Water Management Model) software using the Green-Ampt infiltration method. Advanced surveying techniques are used to model this in order to gather the information required to replicate actual drainage conditions and to further analyze flood scenarios using various rainfall return periods. This also involved processing the incoming data using a geographic information system (GIS). The actual flooding conditions were replicated using the high rainfall event. Previous flood data was then used to modify the model's parameters. In order to prevent floods from periods of intense rainfall while taking field conditions into account, a modified drainage system was implemented. This effort is part of a larger research project that uses both structural and non-structural methods to investigate creative ways to reduce urban floods. The findings will shed crucial light on how well urban flood modelling works to lower the danger of flooding. The foundation for further research in this important field is laid by this study, which will eventually result in more sustainable and resilient urban settings.

Keywords: PCSWMM · Urban Flood Modeling · Advanced Surveying · Flood Mitigation

1 Introduction

As the world becomes increasingly urbanized, over 54% of the population lives in cities, projected to increase to 66% by 2050 [1]. Coastal cities, in particular, face the brunt of climate change, manifesting in frequent floods, water shortages, and erosion [2]. As climate change intensifies, heavy rainfall events of heightened frequency and intensity cause considerable damage to lives, property, and urban infrastructure [3]. Urban flood management is a pressing challenge in India due to the lack of sustainable water systems. Authorities are modernizing existing infrastructure, though effectiveness varies across regions depending on localized planning [4].

Effective adaptation and mitigation strategies can be achieved through integrated sustainable urban water management approaches that incorporate climate change data

H. P. Dasari et al. (Eds.): NITK-CREST 2025, PTT, pp. 229–238, 2026.
https://doi.org/10.1007/978-981-95-3218-6_16

into the decision-making process. Flood forecasting, as a non-structural mitigation technique, plays a vital role in predicting flood events, minimizing their impact, and reducing the risk of loss of life [5, 6].

It uses an integrated hydrologic and hydraulic modelling method. The example of Mangalore, an Indian coastal city, exemplifies the difficulties experienced by emerging nations. Floods brought on by monsoon rains and cyclonic activity regularly threaten Mangalore. Since floods have been happening more often lately, there have been several recorded flood events in Mangalore. To address the impacts of climate change on urban wellbeing, local governments have put in place a number of adaptation and mitigation strategies, such as improved drainage systems and flood forecasts. However, given climate change, flood control and sustainable urban water management continue to be difficult tasks. Decision making can be significantly improved by incorporating climatic data into sustainable urban water management plans. Flood forecasting is one of the non-structural flood mitigation strategies used globally that has shown promise in lessening the negative impacts of urban flooding. As demonstrated in the case of Mangalore, site specific planning that incorporates climate change considerations is crucial for formulating effective adaptation and mitigation measures [1].

To manage urban flooding, both structural and non-structural approaches are explored to identify innovative solutions [7]. The US Environmental Protection Agency created the Storm Water Management Model (SWMM), which is extensively used in urban planning. In addition to modelling urban drainage systems, SWMM is used for a variety of other objectives, including assessing hydrologic impacts, identifying catchment areas, and simulating runoff behavior and water quality over different time frames [8]. However, the absence of integrated spatial functionality presents difficulties when utilizing typical SWMM tools because metropolitan surroundings are complicated. To address this, the commercial version Personal Computer Storm Water Management Model (PCSWMM) has been developed. PCSWMM incorporates standalone GIS capabilities, enabling comprehensive 1D-2D and quasi 2D modeling. Its proven reliability and functionality have led to its adoption in several countries, such as India, the United States, Australia, the Philippines, China, Korea, Thailand, Cambodia, and Norway [9].

2 Materials and Methods

2.1 Site Description

The study area is a 4.75 km^2 subcatchment located within Mangalore City of Karnataka, a state in the southwestern region of India and positioned at a latitude of 12°51.330′N and a longitude of 74°50.656′E. The main stormwater drain length is about 2.84 km from Attavar Locality to Bunder Locality, as shown in Fig. 1. Its elevation ranges from 91 to 13 m above mean sea level. The city's temperatures range from 20 °C to 35 °C, and the annual average rainfall is roughly 3695 mm.

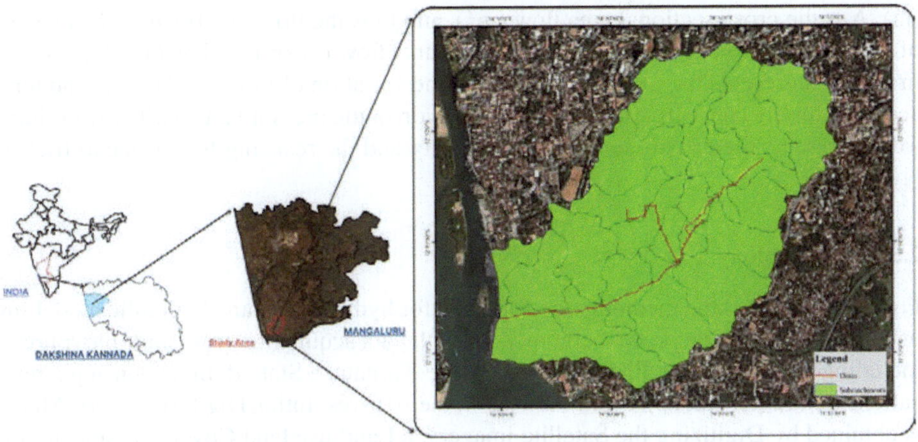

Fig. 1. Geographical Overview of the Study Area

2.2 PCSWMM Model Description

Hydrological modelling in PCSWMM entails simulating the rainfall-runoff process within the catchment region. This process's main components are precipitation, infiltration, evaporation, and runoff. The essential formula used in hydrological modelling is the continuity equation, which can be represented as:

$$\mathrm{P} = \mathrm{I} + \mathrm{E} + \mathrm{R} + \Delta S \tag{1}$$

where P = Precipitation, I = Infiltration, E = Evaporation, R = Runoff = Change in storage (1). In PCSWMM, hydraulic modeling is used to simulate water flow through channels, conduits, and other drainage infrastructure. The shallow water equations also referred to as the Saint-Venant equations, are applied to one-dimensional flow analysis. These equations consist of the Continuity Equation and the Momentum Equation. The Continuity Equation is responsible for representing the principle of mass conservation, which can be represented as:

$$\frac{\partial A}{\partial t} + \frac{\partial Q}{\partial x} = 0 \tag{2}$$

where t = time (s), x = distance along the flow path (m), A = cross-sectional area flow (m^3), and Q = flow rate (m^3/s). The conservation of momentum under unstable flow circumstances is explained by the momentum equation. It takes into consideration the forces operating on the fluid, such as friction, gravity, and pressure forces (2). The following is an expression for the one-dimensional momentum equation:

$$\frac{\partial Q}{\partial t} + \frac{\partial}{\partial x}\left(\frac{Q^2}{A} + gAh\right) = gA\left(S_0 - S_f\right) \tag{3}$$

where g is the acceleration triggered by gravity (m/s^2), h is the flow depth (m), S0 is the bed slope, Sf is the friction slope, t is the time (s), x is the distance along the flow route (m), A is the cross-sectional area flow (m^2), and Q is the flow rate (m^3/s). The balance of momentum in a one-dimensional open-channel flow is expressed by this Eq. (3). The first term represents the local acceleration (time variation of flow), and the second term accounts for the convective acceleration and hydrostatic pressure effects. The right-hand side represents the driving forces due to gravity and the resisting forces due to friction [10].

2.3 Methodology

Figure 2 illustrates the methodology followed for hydrological and hydraulic modelling using PCSWMM. The process commences with data acquisition from multiple sources, including Cartosat-1, satellite imagery, and the Karnataka State Remote Sensing Applications Centre (KSRSAC). From Cartosat-1, the 10 m resolution Digital Elevation Model is obtained by Digitizing the Satellite imagery, a Land use land Cover Classification is Carried out by collecting the Soil Map from the KSRSAC and clipping the study area extent, the Soil map is Digitized which are essential for defining the topography and hydrological characteristics of the study area. Using DGPS (Differential Global Positioning System), a field study was carried out to gather information on the drainage system. Accurate positioning and elevation measurements throughout the drain's path, including its location, dimensions, and construction materials, are achieved by the use of DGPS Survey. The physical precision of the drain properties is guaranteed in this step. The collected data is processed and integrated into PCSWMM during model setup and the Green-Ampt infiltration method used for calculating soil water infiltration. Hydrological and hydraulic parameters are defined, and the drainage network is structured to reflect real world conditions. Once the setup is completed, the model undergoes parameter optimization to refine input variables, followed by sensitivity analysis calibration. This calibration process is essential to ensure the model produces reliable results under varying conditions.

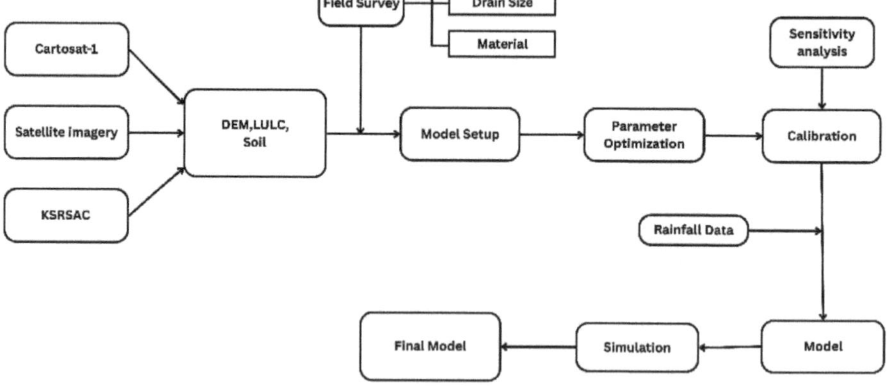

Fig. 2. Methodology for PCSWMM

That is a five-year return time following the simulation of the calibration model using rainfall data. The IDF curve with a 5-year return term is chosen. The rainfall data from Mangalore shows that the Generalised Extreme Value (GEV) distribution is the most accurate fit for the annual maximum daily rainfall, in accordance with the guidelines provided from the Central Public Health and Environmental Engineering Organisation (CPHEEO 2019) Manual on Storm Water Drainage Systems – 2019, and the data from the Panambur rain gauge station used for a 5-year return period [11]. The outcome shows how the model responses to the correct field condition and where the flood can occur within the storm water drain. It is used entirely for decision making and mitigating the flood for future prospective.

3 Results

3.1 Storm Water Drain Modeling

The drainage overflow modelling was carried out in the Study region (Fig. 1). The area facing more flooding during Mansoon time the flood in this region is determined by field surveys and conversations with locals in the study area. In Fig. 3, the Study area Subcatchment is done with an average area of 25 ha. A total of 20 Subcatchments have been delineated with an average percentage of imperviousness about 40.97% and an average percentage of slope in the catchment of 15.69%. The total number of conduits in the catchment is 43, the length of conduits ranges from 6 m to 248 m, and the total number of nodes in the catchment is 43, with one outfall.

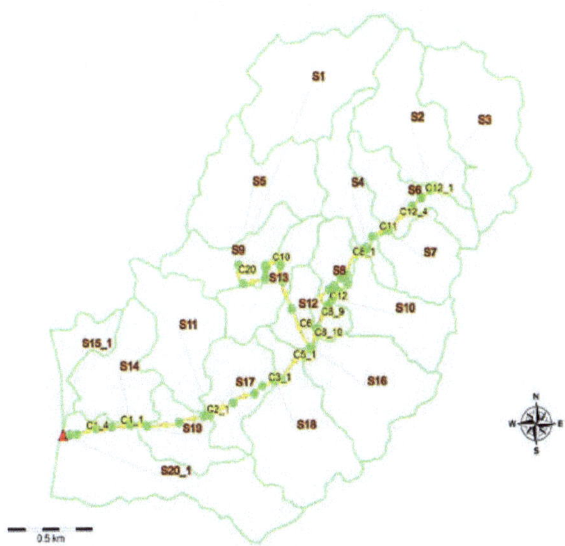

Fig. 3. Catchment discretization for stormwater drain flow modeling.

The flood duration and total flood volume at the drainage network's two specific nodes, J30 and J35. Node J30, located near Mangalore Central Railway, beside the Landmark Infratech Construction site, experienced flooding for 1.84 h, with a total flood volume of 15.654 million liters (ML). In contrast, Node J35, situated beside DJ Sajan Mangalore in Pandeshwar, Mangalore, has a significantly shorter flooding duration of

0.42 h, with a total flood volume of 2.803 ML (Fig. 4). These variations in flood impact across different locations highlight the need for targeted flood mitigation measures, particularly at J30, where flood volume and duration are considerably higher. Understanding these localized flooding patterns is essential for improving Mangalore's urban drainage planning and infrastructure resilience. These findings suggest that the current urban stormwater drainage system lacks sufficient capacity to manage the inflow effectively. Consequently, it is recommended that the system undergo capacity augmentation to enhance its performance and mitigate flooding risks.

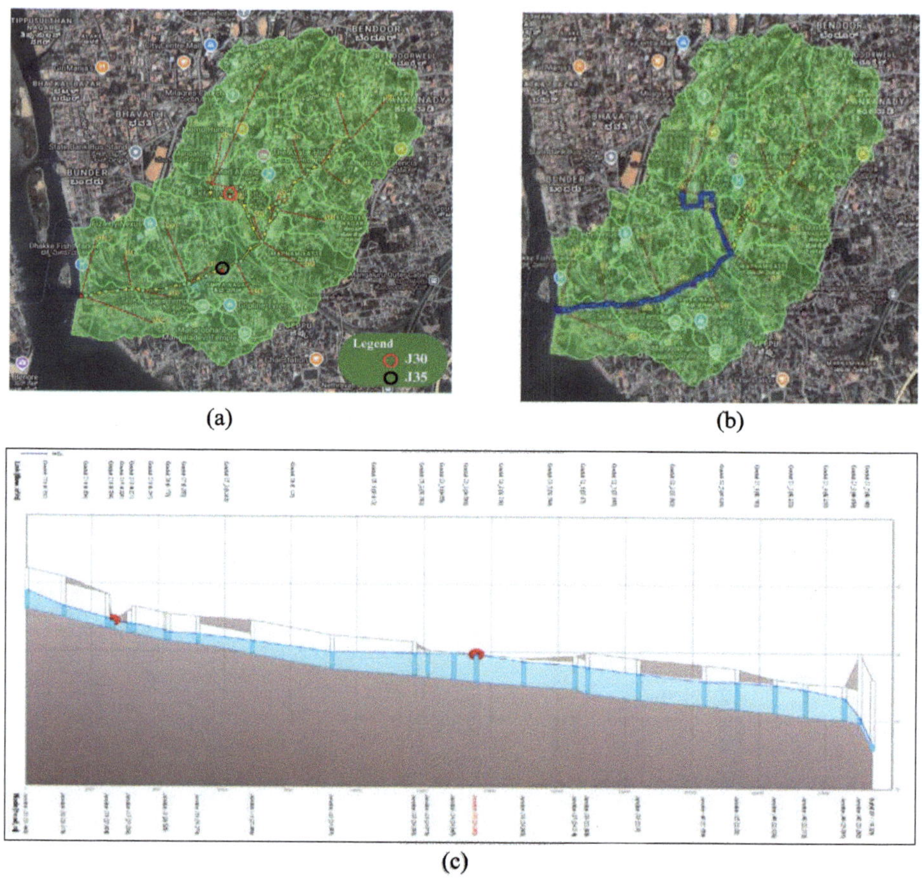

Fig. 4. (a) Flooding Node Location (b) Selected Conduits (c) Cross-Section Profile of Selected Conduits.

3.2 OAT Sensitivity Analysis

Sensitivity analysis determines how altering an independent variable's values under assumptions influences a particular dependent variable. Sensitivity analysis determines the most relevant parameters affecting the hydrological response within the research domain. This is accomplished by systematically adjusting the model input parameters using the One- factor-at-a-time (OAT) technique and assessing their influence on model output [12].

Individual percentage modifications of 10%, 25%, 50%, and 75%, positive and negative, were applied to each input parameter, and their influence on the output values was recorded.

The model input parameters considered for the OAT sensitivity analysis were:

1. Manning's roughness coefficient (N Impv) for impervious surfaces
2. Manning's roughness coefficient (N Perv) for the previous surface
3. Depression storage (DS Impv) for Impervious surfaces
4. Depression storage (DS Perv) for pervious surfaces
5. Subcatchement width (Scw)

The model output parameters considered for the OAT sensitivity analysis were:

1. Percentage change in Flood volume
2. Percentage change in Flood duration

The relative impact of important hydrologic parameters on flood volume and nodal flood duration was evaluated using the OAT method of parameter sensitivity analysis. Subcatchment width (Scw), impervious and pervious depression storage (DS_Impv, DS_Perv), and Manning's roughness coefficients for impervious (N_Impv) and pervious (N_Perv) areas are among the characteristics that are examined. The change in flood volume and duration in response to ±10% to ±75% perturbations in these parameters is depicted in Figs. 5 and 6.

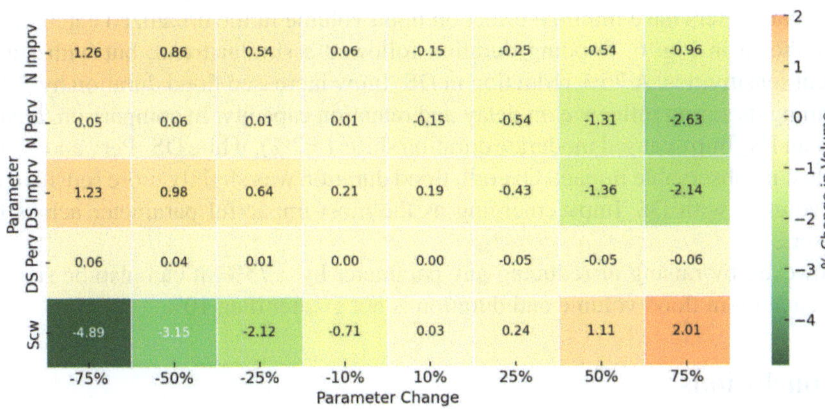

Fig. 5. Percentage change in Flood volume

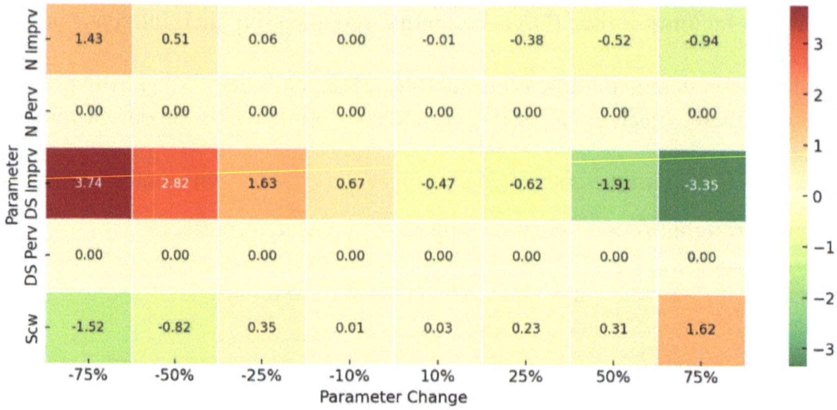

Fig. 6. Percentage change in Flood duration

Figure 5 shows that there is a direct correlation between flow width and runoff volume, with a 75% decrease in subcatchment width (Scw) resulting in a 4.89% decrease in flood volume and a 75% rise in Scw leading to a 2.01% increase. An inverse link between surface storage and runoff is also confirmed by the fact that a 75% drop in impervious depression storage (DS_Impv) results in a 1.23% rise in flood volume, and a 75% increase in it causes a 2.14% decrease. Manning's n for impervious areas (N_Impv), on the other hand, exhibited mildly inverse behaviour, with volume fluctuations of only ±1.26% at a maximum change of ±75%. The N_Perv and DS_Perv parameters exhibited negligible influence, contributing to changes below 0.1%, suggesting that pervious surface parameters have minimal effect on flood volume in the urbanized catchment.

As shown in Fig. 6, flooding duration followed a similar trend, but with slightly different sensitivities. A 75% reduction in DS_Impv increased flood duration by 3.74%, confirming its strong influence on delay and retention capacity. In comparison, changes in Scw and N_Impv caused moderate duration shifts (<2%), while DS_Perv and N_Perv again had no observable impact. Overall, flood duration was slightly more sensitive than flood volume, with DS_Impv emerging as the most impactful parameter across both indicators.

However, by raising or reducing any parameter by ±75%, it can also be seen that the fluctuation in flood volume and duration is not greater than 10%.

4 Conclusions

This study aimed to closely replicate actual field conditions in the Mangalore subcatchment to identify flood prone locations in the stormwater drainage system and support effective mitigation planning. To achieve this goal, field reconnaissance surveys were conducted to map existing storm drains, and a DGPS survey was carried out to obtain accurate geographic positions and elevations (relative to mean sea level) of the drainage network. These data were used to configure a hydraulic model in PCSWMM, which delineated subcatchments and simulated the stormwater system's response to a five-year return period rainfall event. The simulation results revealed critical flood hotspots within the stormwater network. Nodes J30 and J35 were predicted by the model to be severely flooded, with node J30 exhibiting the worst flooding during the design storm. These findings show that the model successfully identified the catchment's key flood risk zones and faithfully replicated field data. To assess the model's robustness, a sensitivity analysis was conducted using the One-Factor-at-a-Time (OAT) method. Five important hydrological parameters were individually changed by $\pm 10\%$, $\pm 25\%$, $\pm 50\%$, and $\pm 75\%$: subcatchment width, depression storage for both surfaces, and Manning's roughness coefficients for pervious and impervious surfaces. The analysis showed that the resulting changes in both flood volume and flood duration remained within a $\pm 10\%$ range for all parameter variations. Among the tested inputs, impervious depression storage (DS_Impv) and subcatchment width (Scw) were found to be the most sensitive, while pervious Manning's n (N_Perv) and pervious depression storage (DS_Perv) had negligible impact on model outputs. These findings indicate that the model is not overly sensitive to input uncertainty and thus provides reliable predictions under realistic conditions.

Acknowledgements. Sincere thanks are extended by the authors to the Minor Irrigation Department authorities. They also want to thank Prof. Gangadharan K V, the coordinator of the Centre for System Design (CSD), a prestigious Centre of Excellence at NITK-Surathkal, for his invaluable assistance and for granting them access to research facilities. Additionally, the authors also extend their appreciation to their colleagues for their support and constructive inputs throughout the course of this work.

References

1. Pai, A., Mahima, N., Pruthviraj, U., Abhijith, P.: Modelling and mitigation of urban floods: a case study in Mangalore. In: Sivakumar Babu, G.L., et al. (eds.) Technologies for Sustainable Transportation Infrastructures: Select Proceedings of SIIOC 2023 Lecture Notes in Civil Engineering, vol. 529, pp. 931–944. Springer, Singapore (2024)
2. Satterthwaite, D.: The implications of population growth and urbanization for climate change. Environ. Urban. **21**(2), 545–567 (2009)
3. Zhou, Q., Leng, G., Su, J., Ren, Y.: Comparison of urbanization and climate change impacts on urban flood volumes: importance of urban planning and drainage adaptation. Sci. Total Environ. **658**, 24–33 (2019)
4. Sen, D.: Flood hazards in India and management strategies. In: Mirza, M.M.Q., Dixit, A., Biswas, A.K. (eds.) Flood Problem and Management in South Asia, pp. 126–146. Springer, Dordrecht (2010)
5. Parkinson, J., Tayler, K., Mark, O.: Planning and design of urban drainage systems in informal settlements in developing countries. Urban Water J. **4**(3), 137–149 (2007)
6. Paule-Mercado, M., Lee, B.Y., Memon, S.A., Umer, S.R., Salim, I., Lee, C.-H.: Influence of land development on Stormwater runoff from a mixed land use and land cover catchment. Sci. Total Environ. **599**, 2142–2155 (2017)
7. Ahiablame, L.M., Engel, B.A., Chaubey, I.: Effectiveness of low impact development practices: literature review and suggestions for future research. Water Air Soil Pollut. **223**, 4253–4257 (2012)
8. Rossman, L.A.: Storm water management model user's manual, version 5.0. National Risk Management Research Laboratory, Office of Research and Development, U.S. EPA, Cincinnati, OH (2010)

9. Liwanag, F., Mostrales, D.S., Ignacio, M.T.T., Orejudos, J.N.: Flood modeling using GIS and PCSWMM. Eng. J. **22**(3), 279–289 (2018)
10. Chow, V.T., Maidment, D., Mays, L.: Applied hydrology McGraw-Hill Series in Water Resources and Environmental Engineering. Tata McGraw-Hill Education, New York (2010)
11. Femin, C.V., Varija, K.: Identification of the best-fit probability distribution and modeling short-duration intensity-duration-frequency curves—Mangalore. In: Narasimhan, M.C., et al. (eds.) Trends in Civil Engineering and Challenges for Sustainability: Select Proceedings of CTCS 2019 Lecture Notes in Civil Engineering, vol. 99, pp. 765–783. Springer, Singapore (2021)
12. Osheen, M., Kansal, L., Bisht, D.S.: Evaluation of an urban drainage system using functional and structural resilience approach. Urban Water J. **20**(10), 1794–1812 (2023)

Assessment and Mitigation of Urban Flooding: A Case Study of a Subcatchment in Mangalore

Samanyu P. Rai(✉), Dhanush Devadiga, and Pruthviraj Umesh

Department of Water Resource and Ocean Engineering, National Institute of Technology Karnataka Surathkal, Mangaluru 575025, India
samanyupanne@gmail.com

Abstract. Urban flooding has emerged as a significant problem in fast-growing urban cities, with uncontrolled growth of impervious surfaces and overwhelmed drainage systems as the primary causes, leading to impaired stormwater management and flood susceptibility. The present study aims to analyze the main stormwater drain from Pumpwell to Jeppinamogaru, Mangalore, which flows into the Nethravathi River and then into the Arabian Sea. This drain is the main conduit for stormwater runoff. Heavy rain events led to extreme flooding in the Mangalore area in 2018 and 2023, with the Pumpwell area being flooded due to the drainage system's low capacity. This paper uses PCSWMM to simulate stormwater flow and Geographic Information Systems (GIS) to pre-process input data. The extreme rainfall events were modeled in PCSWMM to evaluate flood conditions, which were calibrated using historical flooding records from these rainfall events. The stormwater drain was then redesigned in the most critical areas to ensure that the drainage network could manage the stormwater load effectively. Research is included in the broader project to enact structural and non-structural interventions to ease urban stormwater flooding, providing information on good flood mitigation policies.

Keywords: Urban Flooding · Urban Flood Modelling · PCSWMM · Stormwater Drainage

1 Introduction

Flooding is a natural phenomenon that occurs when an area becomes submerged with water, often due to excessive rainfall, river overflow, storm surges, or rapid snowmelt. India is another country that has been frequently affected by floods since Bangladesh [1]. 12.5% of India's geographical area is susceptible to flooding [1]. Flooding is caused by various reasons, ranging from heavy rainfall, poor drainage systems, and land use modifications that change the natural flow of water to other causes. The occurrences have devastating impacts, including property damage, infrastructure collapse, ecological balance, and severe long-term economic consequences. Urban flooding, especially in high-density metropolitan cities, is increasingly becoming a significant problem. Urban areas are most susceptible due to the high impervious surface rate, which causes high

H. P. Dasari et al. (Eds.): NITK-CREST 2025, PTT, pp. 239–250, 2026.
https://doi.org/10.1007/978-981-95-3218-6_17

surface runoff and subjects existing drainage systems to tremendous pressure, often causing flooding during even moderate rainfall. The issue continues to escalate as the number of urban residents increases and urban areas spread to low-lying areas that are naturally more susceptible to flooding. The trend makes it all the more imperative to create robust, dependable, and realistic flood modeling approaches capable of preventing and mitigating the dangers of urban flooding.

Urban flooding is emerging as an important issue in rapidly urbanizing regions, driven by increasing impervious surfaces such as roads, buildings, and concrete parking lots that prevent rain from penetrating the ground naturally. This results in a surplus volume of runoff during high-intensity rainfalls and overloading of drainage facilities not designed to manage such concentrated flow. Effects of urban flooding are far-reaching, varying from damage to properties and disruption of infrastructure to health risks and economic losses. There are a number of causes of urban floods, but one of the main causative factors is rain over a limited time period, i.e., high-intensity rain. It is estimated that 5 billion humans will live in urban zones by 2030, with Africa and Asia's urban populations doubling in a single generation [2, 3].

Urban flooding can be defined as the flooding or inundation of land, particularly in populous areas like cities, primarily due to heavy rain beyond the capability of the stormwater drains. The objective of this project is merely to reduce human, economic, and environmental losses. Integrated strategic action against flood hazard comprises structural (such as dams, dikes, drainage systems, and natural, sustainable means wetlands and natural buffers) and non-structural (such as insurance, flood zoning, and forecasting of floods) measures [4]. Flood Risk Management (FRM) has yielded a wealth of knowledge and today calls for a continuous, comprehensive, and sustainable approach to flood risk reduction. The foundations of contemporary FRM include a thorough comprehension of risks and uncertainties, an awareness of how the system behaves through integrated risk assessment across several scales and dimensions, and using a wide range of tools and measures. When combined, these ideas guarantee robust and successful flood risk reduction plans [5].

Mangalore's sub-catchment area also poses problems that make it impossible to model and predict exactly in the context of flood hazard. The area's topography variability, high-velocity urbanization, and coastal location make it a hydrologically advanced environment that is extremely sensitive to rainfall intensity and land use alteration. Although applied widely elsewhere, existing model approaches cannot adapt to Mangalore's local conditions because of the lack of data, resolution, and infrastructure representation. The models obtained cannot predict flood extent and duration to the desired level of quality for implementing flood mitigation and management.

Urban flood modeling is an important area of research that tries to model and understand the complex interactions among surface runoff, river flow, and urban drainage networks during flooding. With growing urbanization and the area of impervious surfaces in cities, flooding has become a major issue, causing enormous loss and impacting the health and security of inhabitants. Urban flood modeling is necessary for assessing the hazards and impacts of urban floods and developing effective flood management plans to reduce their impact. The hydrological model uses an integrated hydrologic and hydraulic modeling approach, demonstrated in Mangalore, an Indian coastal city prone to monsoon and cyclone-induced floods.

Despite numerous adaptation and mitigation strategies, the frequency of flooding incidents has grown recently, posing a challenge to local authorities. Improvements including improved drainage systems and flood forecasts, which have been put in place to deal with the effects of climate change. Nevertheless, sustainable urban water management is still challenging, given the shifting climate. As demonstrated in Mangalore, integrating climate data into urban planning and forecasts can significantly improve decision-making and lower the danger of flooding [6].

Urban flood modeling uses mathematical models to simulate water flow and predict the severity and extent of urban flooding. The models are based on several parameters, including topography, land use, soil type, and rain intensity, to simulate the complex hydrological processes involved in urban flooding. The models help evaluate the effectiveness of different flood management measures, including green infrastructure, structural and non-structural measures, and emergency response measures. For this research, the catchment area is delineated using high-resolution DEM and aerial photos, and the drainage network is delineated using a detailed field survey to simulate hydrological features well.

To enhance flood forecasting and mitigation, stormwater drains are simulated in PCSWMM to identify flooding nodes and drainage efficiency, thus allowing data-driven urban flood management. The output from this simulation process helps to redesign susceptible areas of the drainage system to increase their capacity and effectiveness with enhanced flood mitigation strategies. From the study [7], it was found that increasing the pipe diameter could decrease flooding in the study area by 81.62%, while modifying the size of the sewer holes reduced flooding by 44.78%. With these measures, the study aims to develop an integrated plan for urban flood management, maximizing stormwater management and improving overall urban resilience against extreme weather.

2 Study Area

The study area, spanning 11.85 km^2 from Pumpwell (12.885443°N, 74.877148°E) to Jeppinamogaru (12.841686°N, 74.858295°E) in the Mangalore catchment, features undulating terrain with urban and semi-urban landscapes. It has a tropical monsoon climate, receiving 3,695 mm of annual rainfall and temperatures ranging from 24 °C to 34 °C. Urban flooding is driven by proximity to the Arabian Sea, rapid urbanization, inadequate drainage, and land use changes. The geographic extent and location of the study area are illustrated in (Fig. 1).

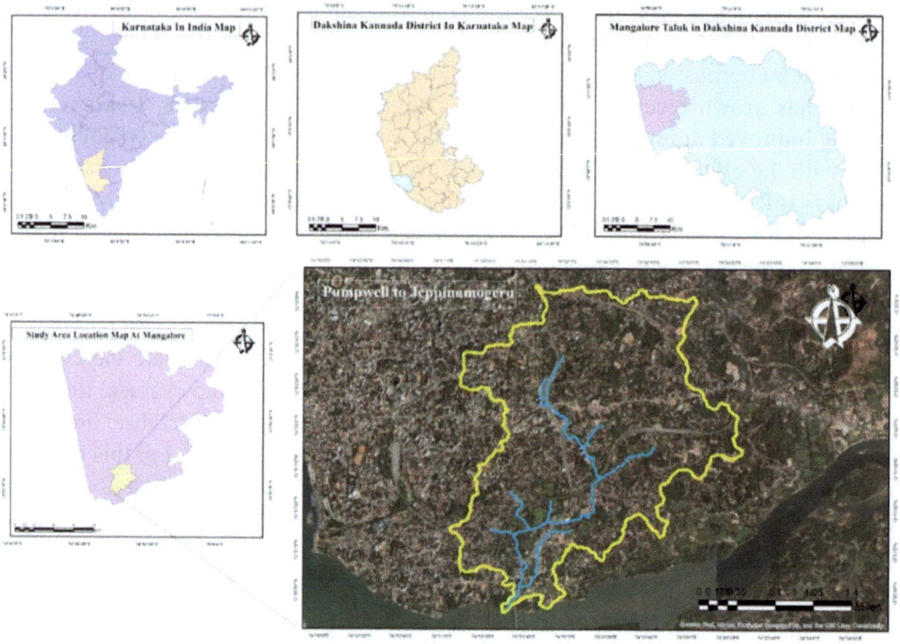

Fig. 1. Location Map of the Study Area (Pumpwell to Jeppinamogaru), Mangalore, Karnataka, India

3 Data Collection

Data from drainage networks, rainfall, soil, terrain, and land use and land cover are all required for the urban rainfall-runoff model. The Panambur rain gauge provides rainfall data, the 10-m resolution CARTOSAT Digital Elevation Model is obtained, Sentinel-2 LULC data is obtained, and the physical attributes of the drain-age network are depth, width, and drainage construction material.

4 PCSWMM Model Description

Hydrological modelling in PCSWMM involves simulating the rainfall-runoff process within the catchment area. The core components of this process include precipitation, infiltration, evaporation and runoff. The key formula used in hydrological modeling is the Continuity Equation, which can be expressed as:

$$P = E + I + R + \Delta S \tag{1}$$

where, P = Precipitation, I = Infiltration, E = Evaporation, R = Runoff, ΔS = Change in storage.

Hydraulic modelling in PCSWMM involves analyzing the flow of water through conduits, channels, and other drainage structures. The software uses the Saint-Venant Equations (also known as the shallow water equations) for one-dimensional flow modelling. These equations consist of the Continuity Equation and the Momentum Equation.

The continuity equation is a fundamental principle in fluid dynamics that expresses the conservation of mass.

$$\frac{\partial A}{\partial t} + \frac{\partial Q}{\partial x} = 0 \quad (2)$$

where, t = time (s), x = distance along the flow path (m), A = cross-sectional area flow (m^3) and Q = flow rate (m^3/s).

For unsteady flow condition, conservation of momentum is described by momentum equation. It accounts for the forces acting on the fluid, including pressure forces, gravity, and friction. The one-dimensional momentum equation can be expressed as:

$$\frac{\partial Q}{\partial t} + \frac{\partial}{\partial x}\left(\frac{Q^2}{A} + gAh\right) = gA(S_0 - S_f) \quad (3)$$

where, Q = Flow rate (m^3/s), A = Cross-sectional area flow (m^2), x = Distance along the flow path (m) t = time (s), g is acceleration due to gravity (m/s^2), h is Flow depth (m), S0 = Bed slope, Sf = Friction slope. This equation expresses the balance of momentum in a one-dimensional open-channel flow. The first term represents the local acceleration (time variation of flow), and the second term accounts for the convective acceleration and hydrostatic pressure effects. The right-hand side represents the driving forces due to gravity and the resisting forces due to friction [8].

5 Methodology

The study starts by demarcating the catchment area using a high-resolution Digital Elevation Model (DEM), which offers fine-scale topographic details. Sentinel-2 multispectral imagery generates a land use/land cover (LULC) map, categorizing the region into pervious and impervious surfaces. Field surveys delineate the drainage network and obtain fundamental physical information, including drain width, depth, bed conditions, lining materials, and blockage level at important junctions and culverts. All these data are vital for judging drainage performance and building a good hydrological model. A Differential Global Positioning System (DGPS) survey is carried out to acquire high-accuracy drain bed elevations with very high spatial precision in the model. Collected data are incorporated into an initial PCSWMM model, simulating the current drainage system and hydrological conditions.

The PCSWMM model integrates hydrological and hydraulic parameters, such as Manning's roughness coefficients, infiltration rate, soil condition and initial flow status. The observed rainfall data of extreme weather events is fed into the model-to-model runoff generation and possible flooding. Simulation is performed for a 5-year return period recommended by the Central Public Health and Environmental Engineering Organization (CPHEEO 2019).

If flooding occurs in any study area region, the drainage system will be modified to meet the requirements, and all subsequent procedures will be followed. The same procedure is repeated until flooding does not occur in the study area due to various rainfall events, and the drainage system is proven to carry rainwater effectively. The complete methodology adopted in this study is illustrated in (Fig. 2).

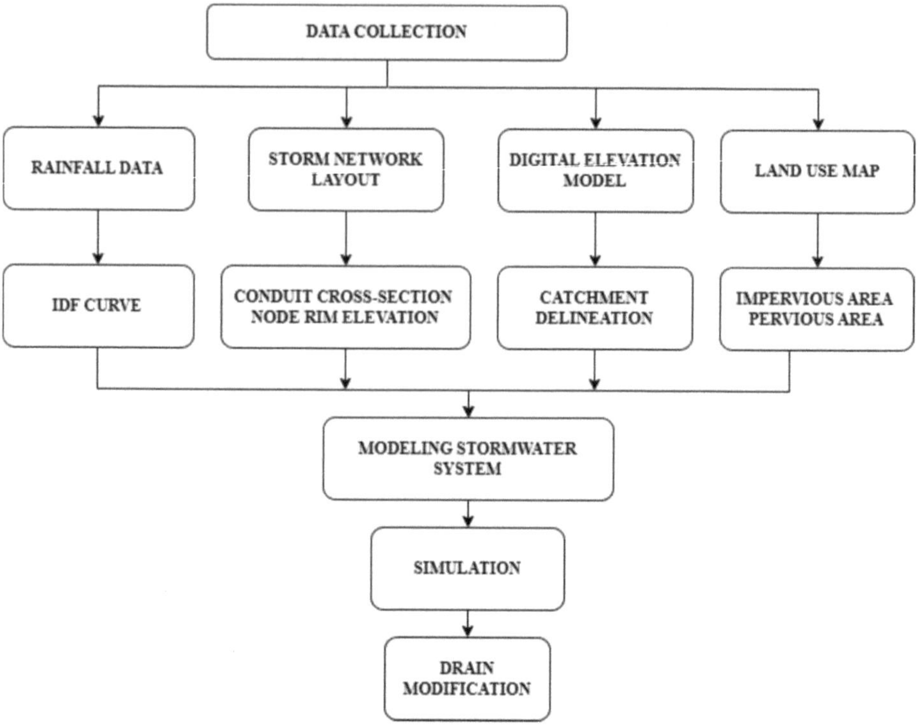

Fig. 2. Methodology of the Study

6 Results

Here in this section, we will be discussing briefly the outcomes which are obtained using this research. The main drain's route, as established by the reconnaissance study, is depicted in Fig. 3 below. A DGPS survey was also conducted for the same area to acquire the digital elevation model and the exact elevation and width of the drains.

Fig. 3. Major channels and the main drain

Field survey data were employed to compile a main drain catchment area. The central drain was digitized by ArcGIS software, which was also engaged to map the subcatchments (Fig. 4) spatially. The total catchment area is 11.85 km^2 and is also mapped into subcatchments of smaller units using 10-m resolution CARTOSAT DEM (Fig. 5). Sentinel 2 Land use and Land Cover data (Fig. 6) assisted in identifying the urban land cover of the catchment area. The proportion of previous and impervious surfaces in every subcatchment was calculated. The Karnataka State Remote Sensing Application Centre (KSRSAC) obtained soil data on the catchment area. Small adjacent drains are not considered because the study is of pilot type. However, the whole catchment of the drain also includes the small adjacent drains, which have been surveyed accordingly.

Fig. 4. Picture showing the main drain and the catchment area.

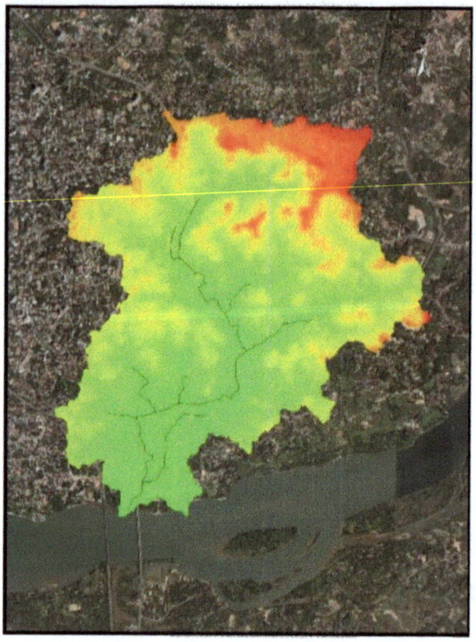

Fig. 5. Digital Elevation Model Map of the study area

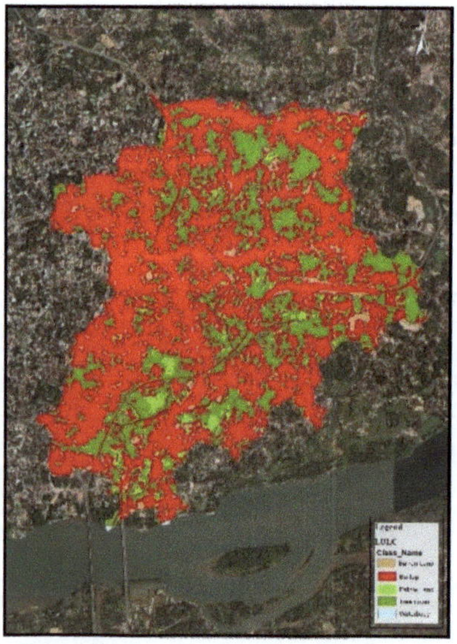

Fig. 6. Spatial Distribution of Land Use Land Cover of the study area

Every junction and outfall's rim (surface) and invert height were determined using a 10-m resolution Digital Elevation Model (DEM) acquired from CARTOSAT, with an elevation between 92.193 m and −1.807 m. The DEM contained key topographic information required to establish the hydraulic gradient throughout the study area accurately.

The elevation data was also tested by field surveys to provide a reliable input to be used for future hydrologic and hydraulic modeling.

The land use and land cover (LULC) map, derived from Sentinel-2 satellite imagery, is classified into built-up areas, tree cover, fallow land, water bodies and barren land. This classification helps to interpret the spatial pattern of different land types.

A referenced study [9] conducted frequency analysis on the yearly maximum daily rainfall from four rain gauge stations in Mangalore (Bajpe ACRO, Panambur, MAQ Railway, and MAQ KSNDMC) in order to estimate design storms. Using goodness-of-fit tests (Kolmogorov-Smirnov, Anderson-Darling, and Chi-squared), it compared the log-normal, log-Pearson, and GEV distributions and concluded that GEV suited the data the best. IDF curves with return durations ranging from two to fifty years were developed based on this. This study adheres to the 5-year return period that CPHEEO (2019) recommends for city catchments.

The 5-year return period chosen for this study aligns with the CPHEEO (2019) recommended design standard for urban catchments.

The Panambur rain gauge station's 5-year return period rainfalls, with the GEV distribution parameters shape = −0.009, scale = 6.678, and location = 24.324, have been used for the current study. In the same research, it has also been noticed that the existing and future IDF curves for the 5-year return period are entirely overlapping. Hence, rainfall characteristics are likely to remain the same up to the projected time until 2035.The corresponding IDF curve for the Panambur station is presented in (Fig. 7).

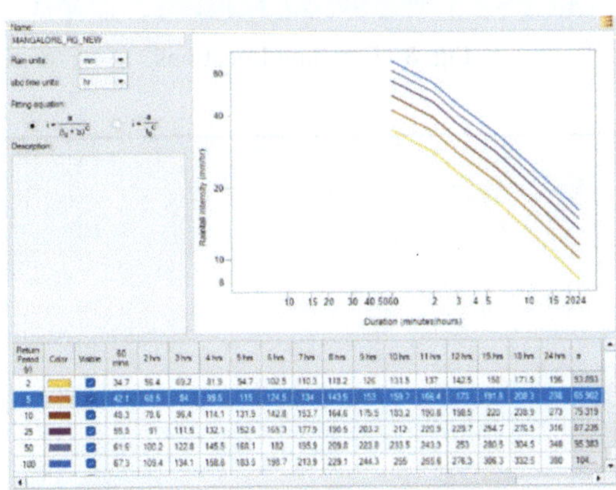

Fig. 7. Intensity Duration frequency Curve for Panambur Rain Gauge Station

6.1 Stormwater Model Simulation

A total of 19 subcatchments have been delineated with an average percentage of imperviousness about 54.822% and an average percentage of slope in the catchment of 15.69%. The total number of conduits in the catchment is 85, the length of conduits ranges from 10.989 m to 341.761 m, and the total number of nodes in the catchment is 85, with one outfall.

Based on the simulation result total 8 nodes were flooding that is J2 and J6 which will come near Maroli, node J5 and J15 will come near Ujjodi Pumpwell, node J16 comes near Nekkare Maar, J24 near Alape, J70 near Jeppinamogaru and node J73 near Jeppu Bappal. It is observed the total flood volume is about 493.266 ML and maximum hours of flooding is about 4.9 h. The identified flooding nodes are illustrated in (Fig. 8), while the vertical profile of the major drain is depicted in (Fig. 9).

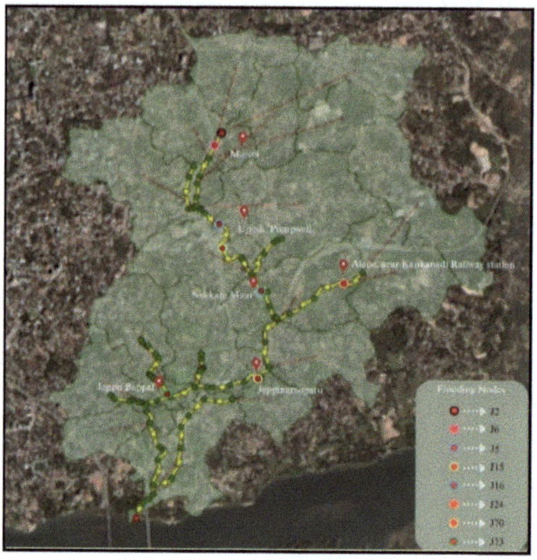

Fig. 8. Flooding Locations

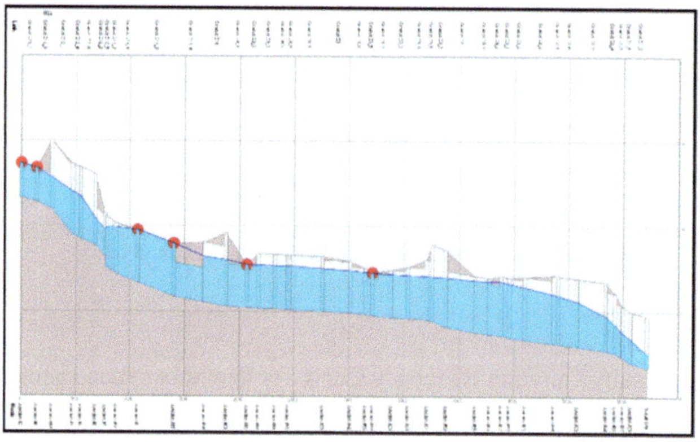

Fig. 9. Vertical Profile of the Major Drain

7 Discussions

Field inspections showed that floating material jams various points in the stormwater drain, diminishing the effective width of flow. In addition, silt, decomposed sewerage, and fine material have accumulated along the course of the main drain over the years, with deposition levels varying up to about 30 cm. Silt layer cannot be precisely measured since the drain base is composed of graded ground, Minor Irrigation, GoK, officials stated. Fixing it can be envisioned as maintenance work, with routine excavation of silt being an available soft option.

The drain has been considered for effective modeling because silt clearance has already been done to avoid overestimating the width needs, mainly because the area is within a highly urbanized environment. Additionally, parts of the retaining walls along the drain have collapsed or are in disrepair. Rehabilitation of these structures would assist in improving the functionality of the current drainage system.

Iterative simulations of flood-prone areas were used to calculate the appropriate width for a five-year return period of rainfall. The following observations were made in light of this simulation.

- Maroli to Pumpwell Circle drainage section of 680.76 m in length is currently 3 m wide and 2 m deep. To enhance its capacity, the changes proposed are to deepen it to 3.1 m and to widen it to 4.5 m.
- The drainage section from Pumpwell Circle to Indiana Hospital & Heart Institute, Ujjodi, 805.4 m long, is 6.5 m wide and 3 m deep. The proposed changes suggest widening to 7 m and deepening to 4 m to improve water conveyance.
- For the section of 565.504 m from Dhoomappa Residency to Nekkaremar Pademane Nagabana, the current width is 10 m, and the depth is 2.5 m. The width remains the same at 10 m but is being deepened to 4 m to augment the drainage capacity.
- The Kudlas Supermarket to the Indiana Convention Center portion of Jeppinamogaru drainage, which is 948.516 m in length, currently is 8.5 m in width and 3 m in depth. The upgrade will have it made 9 m in width and 4.5 m in depth.
- Lastly, the Grace Ice-cream Factory drainage line to Zorion Apartments at Jeppu Bappal has a length of 251.613 m, a width of 3 m, and a depth of 2.5 m. The proposed modifications will make its width 5.3 m and depth 3.7 m.

These proposed upgrades will help alleviate urban flooding as well as help ensure the efficient functioning of the drainage system.

8 Conclusions

This study on urban flood modeling in the subcatchment in Mangalore aims to simulate real-field conditions in the model and identify flood-prone locations along storm drains to formulate effective mitigation measures. A reconnaissance survey was conducted for drain location mapping, followed by a DGPS survey for position and elevation recording against Mean Sea Level (MSL). PCSWMM software delineated the subcatchment and analyzed land use and land cover (LULC). GIS software was also used to digitize satellite imagery and soil maps from the Karnataka State Remote Sensing Applications Centre (KSRSAC). Each junction and outfall's rim and inversion heights were assigned by the model using the DGPS data. Simulation of a 5-year return period storm drain in PCSWMM revealed that the existing drainage system cannot accommodate runoff. The analysis revealed eight critical flood-prone nodes: J2, J15, J16, J24, J5, J6, J70, and J73. The findings suggest the system's limitations in accommodating peak inflows, suggesting the need for targeted flood mitigation at vulnerable locations. The analysis revealed eight critical flood-prone nodes requiring capacity augmentation at essential sections of drainage to avoid flooding. Upgrades involve widening and deepening drains on the Maroli to Jeppu Bappal section to improve water conveyance and ensure adequate drainage. These upgrades are expected to significantly minimize urban flooding and enhance the resilience of the stormwater management system.

Acknowledgements. The authors like to convey their heartfelt gratitude to the Minor Irrigation Department personnel for their assistance. They also thank Prof. Gangadharan K V, the Centre for System Design (CSD) coordinator, an approved Center of Excellence at NITK Surathkal, for the facilities and research assistance. Additionally, they expressed gratitude for the encouragement and perceptive feedback they got during the study.

References

1. Shaikh, M.P., Yadav, S.M., Manekar, V.L.: Assessment of the empirical methods for the development of the synthetic unit hydrograph: a case study of a semi-arid river basin. Water Pract. Technol. **17**(1), 139–156 (2022)
2. Montgomery, M.: United nations population fund: State of world population 2007: Unleashing the potential of urban growth. Popul. Dev. Rev. **33**(3), 639–641 (2007)
3. Surwase, T., Manjusree, P.: Urban flood simulation – a case study of Hyderabad city. In: The National Conference on Flood Early Warning for Disaster Risk Reduction, vol. 30, (2019)
4. Parkinson, J., Tayler, K., Mark, O.: Planning and design of urban drainage systems in informal settlements in developing countries. Urban Water J. **4**(3), 137–149 (2007)
5. Sidek, L.M., Chua, L.H.C., Azizi, A.S.M., Basri, H., Jaafar, A.S., Moon, W.C.: Application of pcswmm for the 1-d and 1-d–2-d modeling of urban looding in damansara catchment, Malaysia. Appl. Sci. **11**(19), 9300 (2021)
6. Pai, A., Mahima, N., Pruthviraj, U., Abhijith, P.: Modelling and mitigation of urban floods: A case study in Mangalore. In: International Conference on Sustainable Infrastructure: Innovation, Opportunities and Challenges. Springer (2023)
7. Giron'as, J., Roesner, L.A., Rossman, L.A., Davis, J.: A new applications manual for the storm water management model (swmm). Environ. Model. Softw. **25**(6), 813–814 (2010)
8. Te Chow, V., Maidment, D., Mays, L.: Applied Hydrology McGraw-Hill series in water resources and environmental engineering. Tata McGraw-Hill Education (2010)
9. Femin, C.V., Varija, K.: Identification of the best-fit probability distribution and modeling short-duration intensity-duration-frequency curves mangalore. Trensa Civil Eng. Challen. Sustain. Select Proc. CTCS. **2019**, 765–783 (2021)

Preperation, Characterization and Adsorption Properties of Black Cotton Soil, Sand Particles and Sugarcane Bagasse for Environmental Applications

Ishwar Chandra[1(✉)], N. Ramesh[2], and Anima Upadhyay[1]

[1] Sir M. Visvesvaraya Institute of Technology, Affiliated to VTU, Belagavi, Krishnadeveraya Nagar, Hunasamaranahalli, Bangalore 562157, Karnataka, India
iswar_biotech@sirmvit.edu

[2] School of Applied Sciences, REVA University, Rukmini Knowledge Park, Kattigenahalli, Yelahanka, Bangalore 560 064, Karnataka, India

Abstract. This study examines the properties of Black Cotton Soil (BCS), Sand Particles (SP), and Sugarcane Bagasse (SB) through various analytical techniques. It involves sequential preparation steps like washing, drying, and grinding. Isotherm analysis reveals Type IV adsorption/desorption characteristics for all samples. BET analysis determines monolayer volumes (V_m) as 12.849, 1.3294, and 0.5768 cm^3 (STP)g^{-1} for BCS, SP, and SB, with energy constants (C) of 2535.1, 49.883, and 6.6705, respectively. Mean pore diameters (d_p) are 4.0273, 7.2504, and 6.0617 nm. The specific surface areas ($a_{s,BET}$) are 55.925, 5.7861, and 2.5105 m^2g^{-1}. BJH plots reveal pore volumes (V_p) of 0.038943, 0.010058, and 0.004079 cm^3g^{-1}, and pore specific surface areas (a_p) of 19.746, 5.2282, and 2.6624 m^2g^{-1}. Mesopore distribution peaks ($d_{p,peak}$) are consistently 2.43 nm. MP plot analysis indicates total specific surface areas (a_1) and external surface areas (a_2) of 66.103, 3.7087, 0.3787 m^3g^{-1}, and 11.498, 5.1618, 0.8742 m^3g^{-1}, respectively. Total V_p for BCS, SP, SB are 0.024967, 0.0006499, and 0.0011581 cm^3g^{-1}, with micropore distribution peaks ($d_{p,peak}$) at 0.60, 0.70, and 1.10 nm. FTIR analysis identifies hydroxyl, methyl, nitrile, carboxylic acids, aromatic, ether, and polysaccharides, indicating diverse adsorption applications. Energy dispersive X-ray diffraction identifies BCS as containing C, O, Al, Si, and Ca, SP as containing C, O, Al, Si, K, and Br, and SB as containing C and O with fiber structures. This analysis underscores the materials structural, chemical, and adsorption properties, highlighting their potential for wastewater treatment applications.

Keywords: Adsorption · Bagasse · EDX Analysis · FTIR · Sand · Soil

1 Introduction

Rapid industrial growth and urban expansion have contributed significantly to water pollution, primarily through the release of substantial amounts of contaminated wastewater into natural aquatic systems. Among various industrial effluents, biomethanated distillery spent wash (BDSW) is particularly problematic due to its high colour, chemical

H. P. Dasari et al. (Eds.): NITK-CREST 2025, PTT, pp. 251–261, 2026.
https://doi.org/10.1007/978-981-95-3218-6_18

oxygen demand (COD), and biological oxygen demand (BOD) levels [1, 2], which pose serious environmental hazards if left untreated [3]. To lessen the effects of BDSW on aquatic ecosystems and human health, efficient treatment techniques are required [4].

Adsorption is a widely recognized method for the removal of pollutants from wastewater due to its ease of use, affordability, and effectiveness in eliminating complex organic compounds [5–8]. Several cost-effective adsorbents have been investigated for their potential to remove organic pollutants and colour from industrial efflents, including sugarcane bagasse (SB), sand particles (SP), and black cotton soil (BCS) [9–14]. Because of their large porosity and surface area, these adsorbents are efficient at removing contaminants from aqueous solutions.

Because BCS contains montmorillonite clay particles, which offer adsorption active sites, it has been shown to have a high cation exchange capacity and substantial adsorption potential [15]. Sand particles, largely made of quartz, exhibit moderate adsorption potential, are largely used as a support material in filtration systems [16–18]. Sugarcane bagasse, a lignocellulosic natural material, contains various functional groups such as hydroxyl and carboxyl, which elevates its ability to remove pollutants via surface adsorption mechanisms [19–23].

A number of variables, including surface area, pore size distribution, and chemical composition, affect these material's adsorption efficiency [24–25]. According to studies, materials with a combination of micro- and mesopores show superior adsorption capacity, and mesoporous structures improve the adsorption of organic pollutants. Nitrogen adsorption-desorption isotherms, Barrett-Joyner-Halenda (BJH) pore size distribution studies, and Brunauer-Emmett-Teller (BET) surface area are frequently used to assess the textural characteristics of adsorbents [26–27].

This study aims to assess the adsorption efficiency of BCS, SP, and SB for bioremediation application. The research will involve characterization techniques such as BET analysis, Fourier-transform infrared spectroscopy (FTIR), scanning electron microscopy (SEM) coupled with energy-dispersive X-ray (EDX) evaluate the structural and chemical properties of the selected adsorbents. The outcomes of this research offer valuable understanding of how these materials could be effectively utilized in wastewater treatment applications [28].

2 Materials and Methods

2.1 Collection and Size Reduction of Adsorbents

All of BCS, SP, and SB's absorbent raw materials were gathered from the college grounds. Before being used, the adsorbents were manually washed and well dried. Before being used, all adsorbents were ground to a very fine size. A shaker sieve equipement (Universal Engineering, Bangalore) was used to calculate the particle size distribution. The adsorbent powder was passed through a stack of sieves with varying mesh sizes and mechanically shaken for 15 min. The agitation separated particles based on size, which were then collected and weighed using a precision balance. The sieve stack was arranged from smallest (bottom) to largest (top) with a collection pan underneath to ensure accurate particle distribution analysis.

2.2 FTIR Analysis of BDSW and Adsorbents

Molecular absorption and transmission in BDSW and adsorbents were recorded using FTIR (Bruker-alpha). Spectra were recorded in the range of 500–4000 cm^{-1} using a diamond Attenuated Total Reflectance (ATR) window. The OPUS software (Bruker) was used for data analysis. The study was conducted at the Nanotechnology Department, Visvesvaraya Technological University (VTU), Bengaluru.

2.3 SEM-EDX Analysis

The surface morphology of adsorbents was examined using SEM (JEOL JSM-7100F, Singapore). The setup included a hot electron gun for imaging and a 129 eV resolution silicon drift detector for EDX analysis to determine elemental composition. The study was performed at the Nanotechnology Department, VTU, Bengaluru.

2.4 Surface Area and Pore Size Distribution Analysis

Porosity of the adsorbents was evaluated through nitrogen adsorption measurements using the BELSORP-MINI II instrument at 77 K, with experiments carried out at PPISR, Bengaluru. Prior to analysis, samples underwent vacuum pre-treatment at 393 K for six hours. Specific surface area was calculated using the BET method, while pore size distribution was assessed via the BJH technique. The resulting adsorption isotherms were utilized to examine adsorption characteristics and performance of the adsorbents.

3 Results and Discussion

3.1 Nitrogen Adsorption/Desorption Isotherms

The nitrogen adsorption/desorption isotherms for BCS, SP, and SB are presented in Fig. 1a, b, and c, respectively. They display how the volume of adsorbed nitrogen (V_a) changes with relative pressure (p/p_o). The adsorption capacity of BCS is significantly higher compared to SP and SB, as indicated by the larger volume of adsorbed nitrogen. This suggests that BCS has a higher surface area or more available adsorption site. SP exhibits moderate adsorption capacity, while SB shows the lowest adsorption capacity among the three samples.

Hysteresis loops observed in the adsorption/desorption isotherms of BCS, SP, and SB suggest that these materials possess mesoporous architectures. The isotherm for BCS (Fig. 1a) corresponds to a type H2 hysteresis loop, suggesting the presence of ink-bottle pores with pore-blocking effects. Sand particles (Fig. 1b) and sugarcane bagasse (Fig. 1c) show H3 hysteresis loops, which are typically associated with slit-shaped pores, indicating different pore structures and adsorption mechanisms compared to BCS.

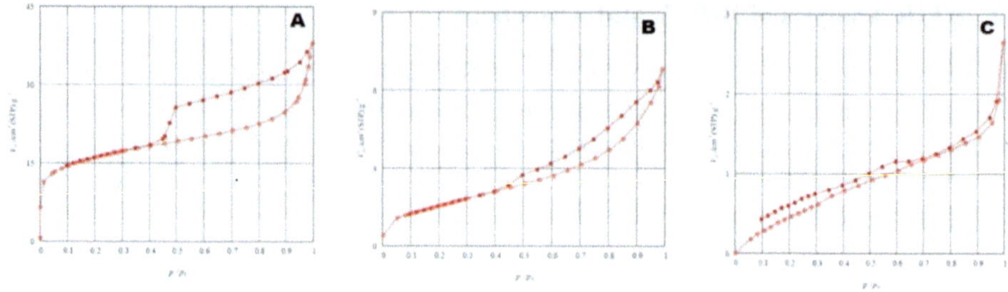

Fig. 1. Adsorption/desorption isotherm of a) BCS, b) SP, c) SB.

3.2 BET Analysis

The BET analysis indicates notable variations in the specific surface areas among the materials. Fig. 2a, b, and c display the BET analytical outcomes for BCS, SP, and SB, respectively. The parameters derived from the BET plots are summarized in Table 1. BCS shows the highest BET-specific surface area ($a_{s,BET}$) of 55.925 m^2/g, implying more surface is accessible for adsorption processes. The C value, which represents the energy constant, is also highest for BCS (2535.1), suggesting stronger interactions between the adsorbent and the adsorbate. SP and SB have lower specific surface areas and energy constants, which correlates with their lower adsorption capacities. BCS shows the largest surface area along with the highest monolayer volume (V_m). This improvised surface area is primarily due to its fine particle size and the presence of montmorillonite clay, which contributes additional active sites for adsorption. Conversely, SP primarily made of quartz, exhibits a smaller surface area, attributed to its larger particle size and minimal porosity. And SB pore structure is likely influenced by its organic nature.

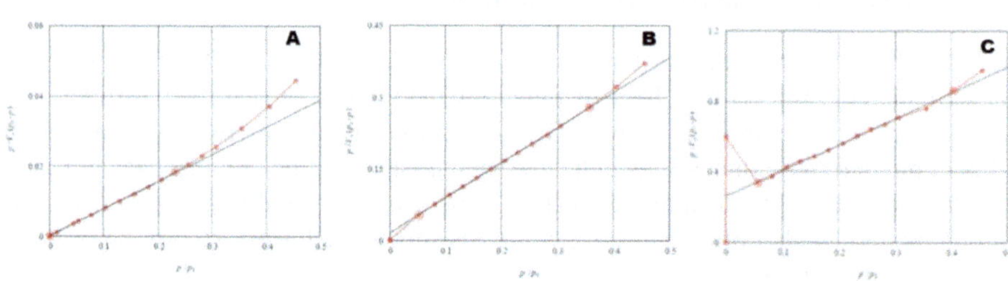

Fig. 2. BET plot of a) BCS, b) SP, c) SB.

Table 1. BET analysis results for BCS, SP, and SB.

Sample	V_m (cm^3(STP)/g)	$a_{s,BET}$ (m^2/g)	C	Total Pore Volume (cm^3/g)	Mean Pore Diameter (nm)
BCS	12.849	55.925	2535.1	0.056307	4.0273
SP	1.3294	5.7861	49.883	0.010488	7.2504
SB	0.5768	2.5105	6.6705	0.0038046	6.0617

Table 2. BJH pore size distribution for BCS, SP, and SB.

Sample	V_p (cm^3/g)	$d_{p,peak}$ (Area) (nm)	a_p (m^2/g)
BCS	0.038943	2.43	19.746
SP	0.010058	2.43	5.2282
SB	0.0040794	2.43	2.6624

Table 3. MP analysis of BCS, SP, and SB.

Sample	a_1 (m^2/g)	a_2 (m^2/g)	V_p (cm^3/g)	$d_{p,peak}$ (nm)
BCS	66.103	11.498	0.024967	0.60
SP	3.7087	5.1618	0.00064992	0.70
SB	0.3787	0.8742	0.0011581	1.10

3.3 BJH Pore Size Distribution

The BJH pore size distribution profiles for BCS, SP, and SB are shown in Fig. 3a, b, and c, respectively. Table 2 provides information on the relevant analytical data. BCS has a significant mesopore distribution peak centered at 2.43 nm and the largest total pore volume (V_p) at 0.038943 cm^3/g. This implies that BCS has a higher mesoporous content, which improves its adsorption effectiveness. In comparison to BCS, SP and SB have lower pore volumes and specific surface areas, which further supports their inferior adsorption capabilities. Additionally, the BJH analysis shows that BCS has a heterogeneous porous architecture with a wide and varied pore size distribution. Given that SP is primarily granular, its distribution is comparatively homogeneous and restricted. SB shows a multimodal pore size distribution, suggesting a highly complex pore network due to its fibrous structure.

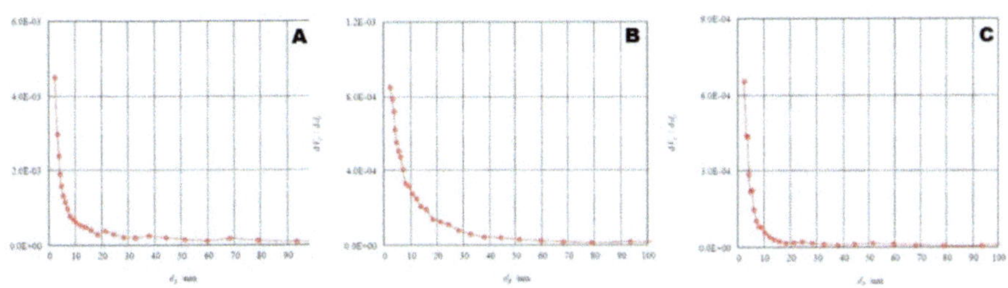

Fig. 3. BJH-Plot of a) BCS, b) SP, c) SB.

3.4 MP Distribution

The microporous character of the samples is revealed by the MP-plot analysis. The MP pore size distribution for BCS, SP, and SB is shown in Fig. 4a, b, and c, respectively.

Table 3 provides a summary of the associated parameters. With a pronounced micropore peak at 0.60 nm and the largest micropore-specific surface area (a_1) of 66.103 m^2/g among the materials, BCS shows that micropores play a major role in its overall adsorption performance. This implies that micropores are essential for increasing BCS's total adsorption capacity. Among the three samples, BCS is shown to be the most effective material for nitrogen adsorption; this is further supported by the fact that SP and SB show much lower micropore specific surface areas and micropore volumes.

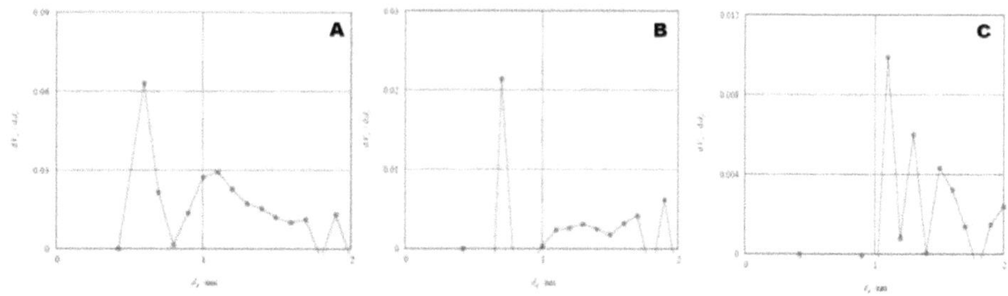

Fig. 4. MP-Plot of a) BCS, b) SP, c) SB.

3.5 FTIR Analysis of BCS, SP, and SB

The BCS, SP, and SB FTIR spectra's offer important information about the functional groups and chemical makeup of each sample. According to these findings, BCS (Fig. 5) has silicate minerals, water molecules, organic content, and hydroxyl groups, all of which support its adsorption capabilities. FTIR spectrum of SP identifies water molecules, organic matter, carbonates, and silicate minerals, all of which are common constituents of sand (Fig. 6). The FTIR spectral data (Fig. 7) in SB reveals the presence of cellulose, hemicellulose, and lignin structurally, as well as hydroxyl functionalities and organic content.

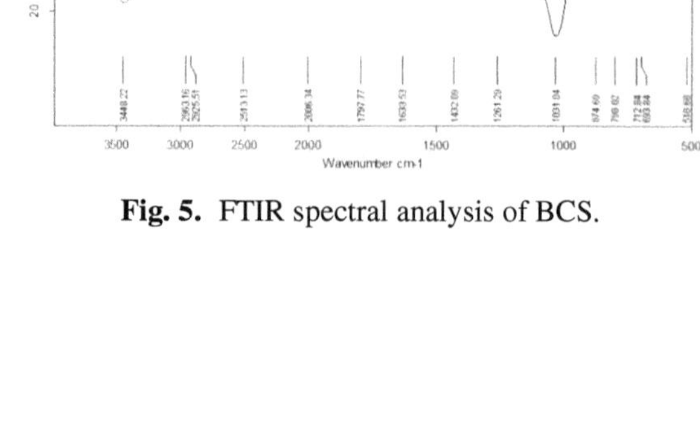

Fig. 5. FTIR spectral analysis of BCS.

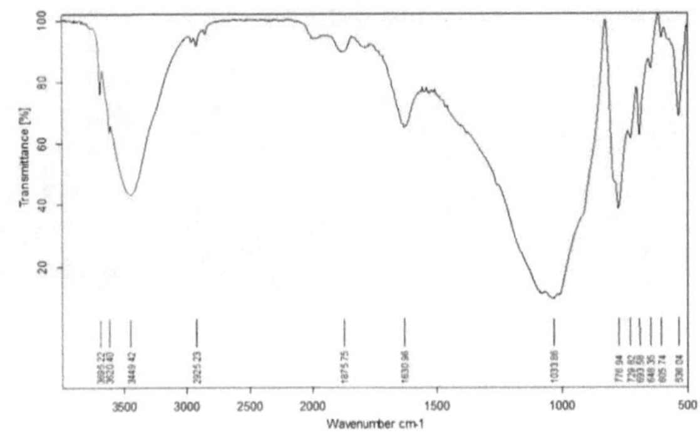

Fig. 6. FTIR spectral analysis of SP.

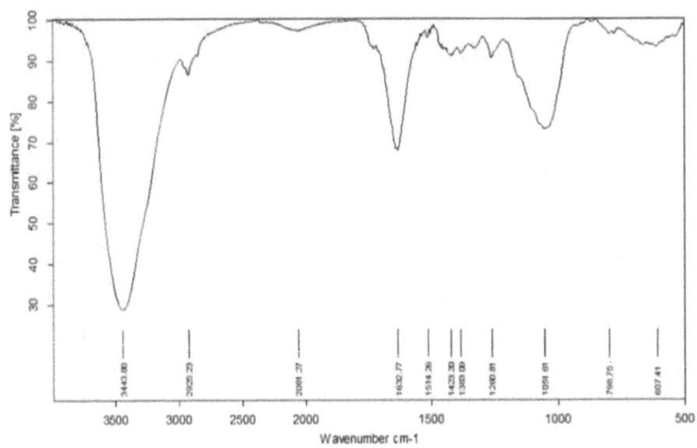

Fig. 7. FTIR spectral analysis of SB.

3.6 SEM Analysis

The SEM coupled with EDX analysis offers detailed information regarding the surface morphology and elemental profiles of BCS, SP, and SB. The SEM images of BCS (Fig. 8a) reveal a heterogeneous surface morphology with irregularly shaped particles. The surface exhibits a high degree of roughness and porosity, which is indicative of potential adsorption sites. Presence of aggregated clusters suggests interparticle interactions and a tendency to retain moisture. The presence of these irregular particles and porous structure contributes to the high adsorption capacity of BCS, as observed in the nitrogen adsorption studies. The SEM images of SP (Fig. 8b) show relatively rounded smooth particles, granular, angular, and well-defined particles, characteristic of quartz-rich sand. The surface morphology is less rough compared to BCS, indicating lower porosity and potentially lower adsorption capacity. The smoother surface and lack of significant porosity suggest that SP may not be as effective as BCS for adsorption applications. The SEM images of SB (Fig. 8c) display a fibrous structure with visible fiber cells, indicative of lignocellulosic material. The presence of hollow channels and large pore spaces suggests good permeability and adsorption potential. The observed network-like arrangement of fibers contributes to its potential as a bio-adsorbent.

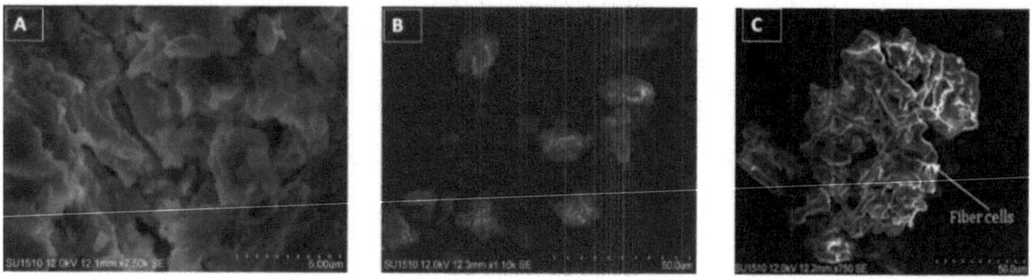

Fig. 8. SEM images of a) BCS, b) SP, c) SB.

3.7 EDX Analysis

BCS shows major elements, Carbon (C), Oxygen (O), Aluminum (Al), Silicon (Si), and Calcium (Ca), detail composition at selected points is shown in Table 4. The presence of Al and Si confirms the dominance of aluminosilicate minerals. The high oxygen content suggests significant moisture retention and the presence of hydroxyl functional groups. The presence of carbon and oxygen indicates organic matter, while silicon, aluminum, and calcium suggest the presence of clay minerals and other inorganic components. This composition is consistent with the adsorption properties of BCS, as the organic matter and clay minerals provide active sites for adsorption. SP shows major elements, Silicon (Si), Oxygen (O), Aluminum (Al), Potassium (K), and traces of Bromine (Br), detail composition at selected points is shown in Table 5. The dominance of Si and O confirms the presence of quartz (SiO_2), which is consistent with the smooth and less porous surface observed in the SEM images. Small amounts of K and Al suggest the possible presence of feldspar minerals.

Table 4. EDX analysis of BCS composition.

Point	C (%)	O (%)	Al (%)	Si (%)	Ca (%)
1	57.05	42.95	–	–	–
2	47.71	15.10	–	37.18	–
3	1.92	29.62	6.55	14.83	47.08

Table 5. EDX analysis of SP composition.

Point	C (%)	O (%)	Al (%)	Si (%)	K (%)	Br (%)
1	43.12	11.49	–	31.78	13.62	–
2	7.46	50.10	11.11	31.33	–	–
3	5.06	42.30	–	31.15	–	21.49

Carbon (C) and Oxygen (O) are the major elements in SB. The high carbon content (40–45%) confirms its organic nature, primarily composed of cellulose, hemicellulose, and lignin, detail composition at selected points is shown in Table 6. The presence of oxygen (~55%) suggests hydroxyl functional groups. This composition supports the fibrous and porous structure observed in the SEM images, contributing to the adsorption potential of SB.

Table 6. EDX analysis of SB composition.

Point	C (%)	O (%)
1	44.51	55.49
2	42.42	57.58

4 Conclusion

The findings demonstrated that BCS exhibited the highest V_m (cm^3(STP)/g), $a_{s,BET}$ (m^2/g), and adsorption capacity due to its montmorillonite clay minerals, heterogeneous morphology, and the presence of active functional groups. SP, while exhibiting lower adsorption potential, offered unique characteristics such as mesoporosity and quartz-dominant composition, making it a viable support material in filtration systems. SB, characterized by its fibrous structure and lignocellulosic composition, presented moderate adsorption capabilities, indicating its suitability as a bio-adsorbent for specific applications. The comparative analysis underscores the potential of BCS as a highly effective material for wastewater treatment and environmental remediation, while SP and SB can be optimized further to enhance their adsorption performance. These findings highlight the significance of utilizing natural and agricultural materials as low-cost, sustainable adsorbents for addressing environmental pollution challenges. Future studies could explore the functionalization and modification of these materials to improve their adsorption performance for specific pollutants, as well as their regeneration and reusability in practical applications. The integration of these materials into large-scale wastewater treatment systems holds promise for developing cost-effective and eco-friendly solutions for environmental sustainability.

References

1. Singh, Y.: Distillery spent wash and its utilisation in agriculture.
2. Shukla, S.K., Mishra, P.K.: Bioremediation and Decolourisation of Biomethanated Distillery Spent Wash. In: Algae and Environmental Sustainability, pp. 107–117. Springer India (2015)
3. Chowdhary, P., Yadav, A., Kaithwas, G., Bharagava, R.N.: Distillery wastewater: a major source of environmental pollution and its biological treatment for environmental safety. In: Green Technologies and Environmental Sustainability, pp. 409–435. Springer International Publishing (2017)
4. Kumar, V., Sharma, D.C.: Distillery effluent: pollution profile, eco-friendly treatment strategies, challenges and future prospects. In: Microbial Metabolism of Xenobiotic Compounds, pp. 337–357 (2019)

5. Kang, K., Li, R., Wang, S., Huang, Z., Li, J.: Adsorption sites and interactions of pigments in molasses-based distillery effluent on starch-based composites: ternary competitive adsorption and theoretical calculations. J. Hazard. Mater. **480**, 136137 (2024)
6. Tyagi, U.: Sustainable and low-cost biomass derived adsorbents for the removal of toxic contaminants from wastewater: approaches and future perspective. Waste Manag. Bull. **2**, 308–325 (2024)
7. Pyrzynska, K.: Removal of cadmium from wastewaters with low-cost adsorbents. J. Environ. Chem. Eng. **7**(1), 102795 (2019)
8. Ahmaruzzaman, M.: Industrial wastes as low-cost potential adsorbents for the treatment of wastewater laden with heavy metals. Adv. Colloid Interf. Sci. **166**(1–2), 36–59 (2011)
9. Arya, P., Patel, S.B., Bharti, G., Shukla, B.K., Hurukadli, P.: Impact of using a blend of bagasse ash and polyester fiber in black cotton soil for improvement of mechanical and geotechnical properties of soil. Mater Today Proc. **78**, 738–743 (2023)
10. Goli, J., Sahu, O.: Effect of distillery industry effluent on fertility of soil and crops. Int. J. Surg. Case Rep. **2**(3), 039–045 (2014)
11. Kumar, B., Bhatt, A., Rawat, A., Dimri, A.: Treatment of distillery spent wash for irrigation purpose by using soil as adsorbent. Int. J. Novel Res. Life Sci. **3**(6), 78–82 (2016)
12. Adikane, H.V., Dange, M.N., Selvakumari, K.: Optimization of anaerobically digested distillery molasses spent wash decolorization using soil as inoculum in the absence of additional carbon and nitrogen source. Bioresour. Technol. **97**(16), 2131–2135 (2006)
13. Hamoda, M.F., Al-Ghusain, I., Al-Mutairi, N.Z.: Sand filtration of wastewater for tertiary treatment and water reuse. Desalination. **164**(3), 203–211 (2004)
14. Fito, J., Tefera, N., Van Hulle, S.W.: Adsorption of distillery spent wash on activated bagasse fly ash: kinetics and thermodynamics. J. Environ. Chem. Eng. **5**(6), 5381–5388 (2017)
15. Miao, S., Shi, J., Sun, Y., Zhang, P., Shen, Z., Nian, H., et al.: Mineral abundances quantification to reveal the swelling property of the black cotton soil in Kenya. Appl. Clay Sci. **161**, 524–532 (2018)
16. Chabi, K., Li, J., Ye, C., Kiki, C., Xiao, X., Li, X., et al.: Rapid sand filtration for< 10 μm-sized microplastic removal in tap water treatment: Efficiency and adsorption mechanisms. Sci. Total Environ. **912**, 169074 (2024)
17. Wang, T., Cao, W., Wang, Y., Qu, C., Xu, Y., Li, H.: Surface modification of quartz sand: a review of its progress and its effect on heavy metal adsorption. Ecotoxicol. Environ. Saf. **262**, 115179 (2023)
18. McDowell-Boyer, L.M., Hunt, J.R., Sitar, N.: Particle transport through porous media. Water Resour. Res. **22**(13), 1901–1921 (1986)
19. Sarker, T.C., Azam, S.M.G.G., El-Gawad, A.M.A., Gaglione, S.A., Bonanomi, G.: Sugarcane bagasse: a potential low-cost biosorbent for the removal of hazardous materials. Clean Techn. Environ. Policy. **19**, 2343–2362 (2017)
20. Halysh, V., Sevastyanova, O., Pikus, S., Dobele, G., Pasalskiy, B., Gun'ko, V.M., Kartel, M.: Sugarcane bagasse and straw as low-cost lignocellulosic sorbents for the removal of dyes and metal ions from water. Cellulose. **27**, 8181–8197 (2020)
21. Patel, H.: Characterization and adsorptive treatment of distillery spent wash using bagasse fly ash. Arab. J. Sci. Eng. **47**(5), 5521–5531 (2022)
22. Hiremath, S.G., Joshi, S.G.: Roadmap to distillery spent wash treatment and use of soft computing techniques. Evol. Intel. **15**(2), 1279–1293 (2022)
23. Mane, J.D., Modi, S., Nagawade, S., Phadnis, S.P., Bhandari, V.M.: Treatment of spentwash using chemically modified bagasse and colour removal studies. Bioresour. Technol. **97**(14), 1752–1755 (2006)
24. Kumar, P.S., Korving, L., Keesman, K.J., van Loosdrecht, M.C., Witkamp, G.J.: Effect of pore size distribution and particle size of porous metal oxides on phosphate adsorption capacity and kinetics. Chem. Eng. J. **358**, 160–169 (2019)

25. Kuila, U., & Prasad, M.: Specific surface area and pore-size distribution in clays and shales. Geophysical prospecting, 61(2-Rock Physics for Reservoir Exploration, Characterisation and Monitoring), 341–362 (2013).
26. Wang, S., Zhang, L., Long, C., Li, A.: Enhanced adsorption and desorption of VOCs vapor on novel micro-mesoporous polymeric adsorbents. J. Colloid Interface Sci. **428**, 185–190 (2014)
27. El Naeem, G.A., Abd-Elhamid, A.I., Farahat, O.O., El-Bardan, A.A., Soliman, H.M., Nayl, A.A.: Adsorption of crystal violet and methylene blue dyes using a cellulose-based adsorbent from sugercane bagasse: characterization, kinetic and isotherm studies. J. Mater. Res. Technol. **19**, 3241–3254 (2022)
28. Rodríguez-Díaz, J.M., García, J.O.P., Sánchez, L.R.B., da Silva, M.G.C., da Silva, V.L., Arteaga-Pérez, L.E.: Comprehensive characterization of sugarcane bagasse ash for its use as an adsorbent. Bioenergy Res. **8**, 1885–1895 (2015)

Assessing the Feasibility of Rooftop Rainwater Harvesting: A Sustainable Solution for Water Conservation and Urban Resilience in a University Campus

Sanjith S. Anchan(✉), Indumathi Shivamallappa, M. Jayanth, B. Kiranpreethi, and C. P. Nesara

The National Institute of Engineering, Mysuru, Karnataka 570008, India
sanjithsj@gmail.com

Abstract. Water scarcity is becoming a serious problem worldwide, due to climate change, increased population, and overuse of groundwater. This paper proposes Rainwater Harvesting (RWH) system for a university campus in Mysuru, Karnataka, India, to reduce dependency on external and groundwater sources. The system collects, stores, and monitors rainwater to use for gardening and sanitation. Mysuru's decent rainfall and climate make it a great spot for RWH system, with the university campus having the total rooftop area of 32.124 acers, it can collect 335.26 million litres annually, which is sufficient to meet the 215.5 million litres of non-potable water needs. The study uses GIS-based rooftop mapping and demand analysis, to evaluate the RWH feasibility across eight campus zones. The analysis show that Zones 1, 2, and 6 are ready for RWH, while minor changes in other zones can improve system efficiency.

This plan offers campus planners a smart way to address water scarcity and sets a model for sustainable water use that other campuses could follow.

Keywords: Feasibility · Rainwater Harvesting (RWH) for University Campus · Sustainable Water Management · University Campus Infrastructure · Water Conservation Strategies

1 Introduction

Water scarcity has become a serious global issue, due to unpredictable climatic variations and over-dependency on groundwater resources. With increasing population and changing climate regime, water supply systems in many cities of the world are under stress. To tackle this problem, water authorities are adopting several measures including demand management and identifying alter-native water sources such as storm water harvesting, greywater and wastewater reuse and desalination. Among all the alternative water sources, storm water harvesting perhaps has received the most attention (Imteaz et al., 2011).Urban centres like Mysuru, face growing challenges in meeting water demands due to rapid urbanization and unpredictable rainfall patterns. As hubs of academic and

H. P. Dasari et al. (Eds.): NITK-CREST 2025, PTT, pp. 262–288, 2026.
https://doi.org/10.1007/978-981-95-3218-6_19

residential activity, universities present a unique opportunity to implement sustainable water management solutions that reduce costs, minimize environmental impact, and ensure consistent water availability. Much of the water used in the buildings can be replaced by rainwater. This water is a natural resource that can be easily collected and consumed in various domestic, commercial and industrial applications especially where the drinking water is not required (Anchan & Shiva Prasad, 2021). However, the use of rainwater as a resource depends on the local precipitation regime, especially on the precipitated volumes and their temporal variability. Equally important are the storage capacity and the availability of surfaces for effectively collecting rainwater, such as roofs or terraces (Ipek Sahin & Manioglu, 2019). Rain water harvesting is a technique through which the rainwater is captured, stored and reused for various purposes (Stec and Kordana, 2015). Various studies has suggested to use harvested storm water for domestic and land-scape purpose (Abdulla and Al-Shareef, 2009). The investigation conducted by (Sturm et al., 2009), presented the economic and technical feasibility of rain water harvesting system (Sturm et al., 2009). Marlow et al. (2013), reviewed a critical reassessment of sustainable urban water resource management by considering its advantages and limitations. Among these solutions, RWH offers a reliable, eco-friendly method to address water needs while promoting sustainability (Fig. 1).

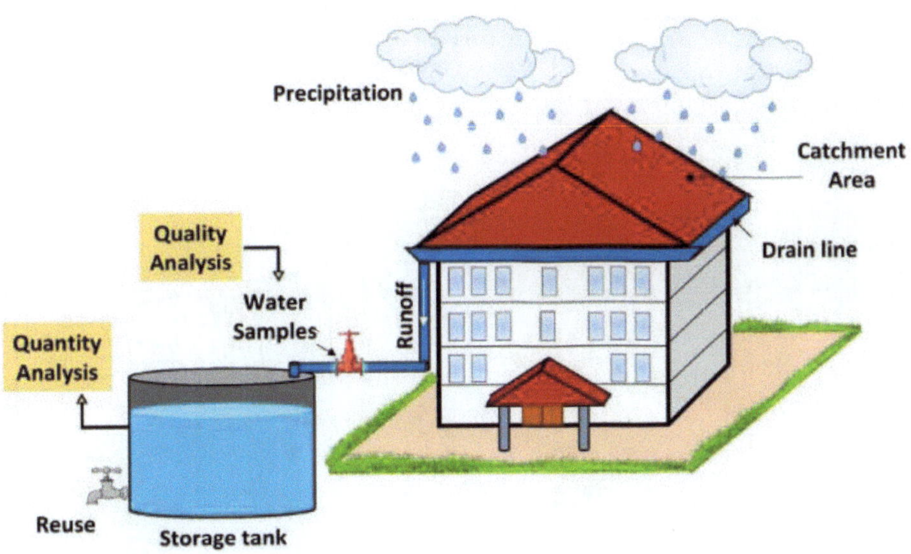

Fig. 1. Schematic representation of the rainwater harvesting process (Anchan & Shiva Prasad, 2021)

2 Methods and Methodologies

2.1 Study Area

The University of Mysore, located in Mysuru, Karnataka, spans a vast, well-planned campus with significant potential for rainwater harvesting due to its geographical and infrastructural features. Situated at a latitude of 12.30° N and longitude 76.64° E, the campus experiences a tropical savanna climate with an average annual rainfall of 1339 mm, primarily during the southwest and northeast monsoons.

The campus being approximately 739 acres (Fig. 2), includes large academic buildings, residential quarters, a lake, and administrative offices, with diverse rooftop types (sloped metal/polycarbonate and flat concrete roofs) serving as ideal catchments for rainwater collection. Open spaces such as playgrounds, roads, and parking lots contribute to surface runoff, which can be harvested through infiltration wells and storage tanks. With a population of over 10,000 students, faculty, and staff, the daily water demand exceeds 1.35 million litres, putting immense pressure on groundwater and external supply sources. Implementing rainwater harvesting systems across the campus will address water scarcity, reduce dependence on groundwater, and promote sustainability.

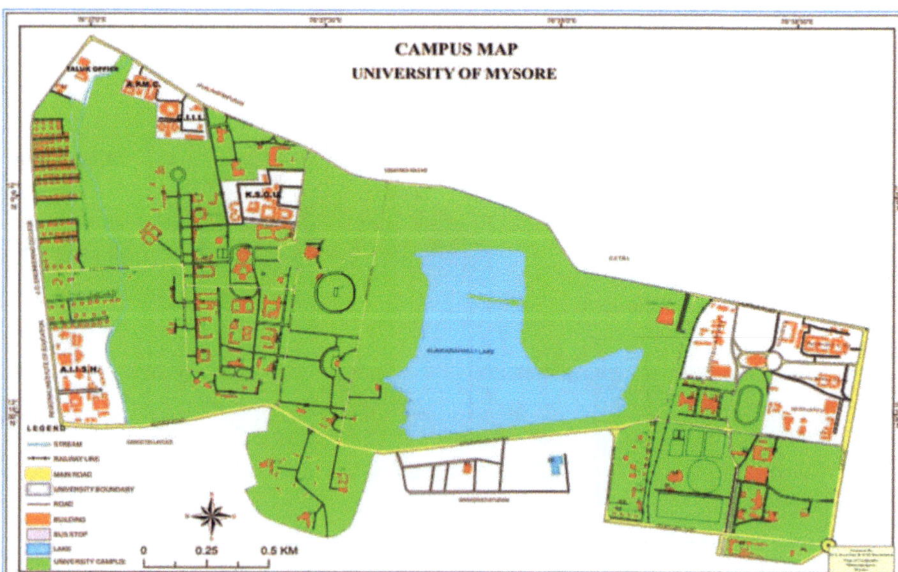

Fig. 2. Layout of University of Mysuru.

The detailed study of rooftop catchment areas, runoff efficiency, and water quality will help optimize the design of storage systems to fulfil campus water demands effectively throughout the year.

3 Methodologies

Figure 3 represents the flowchart of the methodology, where rainfall data is collected, rooftop areas are calculated, the quality of harvested rainwater is assessed, and population counts are utilized to analyze per capita demand, thereby understanding the overall utilization of rainwater.

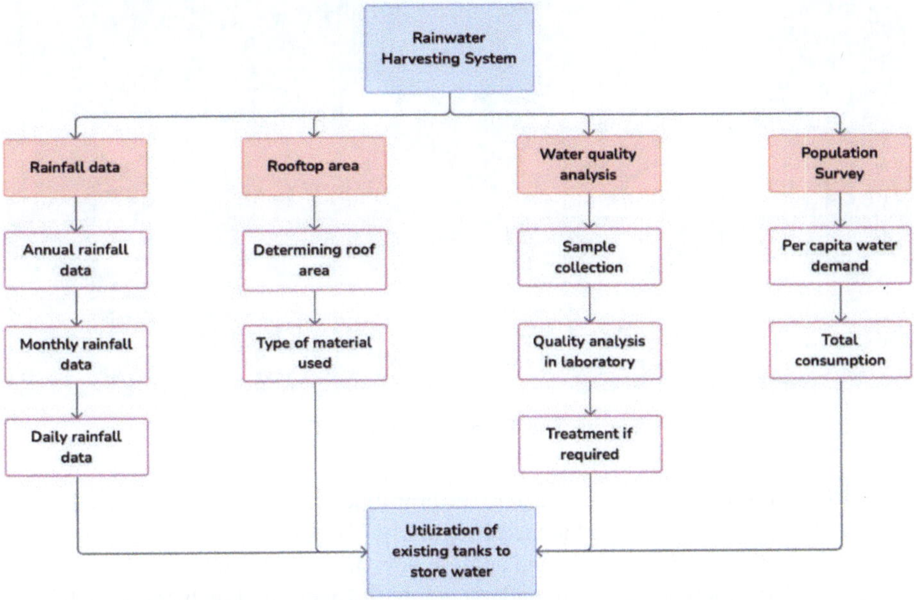

Fig. 3. Flowchart representing the overview of methodologies adopted.

3.1 Zone Classification

The University of Mysore campus was systematically divided into 9 zones (Fig. 4) to facilitate a focused and detailed analysis of its rainwater harvesting potential. The campus was divided into different zones based on the types of buildings, their structure, and how they are used. This helped in clearly identifying which areas could collect rainwater. By organizing the campus this way, it became easier to measure the rooftop areas that contribute to rainwater and to understand how much water is needed in each zone. This step-by-step approach made sure that all possible rainwater collection spots were properly studied.

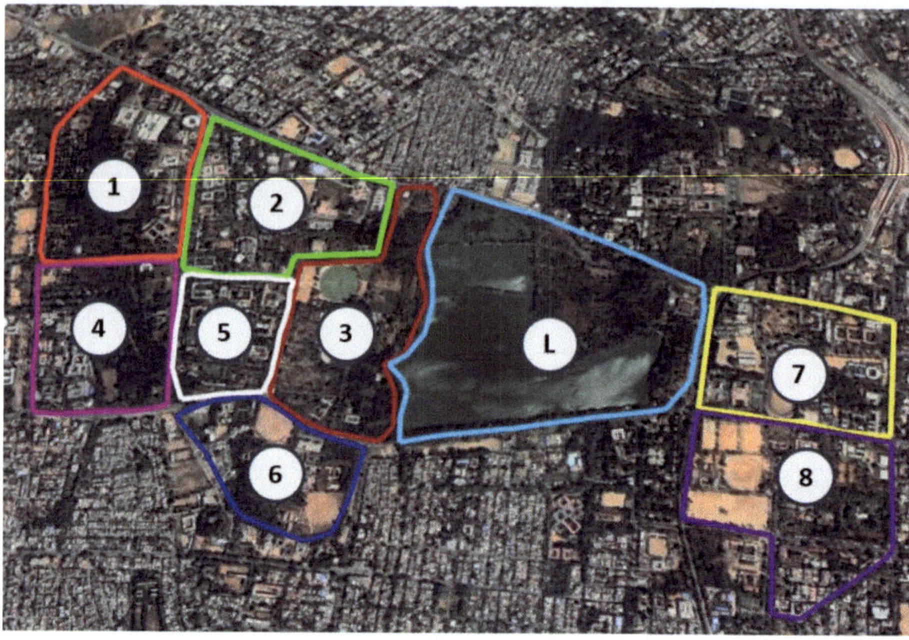

Fig. 4. Zonal Classification of UOM

Residential rooftops were examined to see how much rainwater they could collect, as they have large surface areas and are used regularly. The study divided the campus into different zones to properly identify and analyse how much rainwater could be harvested in each area. This method helped in understanding the needs and features of each zone. Based on this, an effective rainwater management system was planned to suit the different buildings and landscapes across the University of Mysore.

3.2 Rainfall Data Collection

The rainfall data collection was collected to estimate the campus rainwater harvesting potential based on historical precipitation records. Daily and monthly rainfall data for Mysuru city, spanning the last 10–15 years, was obtained from the CHRS. Below is the graphical representation of yearly rainfall data from the year 2010–23, where the highest rainfall found was 2950 mm in 2011 and the least rainfall was 735 mm in 2016 (Fig. 5), and its mean rainfall data is found to be 1339 mm (Fig. 6).

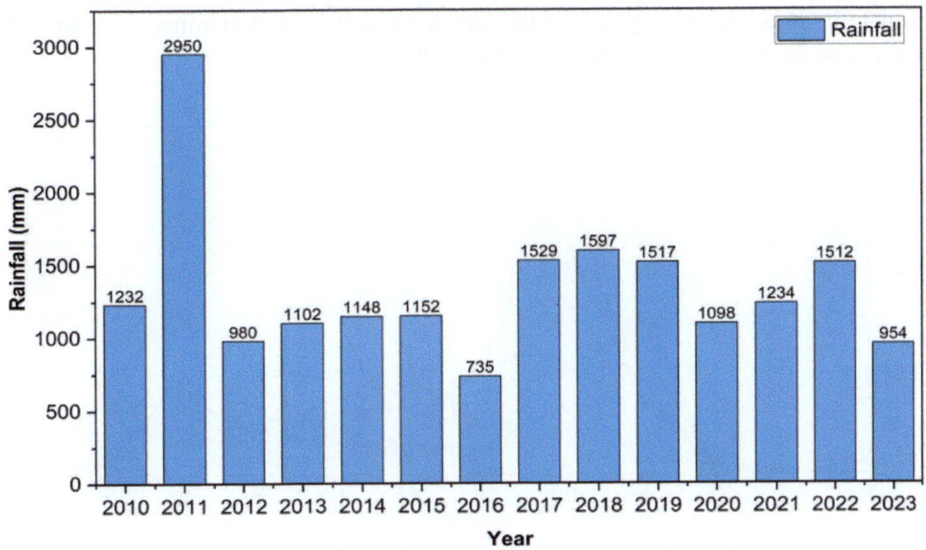

Fig. 5. Rainfall data (2010–2023)

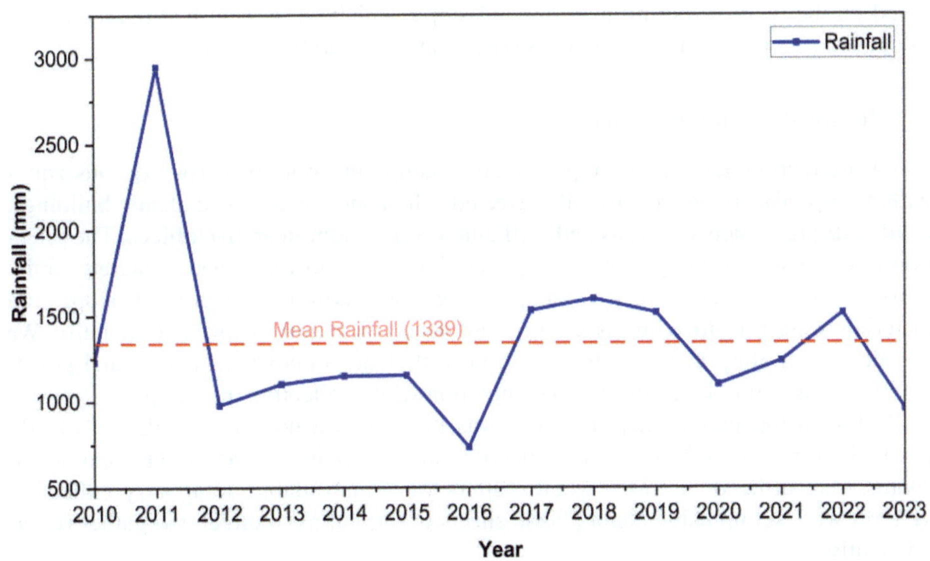

Fig. 6. Rainfall data (2010–2023) and mean rainfall data representation.

Figure 7 is a graphical representation of monthly rainfall data from 2010–2023, showing major peaks in May and minor peaks in February. This data analysis helps to identify the average annual rainfall, peak rainfall months, and variability over the years, seasonal distributions- monsoon and non-monsoon rainfall periods to assess water availability throughout the year, and to Identify the periods with minimal or no rainfall to estimate the storage needs during such months.

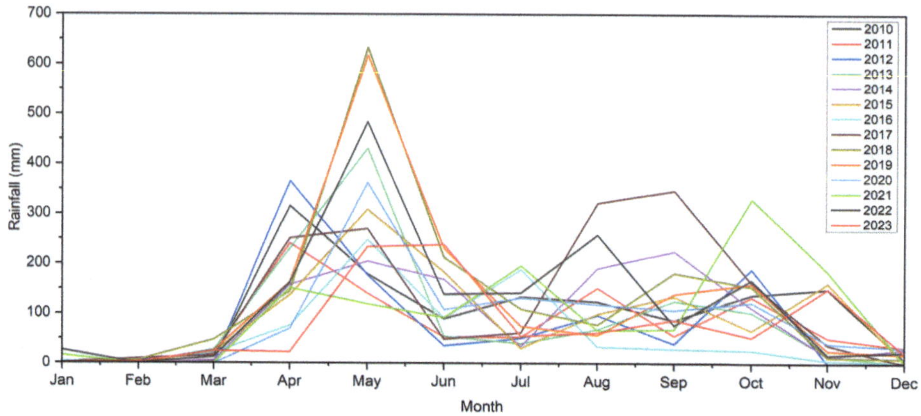

Fig. 7. Monthly rainfall data for each year from 2010–2023.

By using these rainfall data and the catchment area for each zone, the runoff potential was calculated using the standard equation, where the runoff coefficient was determined based on the type of roofing material and slope, which were identified manually and calculated accordingly using the mentioned standards and the formula.

3.3 Rainwater Sample Collection

To ensure comprehensive and representative sampling, rooftops across various zones within the study area were carefully selected. The zones included academic buildings, residential areas such as hostels and staff quarters, and administrative offices. The selection was based on building type, rooftop material (metal, polycarbonate, concrete, or flat gravel roofs), and accessibility for sample collection. Rooftops with different slopes and materials were prioritized to assess their influence on runoff quality and quantity. We got all the sampling materials like clean polyethylene containers, funnels, and gloves ready, to make sure there was no contamination while collecting the samples.

Before using them, each container was properly washed, sterilized, and clearly labelled. To guide rooftop runoff to specific sampling points, clean gutters, and downspouts were utilized. In cases where gutters were unavailable, temporary collection systems were set up using clean plastic sheets or tarpaulins to direct and gather runoff efficiently.

The collection process began with implementing the first flush technique to improve sample quality. During each rainfall event, the initial runoff flow from the rooftop (5–10 min) was allowed to drain out, as it often contains accumulated debris, dust, and pollutants from the roof surface. This step significantly reduced contamination and ensured a more accurate representation of the actual rainwater quality. Then, runoff samples were collected directly from designated points, such as clean downspouts or flow systems, using sterilized containers. To maintain consistency, 3–4 samples were collected, the collected samples were immediately sealed to prevent any external contamination, and each container was labelled with essential information, including the sampling location, date, time, and type of rooftop.

After the samples were collected across various buildings including different roof types and sealed (Fig. 8), they were stored in coolers to maintain temperature stability and prevent microbial growth during transportation. This step was crucial for preserving the integrity of the water samples, particularly for microbial analysis. Samples were transported to the laboratory within a short time frame (within 6–12 h) to ensure reliable and accurate testing. By adhering to this timeline, the analysis could reflect the true physical, chemical, and biological characteristics of the rainwater. Proper handling and timely transportation of samples minimized the risk of degradation or contamination, ensuring that the results obtained from laboratory testing was credible and representative of the collected runoff.

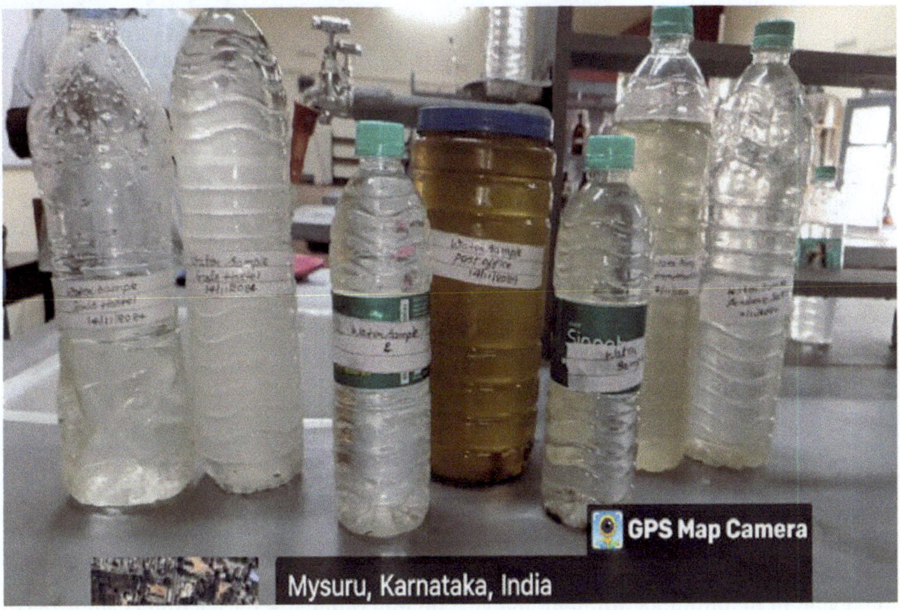

Fig. 8. Rainwater sample collected to store

3.4 Rooftop Area Calculations

Google Earth Pro software was utilized due to its high-resolution satellite imagery and precise measurement tools to calculate the rooftop areas of each building and the garden areas of the University of Mysore campus. The process began by navigating to the campus location and outlining its boundaries using the polygon tool, ensuring the measurements were confined to university premises. Individual buildings were systematically selected, and their rooftop perimeters were marked with the polygon tool to calculate their areas in square meters. Similarly, garden spaces, lawns, and other vegetation zones were identified and measured using the same method. All measurements were carefully recorded for further analysis.

Figure 9 shows the rooftop area calculations of each building in ach zones, using Google Earth Pro software. The collected data, including building names, rooftop areas, and garden spaces, was organized in a spreadsheet to allow easy comparison and calculation of rainwater harvesting potential. To ensure accuracy, random measurements were cross-verified, and any discrepancies were rechecked and resolved.

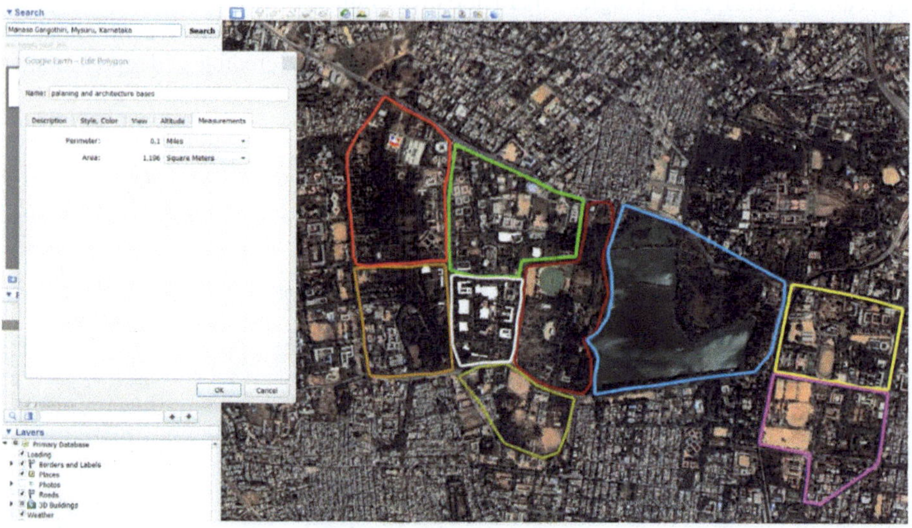

Fig. 9. Rooftop area calculations in Google Earth Pro.

Though the rooftop areas were calculated using the software each area was verified with the data from the engineering department. The rooftop areas were used to estimate rainwater collection capacities, while the garden areas helped to analyze, surface runoff and water utilization. This methodology provided reliable measurements and formed a critical step in planning an efficient and sustainable rainwater harvesting system for the campus (Fig. 10).

Fig. 10. The different rooftops in UoM

These rooftop materials are also very important, as they impact run off efficiency, based on the material type and based on each material the runoff coefficient values are obtained as per standards shown in Table 1.

Table 1. Run off coefficient values

Sl. no	Type of Roof	Runoff Coefficient
1.	Galvanized Sheets	0.90
2.	Asbestos Sheets	0.80
3.	Polycarbonate Sheets	0.80
4.	Tiled Roof	0.75
5.	Concrete Roof	0.70

After calculating the rooftop area, the area was used to calculate the total volume of water consumption using the formula,

$$V = C \times P \times A \quad (1)$$

where,

V: volume of rainwater in the reference period (L)
C: runoff coefficient (dimensionless)
P: average precipitation accumulated at the site (mm)
A: catchment area (roof) (m^2).

And the water demand per day was calculated using,

$$\text{Water Demand (L/day)} = P \times \text{Per Capita demand (L/person/day)} \quad (2)$$

where,

P = Number of people in the area
Per Capita demand = Average daily water use per person (in liters)

3.5 Rainwater Quality Analysis

The collected RW quality analysis was conducted for 4 different samples using standard procedures ensuring the accuracy and reliability of the results for each parameter:

1. pH Measurement: The pH was measured using a digital pH meter. The probe was immersed in the water sample, and the reading was noted. The pH values exceeded the permissible range (5.0–6.5), indicating alkalinity in most samples.
2. Conductivity Test: Conductivity was assessed using a conductivity meter, which measures the ability of water to conduct electrical current. The values observed were within permissible limits (10–100), except for slight elevations in the Post Office sample.
3. Total Dissolved Solids (TDS): TDS was determined by evaporating a known volume of water and weighing the residue or using a TDS meter. Results showed values within the permissible range (20–100 mg/L).
4. Total Alkalinity: Alkalinity was measured through titration using a strong acid, typically hydrochloric acid, with phenolphthalein and methyl orange as indicators. Some samples showed levels below the permissible range (10–50), suggesting low buffering capacity.
5. Total Hardness and Calcium/Magnesium Hardness: These were measured through complexometric titration using EDTA (Ethylene Diamine Tetra acetic acid) with an Eriochrome Black T indicator. Calcium and magnesium hardness were calculated separately. Results showed values mostly near or below permissible limits (<50 mg/L), indicating moderate mineral content.
6. Sulphate and Nitrate: Sulphate and Nitrate were analyzed using UV spectrophotometry. Nitrate levels were within safe limits (<10 mg/L), but sulphates were nil.
7. Total Solids and Suspended Solids: Total solids were measured by evaporating the water and weighing the residue, while suspended solids were calculated by filtering the water and measuring the residue. Results were compliant with permissible standards (<100 mg/L for TS and <10 mg/L for TSS).The detailed values of all tests conducted for different samples are shown in Table 2.

3.6 Area Calculation

The rooftop area and garden area were calculated using Google Earth Pro for accurate and precise calculations of individual buildings in each zone and for calculation of the total garden area. Figure 11 is a representation of the rooftop area of each building under the zone-1 classification (Tables 3, 4, 5, 6, 7, 8, 9, 10 and 11).

Table 2. Tabulation of results of tested samples.

Sl No	Parameters	AB	BS	GH	PO	Units	Permissible Range
1	pH	8.34	6.88	6.74	6.57	=	5.0–6.5
2	Conductivity	67	97	86	184	μS/cm	10–100
3	Turbidity	45	63	57	127	mg/l	20–100
5	Total Alkalinity	3	5	2	49	mg/l	10–50 mg/L as $CaCO_3$
6	Total Hardness	53	35	33	49	mg/l	< 50 mg/L as $CaCO_3$
7	Ortho Phosphate	- nil	- nil	0.7	2.2	mg/l	< 0.1
8	Sulphate	- nil	- nil	2	- nil	mg/l	< 50
9	Nitrate	6.7	6.1	6.3	3.8	mg/l	< 10
10	Ca as $CaCO_3$	40	42	32	49	mg/l	< 50
11	Ca as Ca^{++}	16.16	14.589	12.928	19.796	mg/l	< 20
12	Mg as $CaCO_3$	13	Nil	1	nil	mg/l	< 50
13	Mg as Mg^{++}	3.12	Nil	0.24	nil	mg/l	< 20
14	TS	100	100	100	100	mg/l	< 100
15	TDS	100	100	100	100	mg/l	20–100
16	TSS	0	0	0	0	mg/l	< 10

Total Suspended Solids (TSS), Total Dissolved Solids (TDS), Administrative Building (AB), Building Sample (BS), Girls Hostel (GH), Post Office (PO), Total Solids (TS)

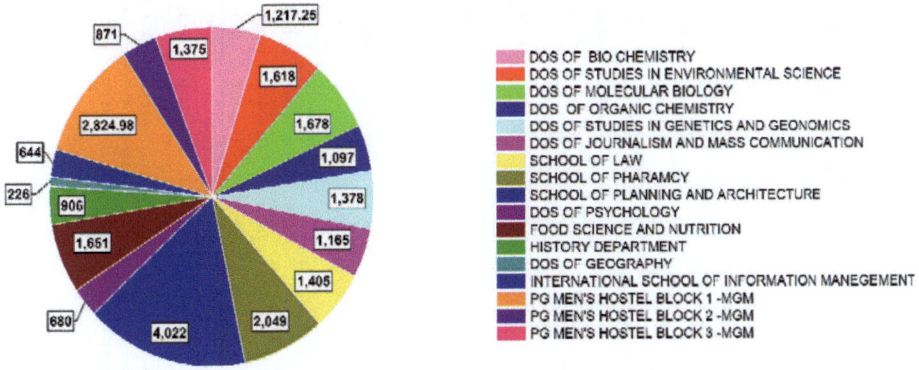

Fig. 11. Rooftop area of each building in Zone-1 (units in m^2)

The pie charts show the garden area of each zone in University of Mysuru (Fig. 12) and its water demand for gardening purposes (Fig. 13).

Table 3. Calculated data of water consumption and volume of rainwater of Zone-1

SL No.	Name of the building	Roof top area (Sq Mtr)	Water Consumption (in litres/day)	Water Consumption (in litres/month)	Water Consumption (in litres/year)	Volume of Rain Water (in litres)											
						JAN	FEB	MAR	APR	MAY	JUN	JUL	AUG	SEP	OCT	NOV	DEC
Average Rainfall: 2010–2023 (in mm)						3.93	3.28	16.50	177.93	315.50	125.07	93.36	123.57	122.14	137.50	62.71	13.43
1	DOS-BC	1217.25	4050.00	89100.00	891000.00	4783.79	3992.58	20084.63	216585.29	384042.38	152241.46	113642.46	150415.58	148674.92	167371.88	76333.75	16347.67
2	DOS-ES	1618.00	1775.00	39050.00	390500.00	6358.74	5307.04	26697.00	287890.74	510479.00	202363.26	151056.48	199936.26	197622.52	222475.00	101464.78	21729.74
3	DOS-MB	1678.00	2475.00	54450.00	544500.00	6594.54	5503.84	27687.00	298566.54	529409.00	209867.46	156658.08	207350.46	204950.92	230725.00	105227.38	22535.54
4	DOS-OC	1097.00	3000.00	66000.00	660000.00	4311.21	3598.16	18100.50	195189.21	346103.50	137201.79	102415.92	135556.29	133987.58	150837.50	68792.87	14732.71
5	DOS-GG	1378.00	1705.00	37510.00	375100.00	5415.54	4519.84	22737.00	245187.54	434759.00	172346.46	128650.08	170279.46	168308.92	189475.00	86414.38	18506.54
6	DOS-JM	1165.00	3455.00	76010.00	760100.00	4578.45	3821.20	19222.50	207288.45	367557.50	145706.55	108764.40	143959.05	142293.10	160187.50	73057.15	15645.95
7	SL	1405.00	2295.00	50490.00	504900.00	5521.65	4608.40	23182.50	249991.65	443277.50	175723.35	131170.80	173615.85	171606.70	193187.50	88107.55	18869.15
8	SP	2049.00	2345.00	51590.00	515900.00	8052.57	6720.72	33808.50	364578.57	646459.50	256268.43	191294.64	253194.93	250264.86	281737.50	128492.79	27518.07
9	SPA	4022.00	2104.00	46288.00	462880.00	15806.46	13192.16	66363.00	715634.46	1268941.00	503031.54	375493.92	496998.54	491247.08	553025.00	252219.62	54015.46
10	SPY	680.00	3100.00	68200.00	682000.00	2672.40	2230.40	11220.00	120992.40	214540.00	85047.60	63484.80	84027.60	83055.20	93500.00	42642.80	9132.40
11	FN	1651.00	4015.00	88330.00	883300.00	6488.43	5415.28	27241.50	293762.43	520890.50	206490.57	154137.36	204014.07	201653.14	227012.50	103534.21	22172.93
12	HI	906.00	4990.00	109780.00	1097800.00	3560.58	2971.68	14949.00	161204.58	285843.00	113313.42	84584.16	111954.42	110658.84	124575.00	56815.26	12167.58
13	GE	226.00	2475.00	54450.00	544500.00	799.36	8754.08	192589.65	2827.34	30963.16	681189.59	10000.31	109516.71	2409367.59	35371.11	387360.60	8521933.16
14	IM	644.00	3100.00	68200.00	682000.00	1771.64	8528.10	187618.20	4873.79	23460.80	516137.67	13407.80	64540.67	1419894.73	88550.00	40385.24	8648.92
15	PG-MB1	2824.98	52885.00	1163470.00	11634700.00	11102.18	9265.94	46612.20	502649.05	891281.82	353320.50	263740.32	349083.03	345043.30	388435.03	177154.62	37939.51
16	PG-MB2	871.00	51400.00	1130800.00	11308000.00	3423.03	2856.88	14371.50	154977.03	274800.50	108935.97	81316.56	107629.47	106383.94	119762.50	54620.41	11697.53
17	PG-MB3	1375.00	27100.00	596200.00	5962000.00	5403.75	4510.00	22687.50	244653.75	433812.50	171971.25	128370.00	169908.75	167942.50	189062.50	86226.25	18466.25
	Total	**24807.23**	**172269.00**	**3789918.00**	**37899180.00**	**96644.33**	**95796.30**	**775172.18**	**4266852.83**	**7606620.66**	**4191156.87**	**2258188.10**	**3131981.14**	**6752955.83**	**3415290.51**	**1928849.66**	**8852059.10**

DOS-BC: DOS of bio chemistry, DOS-ES: DOS of studies in environmental science, DOS-MB: DOS of molecular biology, DOS-OC: DOS of organic chemistry, DOS-GG: DOS of studies in genetics and geonomics, DOS-JM: DOS of journalism and mass communication, SL: School of law, SP: School of Pharmacy, SPA: School of planning and architecture, SPY: DOS of psychology, FN: Food science and nutrition, HI: History department, GE: DOS of geography, IM: International school of information management, PG-MB1: PG men's hostel block 1 – MGM, PG-MB2: PG men's hostel block 2 – MGM, PG-MB3: PG men's hostel block 3 - MGM

Table 4. Calculated data of water consumption and volume of rainwater of Zone-2

SL No.	Name of the building	Roof top area (Sq Mtr)	Water Consumption (in liters/day)	Water Consumption (in litres/month)	Water Consumption (in litres/year)	Volume of Rainwater (In Litres)											
						JAN	FEB	MAR	APR	MAY	JUN	JUL	AUG	SEP	OCT	NOV	DEC
	Average Rainfall: 2010–2023 (in mm)					3.93	3.28	16.50	177.93	315.50	125.07	93.36	123.57	122.14	137.50	62.71	13.43
1	PG-WB1	2283.00	53460.00	1176120.00	11761200.00	8972.19	7488.24	37669.50	404844.39	720286.50	285534.81	213140.88	282110.31	278845.62	313912.50	143166.93	30660.69
2	PG-WB2	1178.00	44820.00	986040.00	9860400.00	4629.54	3863.84	19437.00	208894.74	371659.00	147332.46	109978.08	145565.46	143880.92	161975.00	73872.38	15820.54
3	PG-WB3	2596.00	41850.00	920700.00	9207000.00	10202.28	8514.88	42834.00	460348.68	819038.00	324681.72	242362.56	320787.72	317075.44	356950.00	162795.16	34864.28
4	CIST	2168.00	2615.00	57530.00	575300.00	8520.24	7111.04	35772.00	384451.44	684004.00	271151.76	202404.48	267899.76	264799.52	298100.00	135955.28	29116.24
5	JVM	13505.33	700.00	15400.00	154000.00	53075.95	44297.48	222837.95	2394900.17	4260931.62	1689111.62	1260857.61	1668853.63	1649541.01	1856982.88	846919.24	181376.58
6	FM	673.00	1200.00	26400.00	264000.00	2644.89	2207.44	11104.50	119343.09	212331.50	84172.11	62831.28	83162.61	82200.22	92537.50	42203.83	9038.39
7	GD	840.00	2450.00	53900.00	539000.00	3301.20	2755.20	13860.00	148957.20	265020.00	105058.80	78422.40	103798.80	102597.60	115500.00	52676.40	11281.20
8	RTC	535.00	8850.00	194700.00	1947000.00	2102.55	1754.80	8827.50	94871.55	168792.50	66912.45	49947.60	66109.95	65344.90	73562.50	33549.85	7185.05
Total		**23778.33**	**155945.00**	**3430790.00**	**34307900.00**	**93448.84**	**77992.92**	**392342.45**	**4216611.26**	**7502063.12**	**2973955.73**	**2219944.89**	**2938288.24**	**2904285.23**	**3269520.38**	**1491139.07**	**319342.97**

PG-WB1: PG Women's Hostel Block 1 –MGM, PG-WB2: PG Women's hostel Block 2 –MGM,, PG-WB3: PG Women's hostel block 3 – MGM, CIST: Centre for Information Science and Technology and Office of Administrative Officer, JVM: Jayalakshmi vilas museum, FM: Folklore museum, GD: Office of garden department, RTC: Round table canteen

Table 5. Calculated data of water consumption and volume of rainwater of Zone-3

Sl No.	Name of the Building	Roof Top Area (Sq Mtr)	Water Consumption (In -Lts/Day)	Water Consumption (In Lts/Mon)	Water Consumption (In Lts/Yrs)	Volume of Rainwater (In Litres)											
						JAN	FEB	MAR	APR	MAY	JUN	JUL	AUG	SEP	OCT	NOV	DEC
	Average Rainfall: 2010–2023 (in mm)					3.93	3.28	16.50	177.93	315.50	125.07	93.36	123.57	122.14	137.50	62.71	13.43
1	SNRWCS	0.00	30000.00	660000.00	6600000.00	0.00	0.00	0.00	0.00	0.00	0.00	0.00	0.00	0.00	0.00	0.00	0.00
2	MIS	1479.00	1325.00	19875.00	291500.00	5812.47	4851.12	24403.50	263158.47	466624.50	184978.53	138079.44	182760.03	180645.06	203362.50	92748.09	19862.97
3	OAT	651.76	1500.00	22500.00	330000.00	2049.13	4716.00	70740.00	6442.48	14827.10	222406.56	20255.14	46616.41	699246.22	63682.17	146562.01	2198430.13
4	UED	676.00	17775.00	391050.00	3910500.00	1859.68	62870.18	1383143.85	6577.67	222371.81	4892179.80	23265.23	786529.09	17303639.94	82289.13	2781953.39	61202974.48
5	MGH	1207.00	1350.00	20250.00	297000.00	4743.51	3958.96	19915.50	214761.51	380808.50	150959.49	112685.52	149148.99	147422.98	165962.50	75690.97	16210.01
6	ASC-HRDC	449.00	1700.00	37400.00	374000.00	1764.57	1472.72	7408.50	79890.57	141659.50	56156.43	41918.64	55482.93	54840.86	61737.50	28156.79	6030.07
7	UoM-OP	280.00	3500.00	77000.00	770000.00	1100.40	918.40	4620.00	49820.40	88340.00	35019.60	26140.80	34599.60	34199.20	38500.00	17558.80	3760.40
	Total	**4742.76**		**1228075.00**	**12573000**	**17329.76**	**78787.38**	**1510231.35**	**620651.10**	**1314631.41**	**5541700.41**	**362344.78**	**1255137.05**	**18419994.27**	**615533.80**	**3142670.04**	**63447268.06**

SNRW-CS: Srikanta Datta Narasimha Raja Wadiyar Cricket Stadium, MIS: Manasa Gangotri Indoor Stadium, OAT: Open Air Theatre, UED: University Engineering Division, MGH: Manasa Gangotri Guest House, ASC-HRDC: Academic Staff College (Human Resource Development Centre), UoM-OP: University of Mysuru Online Program

Table 6. Calculated data of water consumption and volume of rainwater of Zone-4

Sl No.	Name of the Building	Roof Top Area (Sq Mtr)	Water Consumption (In -Lts/Day)	Water Consumption (In Lts/Mon)	Water Consumption (In Lts/Yrs)	Volume of Rainwater (In Litres)											
						JAN	FEB	MAR	APR	MAY	JUN	JUL	AUG	SEP	OCT	NOV	DEC
Average Rainfall: 2010–2023 (in mm)						**3.93**	**3.28**	**16.50**	**177.93**	**315.50**	**125.07**	**93.36**	**123.57**	**122.14**	**137.50**	**62.71**	**13.43**
1	MGH	867.00	4200.00	92400.00	924000.00	3407.31	2843.76	14305.50	154265.31	273538.50	108435.69	80943.12	107135.19	105895.38	119212.50	54369.57	11661.15
2	UHC	429.00	1000.00	22000.00	220000.00	1685.97	1407.12	7078.50	76331.97	135349.50	53655.03	40051.44	53011.53	52398.06	58987.50	26902.59	5770.05
3	DoA	445.00	1185.00	26070.00	260700.00	1748.85	1459.60	7342.50	79178.85	140397.50	55656.15	41545.20	54988.65	54352.30	61187.50	27905.95	5985.25
4	DoS	734.00	2305.00	50710.00	507100.00	2884.62	2407.52	12111.00	130600.62	231577.00	91801.38	68526.24	90700.38	89650.76	100925.00	46029.14	9872.30
5	SBCSC-R&E	1119.00	0.00	0.00	0.00	4397.67	3670.32	18463.50	199103.67	353044.50	139953.33	104469.84	138274.83	136674.66	153862.50	70172.49	15050.55
6	USIC& DoAH:	1132.00	1235.00	27170.00	271700.00	4448.76	3712.96	18678.00	201416.76	357146.00	141579.24	105683.52	139881.24	138262.48	155650.00	70987.72	15225.40
	Total	**4726.00**		**218350.00**	**2183500.00**	**18573.18**	**15501.28**	**77979.00**	**840897.18**	**1491053.00**	**591080.82**	**441219.36**	**583991.82**	**577233.64**	**649825.00**	**296367.46**	**63564.70**

Table 7. Calculated data of water consumption and volume of rainwater of lake zone

Sl No.	Name Of The Building	Roof Top Area (Sq Mtr)	Water Consumption (In -Lts/Day)	Water Consumption (In Lts/Mon)	Water Consumption (In Lts/Yrs)	Volume of Rainwater (In Litres)											
						JAN	FEB	MAR	APR	MAY	JUN	JUL	AUG	SEP	OCT	NOV	DEC
	Avg Rinfall: 2010–2023 (in mm)					3.93	3.28	16.50	177.93	315.50	125.07	93.36	123.57	122.14	137.50	62.71	13.43
1	KK	3385.00	21000.00	315000.00	4620000.00	13303.05	7484.96	37653.00	406036.26	719971.00	285409.74	213047.52	281986.74	278723.48	313775.00	143104.22	30647.26
	Total	**3385.00**		**315000.00**	**4620000.00**	**13303.05**	**7484.96**	**37653.00**	**406036.26**	**719971.00**	**285409.74**	**213047.52**	**281986.74**	**278723.48**	**313775.00**	**143104.22**	**30647.26**

KKA: Karnataka Kalamandira

Table 8. Calculated data of water consumption and volume of rainwater of Zone-5

	Name Of The Building	Roof Top Area (Sq Mtr)	Water Consumption (In -Lts/Day)	Water Consumption (In Lts/Mon)	Water Consumption (In Lts/Yrs)	Volume Of Rainwater (In Litres)											
						JAN	FEB	MAR	APR	MAY	JUN	JUL	AUG	SEP	OCT	NOV	DEC
	Avg Rinfall: 2010–2023 (in mm)					3.93	3.28	16.50	177.93	315.50	125.07	93.36	123.57	122.14	137.50	62.71	13.43
1	DoG	919.00	1180.00	25960.00	259600.00	2889.34	3014.32	15163.50	163517.67	289944.50	114939.33	85797.84	113560.83	112246.66	126362.50	57630.49	12342.17
2	DOS-B&Z	3300.00	7900.00	173800.00	1738000.00	12969.00	10824.00	54450.00	587169.00	1041150.00	412731.00	308088.00	407781.00	403062.00	453750.00	206943.00	44319.00
3	DOS-C	254.00	6465.00	142230.00	1422300.00	998.22	833.12	4191.00	45194.22	80137.00	31767.78	23713.44	31386.78	31023.56	34925.00	15928.34	3411.22
4	DOS-P	1442.00	3950.00	86900.00	869000.00	5667.06	4729.76	23793.00	256575.06	454951.00	180350.94	134625.12	178187.94	176125.88	198275.00	90427.82	19366.06
5	NSH&CKO	645.00	2000.00	44000.00	440000.00	2534.85	2115.60	10642.50	114764.85	203497.50	80670.15	60217.20	79702.65	78780.30	88687.50	40447.95	8662.35
6	DOS-SS	564.00	2790.00	61380.00	613800.00	2216.52	1849.92	9306.00	100352.52	177942.00	70539.48	52655.04	69693.48	68886.96	77550.00	35368.44	7574.52
7	DOS -SA	388.00	690.00	15180.00	151800.00	1524.84	1272.64	6402.00	69036.84	122414.00	48527.16	36223.68	47945.16	47390.32	53350.00	24331.48	5210.84
8	GB	352.20	450.00	9900.00	99000.00	1384.15	1155.22	5811.30	62666.95	111119.10	44049.65	32881.39	43521.35	43017.71	48427.50	22086.46	4730.05
9	KIKS	2952.00	4890.00	107580.00	1075800.00	11601.36	9682.56	48708.00	525249.36	931356.00	369206.64	275598.72	364778.64	360557.28	405900.00	185119.92	39645.36
10	BM-A	419.00	1050.00	23100.00	231000.00	1317.34	3301.20	72626.40	4141.70	10378.97	228337.40	13021.52	32631.49	717892.79	40939.65	102593.41	2257054.93
11	UB	1132.00	350.00	7700.00	77000.00	4448.76	3712.96	18678.00	201416.76	357146.00	141579.24	105683.52	139881.24	138262.48	155650.00	70987.72	15202.76
12	DOS-SMI	1139.00	3625.00	79750.00	797500.00	4476.27	3735.92	18793.50	202662.27	359354.50	142454.73	106337.04	140746.23	139117.46	156612.50	71426.69	15296.77
13	DOS-BT	1909.00	4155.00	91410.00	914100.00	7502.37	6261.52	31498.50	339668.37	602289.50	238758.63	178224.24	235895.13	233165.26	262487.50	119713.39	25637.87
14	SBI	172.00	1500.00	33000.00	330000.00	675.96	564.16	2838.00	30603.96	54266.00	21512.04	16057.92	21254.04	21008.08	23650.00	10786.12	2309.96
15	DOS -CS	1615.00	8075.00	177650.00	1776500.00	6346.95	5297.20	26647.50	287356.95	509532.50	201988.05	150776.40	199565.55	197256.10	222062.50	101276.65	21689.45
16	MPO	227.00	550.00	12100.00	121000.00	892.11	744.56	3745.50	40390.11	71618.50	28390.89	21192.72	28050.39	27725.78	31212.50	14235.17	3048.61
17	DOS-SW	797.00	1550.00	34100.00	341000.00	3132.21	2614.16	13150.50	141810.21	251453.50	99680.79	74407.92	98485.29	97345.58	109587.50	49979.87	10703.71
18	UOM-LI	4112.00	8000.00	176000.00	1760000.00	16160.16	13487.36	67848.00	731648.16	1297336.00	514287.84	383896.32	508119.84	502239.68	565400.00	257863.52	55224.16
19	DOS -SO	2442.00	3515.00	77330.00	773300.00	9597.06	8009.76	40293.00	434505.06	770451.00	305420.94	227985.12	301757.94	298265.88	335775.00	153137.82	32796.06
20	DOS OF STATISTICS	750.00	1425.00	31350.00	313500.00	2947.50	2460.00	12375.00	133447.50	236625.00	93802.50	70020.00	92677.50	91605.00	103125.00	47032.50	10072.50
21	DOS-M	795.00	4820.00	106040.00	1060400.00	3124.35	2607.60	13117.50	141454.35	250822.50	99430.65	74221.20	98238.15	97101.30	109312.50	49854.45	10676.85
22	DOS-C	841.00	12660.00	278520.00	2785200.00	3305.13	2758.48	13876.50	149639.13	265335.50	105183.87	78515.76	103922.37	102719.14	115637.50	52739.11	11294.63
Total	**27166.2**	**81590**	**1794980**	**17949800**	**105711.51**	**91032.02**	**513955.2**	**4763271**	**8449120.57**	**3573609.7**	**2510140.11**	**3337782.99**	**3984795.2**	**3718679.65**	**1779910.32**	**2616269.83**	

DOS-GE: DOS Of Geology, DOS-B&Z: DOS Of Studies In Botonny And Zoology, DOS-C: DOS Of Chemistry, DOS-P: DOS Of Physics, NSH&CKO: National Science History And Classical Kannada Office, DOS-SS: DOS Of Social Studies, DOS-SA: DOS Of Studies In Sanskrit, GB: Gandhi Bhavan, KIKS: Kuvempu Institute Of Kannada Studies, BM-A: B M Sri Auditorium, UB: Unknown Building, DOS-SMI: DOS Of Studies In Microbiology, DOS-BT: DOS In Biotechnology, SBI: State Bank of India, DOS-CS: DOS In Computer Science, MPO: Manasa Gangotri Post Office, DoS-SW: DOS Of Student Welfare, U0M-LI: UOM Library, DOS-SO: DOS Of Sociology, DOS-ST: DOS Of Statistics, DOS-M: DoS Of Mathematics, DoS-C: DOS Of Commerce

Table 9. Calculated data of water consumption and volume of rainwater of Zone-6

Sl No.	Name Of The Building	Roof Top Area (Sq Mtr)	Water Consumption (In -Lts/Day)	Water Consumption (In Lts/Mon)	Water Consumption (In Lts/Yrs)	Volume of Rainwater (In Litres)											
						JAN	FEB	MAR	APR	MAY	JUN	JUL	AUG	SEP	OCT	NOV	DEC
	Avg Rinfall: 2010–2023 (in mm)					3.93	3.28	16.50	177.93	315.50	125.07	93.36	123.57	122.14	137.50	62.71	13.43
1	VHG	0.00	500.00	5000.00	110000.00	0.00	0.00	0.00	0.00	0.00	0.00	0.00	0.00	0.00	0.00	0.00	0.00
2	MB	2282.00	800.00	17600.00	176000.00	8968.26	7484.96	37653.00	406036.26	719971.00	285409.74	213047.52	281986.74	278723.48	313775.00	143104.22	30647.26
3	BRA-H	1112.00	24570.00	540540.00	5405400.00	4370.16	3647.36	18348.00	197858.16	350836.00	139077.84	103816.32	137409.84	135819.68	152900.00	69733.52	14934.16
	Total	**3394.00**		**563140.00**	**5691400.00**	**13338.42**	**11132.32**	**56001.00**	**603894.42**	**1070807.00**	**424487.58**	**316863.84**	**419396.58**	**414543.16**	**466675.00**	**212837.74**	**45581.42**

VHG: Village Hostel Ground, MB: Moulay Bhavan, Dr. B R A H: Dr. B R Ambedkar Hostel

Table 10. Calculated data of water consumption and volume of rainwater of Zone-7

Sl No.	Name of the Building	Roof Top Area (Sq Mtr)	Water Consumption (In -Lts/Day)	Water Consumption (In Lts/Mon)	Water Consumption (In Lts/Yrs)	Volume of Rainwater (In Litres)											
						JAN	FEB	MAR	APR	MAY	JUN	JUL	AUG	SEP	OCT	NOV	DEC
	Avg Rainfall: 2010–2023 (in mm)					3.93	3.28	16.50	177.93	315.50	125.07	93.36	123.57	122.14	137.50	62.71	13.43
1	CH	3812.00	12500.00	275000.00	2750000.00	14981.16	12503.36	62898.00	678269.16	1202686.00	476766.84	355888.32	471048.84	465597.68	524150.00	239050.52	51195.16
2	PH	2009.00	10000.00	220000.00	2200000.00	7895.37	6589.52	33148.50	357461.37	633839.50	251265.63	187560.24	248252.13	245379.26	276237.50	125984.39	26980.87
3	PO	316.97	500.00	11000.00	110000.00	1245.69	1039.66	5230.01	56398.47	100004.04	39643.44	29592.32	39167.98	38714.72	43583.38	19877.19	4256.91
4	SBI	192.11	2000.00	44000.00	440000.00	754.99	630.12	3169.82	34182.13	60610.71	24027.20	17935.39	23739.03	23464.32	26415.13	12047.22	2580.04
5	UOG	0.00	500.00	5000.00	110000.00	0.00	0.00	0.00	0.00	0.00	0.00	0.00	0.00	0.00	0.00	0.00	0.00
6	SP	1050.00	15250.00	335500.00	3355000.00	4126.50	3444.00	17325.00	186826.50	331275.00	131323.50	98028.00	129748.50	128247.00	144375.00	65845.50	14101.50
7	UBBC	0.00	500.00	5000.00	110000.00	0.00	0.00	0.00	0.00	0.00	0.00	0.00	0.00	0.00	0.00	0.00	0.00
	Total	**7380.08**	**41250**	**895500**	**9075000**	**29003.71**	**24206.66**	**121771.33**	**1313137.63**	**2328415.25**	**923026.61**	**689004.27**	**911956.48**	**901402.98**	**1014761.01**	**462804.82**	**99114.48**

CH: Crowfard Hall, PH; Pareeksha Bhavan, PO: Post Office,SBI: State Bank of India, UOG: University Oval Ground, SP: Swimming Pool, UBBC: University Basket Ball Court, M PUC: Maharaja Pu College

Table 11. Calculated data of water consumption and volume of rainwater of Zone-8

	Name Of the Building	Roof Top Area (Sq Mtr)	Water Consumption (In -Lts/Day)	Water Consumption (In Lts/Mon)	Water Consumption (In Lts/Yrs)	Volume of Rainwater (In Litres)											
						JAN	FEB	MAR	APR	MAY	JUN	JUL	AUG	SEP	OCT	NOV	DEC
	Avg Rinfall: 2010–2023 (in mm)					3.93	3.28	16.50	177.93	315.50	125.07	93.36	123.57	122.14	137.50	62.71	13.43
1	UHV	0.00	500.00	5000.00	110000.00	0.00	0.00	0.00	0.00	0.00	0.00	0.00	0.00	0.00	0.00	0.00	0.00
2	UVC	0.00	500.00	5000.00	110000.00	0.00	0.00	0.00	0.00	0.00	0.00	0.00	0.00	0.00	0.00	0.00	0.00
3	GHSPM	856.78	12000.00	264000.00	2640000.00	3367.15	2810.24	14136.87	152446.87	270314.09	107157.47	79988.98	105872.30	104647.11	117807.25	53728.67	11506.56
4	USPV	392.00	6000.00	132000.00	1320000.00	1540.56	1285.76	6468.00	69748.56	123676.00	49027.44	36597.12	48439.44	47878.88	53900.00	24582.32	5264.56
5	PCG	0.00	1000.00	22000.00	220000.00	0.00	0.00	0.00	0.00	0.00	0.00	0.00	0.00	0.00	0.00	0.00	0.00
6	UFG	0.00	700.00	15400.00	154000.00	0.00	0.00	0.00	0.00	0.00	0.00	0.00	0.00	0.00	0.00	0.00	0.00
7	ORI	711.89	1750.00	38500.00	385000.00	2797.73	2335.00	11746.19	126666.59	224601.30	89036.08	66462.05	87968.25	86950.24	97884.88	44642.62	9560.68
8	GP	4760.00	450.00	9900.00	99000.00	18706.80	15612.80	78540.00	846946.80	1501780.00	595333.20	444393.60	588193.20	581386.40	654500.00	298499.60	63926.80
9	YC	2265.00	96565.00	2124430.00	21244300.00	8901.45	7429.20	37372.50	403011.45	714607.50	283283.55	211460.40	279886.05	276647.10	311437.50	142038.15	30418.95
10	MC	2393.00	113575.00	2498650.00	24986500.00	9404.49	7849.04	39484.50	425786.49	754991.50	299292.51	223410.48	295703.01	292281.02	329037.50	150065.03	32137.99
11	MCCH	725.30	8000.00	176000.00	1760000.00	2850.43	2378.98	11967.45	129052.63	228832.15	90713.27	67714.01	89625.32	88588.14	99728.75	45483.56	9740.78
12	VCR	1651.01	20000.00	440000.00	4400000.00	6488.47	5415.31	27241.67	293764.21	520893.66	206491.82	154138.29	204015.31	201654.36	227013.88	103534.84	22173.06
13	UEC	267.00	4260.00	93720.00	937200.00	1049.31	875.76	4405.50	47507.31	84238.50	33393.69	24927.12	32993.19	32611.38	36712.50	16743.57	3585.81
14	CCMC	1111.00	5000.00	110000.00	1100000.00	4366.23	3644.08	18331.50	197680.23	350520.50	138952.77	103722.96	137286.27	135697.54	152762.50	69670.81	14920.73
15	MCC	218.00	8000.00	176000.00	1760000.00	856.74	715.04	3597.00	38788.74	68779.00	27265.26	20352.48	26938.26	26626.52	29975.00	13670.78	2927.74
16	UH	465.00	5000.00	110000.00	1100000.00	1827.45	1525.20	7672.50	82737.45	146707.50	58157.55	43412.40	57460.05	56795.10	63937.50	29160.15	6244.95

(continued)

Table 11. *(continued)*

	Name Of the Building	Roof Top Area (Sq Mtr)	Water Consumption (In -Lts/Day)	Water Consumption (In Lts/Mon)	Water Consumption (In Lts/Yrs)	Volume of Rainwater (In Litres)											
						JAN	FEB	MAR	APR	MAY	JUN	JUL	AUG	SEP	OCT	NOV	DEC
17	UGLYC + MUECL	1200.00	16000.00	352000.00	3520000.00	4716.00	3936.00	19800.00	213516.00	378600.00	150084.00	112032.00	148284.00	146568.00	165000.00	75252.00	16116.00
18	DOC	320.00	1950.00	42900.00	429000.00	1257.60	1049.60	5280.00	56937.60	100960.00	40022.40	29875.20	39542.40	39084.80	44000.00	20067.20	4297.60
19	YCIS	943.00	3000.00	66000.00	660000.00	2964.79	9432.00	207504.00	9321.31	29654.21	652392.58	29306.19	93232.83	2051122.26	92138.65	293124.02	6448728.38
20	MCIS	943.00	3000.00	66000.00	660000.00	2964.79	9432.00	207504.00	9321.31	29654.21	652392.58	29306.19	93232.83	2051122.26	92138.65	293124.02	6448728.38
21	YCPJH	735.00	8750.00	70000.00	1925000.00	2888.55	2410.80	12127.50	130778.55	231892.50	91926.45	68619.60	90823.95	89772.90	101062.50	46091.85	9871.05
22	YCCOE	647.00	420.00	9240.00	92400.00	2542.71	2122.16	10675.50	115120.71	204128.50	80920.29	60403.92	79949.79	79024.58	88962.50	40573.37	8689.21
23	MH	895.00	60680.00	1334960.00	13349600.00	3517.35	2935.60	14767.50	159247.35	282372.50	111937.65	83557.20	110595.15	109315.30	123062.50	56125.45	12019.85
24	YH	614.00	44885.00	987470.00	9874700.00	2413.02	2013.92	10131.00	109249.02	193717.00	76792.98	57323.04	75871.98	74993.96	84425.00	38503.94	8246.02
25	MG	0.00	2000.00	44000.00	440000.00	0.00	0.00	0.00	0.00	0.00	0.00	0.00	0.00	0.00	0.00	0.00	0.00
	Total	**22112.98**		**9193170.00**	**93276700.00**	**85421.62**	**85208.49**	**748753.17**	**3617629.16**	**6440920.61**	**3834573.54**	**1947003.23**	**2685913.58**	**6572767.86**	**2965487.05**	**1854681.95**	**13169105.11**

UHG: University hockey ground, UVC: University volleyball court, GHSPM: Gymnasium hall sports pavillion mysore, USPV: University sports complex, PCG: Pavillion cricket ground, UFG: University football ground, ORI: Oriental research institute, GP: Garden park, YC: Yuvaraja college, MC: Maharaja college, MCCH: Maharaja college centenary hall, VCR: Vice chancellor's residence, UEC: University evening college, CCMC: Computer centre maharaja college, MCC: Maharaja college canteen,UH: University hospital, UGLYC + MUECL: UG library Yuvaraja college + mysore university evening college library, DOC: Department of criminology, YCIS: Yuvaraja college indoor stadium, MCIS: Maharaja college indoor stadium, YCPJH: Yuvaraja college platinum jublie hall, YCCOE: Yuvaraja college controller of examination, MH: Maharaja hostel, YH: Yuvaraja hostel, MG: Maharaja grounds

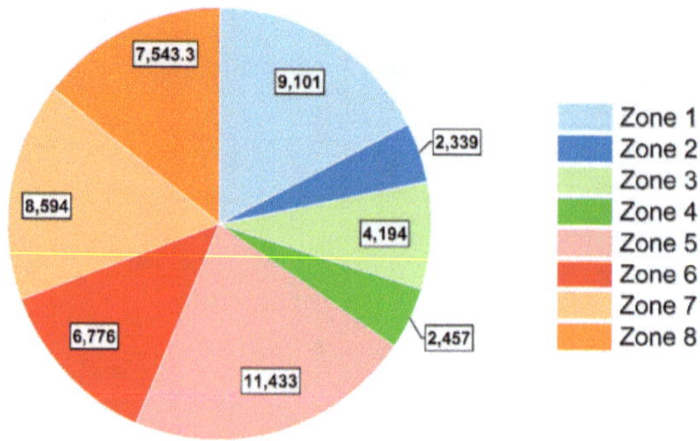

Fig. 12. Garden area of each plotted zone (units in m^2)

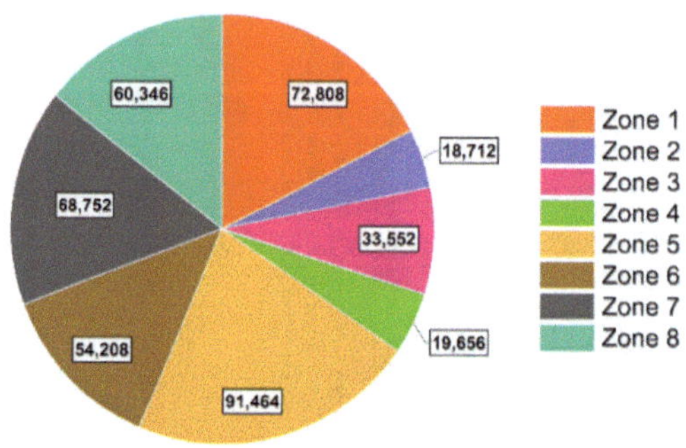

Fig. 13. Water demand for gardening purpose (units in litres)

Figure 14 shows variation in rainfall from the year 2010–23, having peak rainfall in the year 2022; this also represents the average rainfall data of 3.93 mm. And, the Fig. 15 represents the seasonal water demand for zone-1 for various purposes like sanitation, cleaning, etc., where the constant water demand is 38 ML.

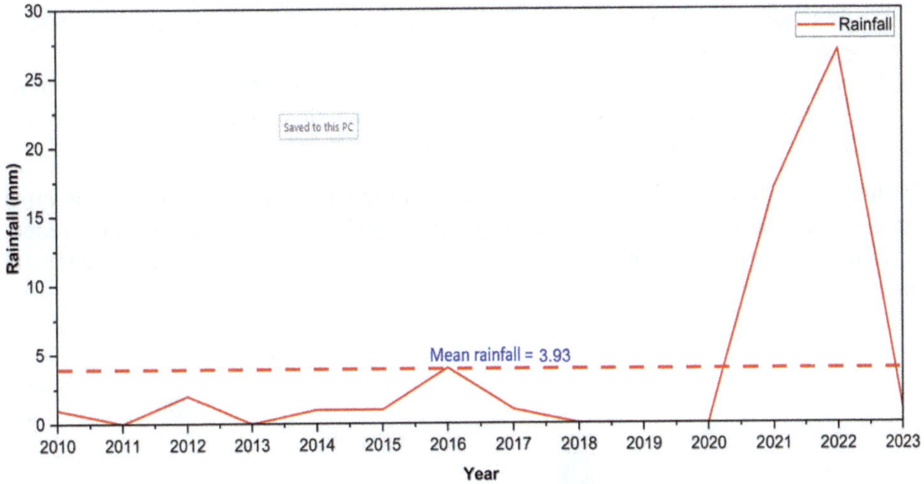

Fig. 14. Average rainfall from 2010–2023 for January

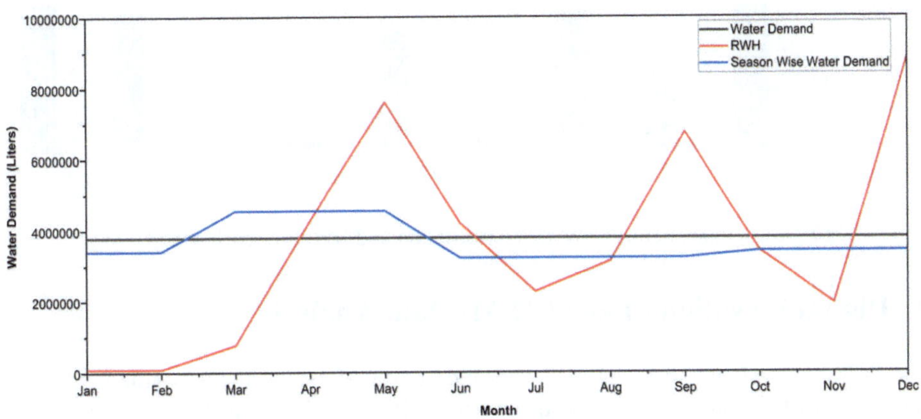

Fig. 15. Seasonal water demand for zone-1 of UoM

The two consecutive bars in the graph (Fig. 16) represent the comparison between the total volume of water consumption and the volume of harvested rainwater in each zone.

- The evaluated water consumption of UoM per year is approximately 215.5 M.
- The total rooftop area of 96 buildings of UoM is estimated to be 32.124 acres.
- The water demand of 12.96 acres of garden area in UoM evaluates to be 100.68 M litres.
- The total estimated rainwater collected is 335.26 M litres, sufficient for non-potable applications - such as gardening, flushing, and cleaning.

- Existing storage tanks can be optimized with minor modifications, such as incorporating filtration systems.
- Through a thorough quality analysis, we can conclude that the water qualifies to be suitable to use for all various purposes like gardening, flushing, and sanitation except for drinking purposes.

The harvested rainwater is more than sufficient to meet this demand, as well as other non-potable requirements such as flushing and cleaning, thereby significantly reducing reliance on external water sources and promoting sustainable water management practices on campus.

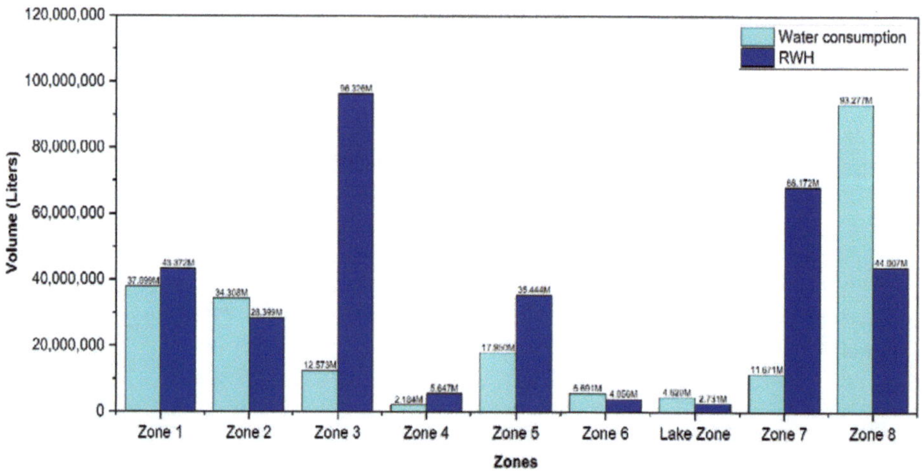

Fig. 16. Comparisons of Water consumption and RWH for each zone UoM

4 Digital Elevation Model (DEM) Data Analysis

The DEM analysis involves using elevation data to understand the topographical characteristics of the University of Mysuru campus. By analysing the slopes and elevation profiles, it is possible to identify areas with optimal gradients for placing storage tanks, ensuring proper water flow through gravity. High-resolution DEM data is processed using GIS software to classify the terrain into different slope categories, highlighting steep, moderate, and flat regions. The analysis prioritizes flat or gently sloping areas at higher elevations to maximize storage capacity and facilitate efficient water distribution. Additionally, the DEM helps identify natural depressions or catchment areas suitable for tank placement, ensuring minimal site alteration and construction feasibility. This systematic evaluation ensures the selection of an optimal location that aligns with hydraulic efficiency, structural stability, and minimal environmental impact. Figure 17 shows that the area has a maximum elevation of 789 m, and a minimum deflection of 756 m which gives us an idea to plan efficient storage tanks.

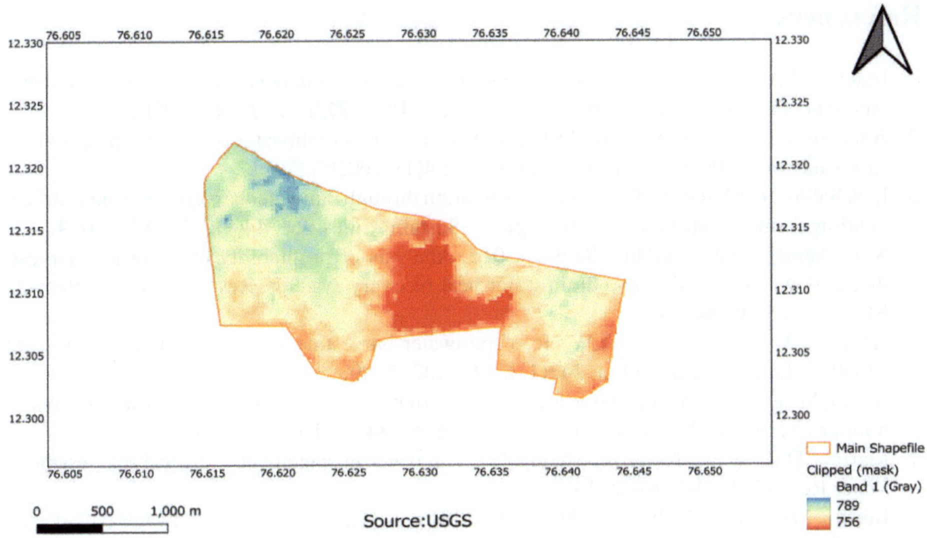

Fig. 17. Map Layout of DEM Analysis.

5 Result

The University of Mysore's annual water consumption is estimated to be 215.5 million litres. The total rooftop area of 96 buildings within the university is 32,124m^2, and the water demand for the 12.96 acres of garden area is 100.68 million litres. The total estimated rainwater that can be collected is around 335.26 million litres, which is more than enough for non-drinking uses like gardening, toilet flushing, and general cleaning. By incorporating few small changes to the existing storage tanks and the addition of proper filtration, water use on campus can be made even more efficient. Quality tests also show that the harvested water is safe for uses like sanitation and gardening, though it's not fit for drinking.

6 Conclusion

This study helps us understand that rain water harvesting isn't just adaptable but an effective sustainable water management. By utilising the naturally available water, we can reduce our dependency on external sources, especially during the peak demands. By adapting the right filtration and proper storage systems in place, the collected rainwater can be sufficiently used for various non-drinking purposes making it ecofriendly. One of the major findings is that the water collected is way more than our current storage tanks available. Hence, by redesigning, cleaning, and using basic filters for few of the existing tanks with concern to the volume of rainfall, we could add the storage capacity, and save the construction cost as well. Overall, using this filtered rainwater helps out reliance on municipal water and reduces water bills on a longer run. Also, since we wouldn't pump water, often from farfetched sources, energy usage also drops-saving the operational costs.

References

1. Imteaz, M.A., Moniruzzaman, M.: Potential impacts of climate change on future rainwater tank outcomes: a case study for Sydney. J. Clean. Prod. **273**(1), 123095 (2020)
2. Anchan, S.S., Shiva Prasad, H.C.: Feasibility of roof top rainwater harvesting potential - a case study of south Indian university. Elsevier. **4**(1), 100206 (2021)
3. Ipek Sahin, N., Manioglu, G.: Water conservation through rainwater harvesting using different building forms in different climatic regions. Sustain. Cities Soc. **44**(1), 367–377 (2019)
4. Stec, Agnieszka & Kordana, Sabina, 2015. Analysis of profitability of rainwater harvesting, gray water recycling and drain water heat recovery systems, Resour. Conserv. Recycl., Elsevier, 105 (1), 84–94.
5. Abdulla, A.A., Al-Shareef, A.W.: Roof rainwater harvesting systems for household water supply in Jordan. Desalination. **243(1-3**, 195–207 (2009)
6. Sturm, M., et al.: Rainwater harvesting as an alternative water resource in rural sites in central northern Namibia. Phys. Chem. Earth Parts A/B/C. **34**(13–16), 776–785 (2009)
7. Marlow, D.R., et al.: Towards sustainable urban water management: a critical reassessment. Water Res. **47**, 7150–7161 (2013)
8. Imteaz, M.A., Shanableh, A., Rahman, A., Ahsan, A.: Optimization of rainwater tank design from large A case study in Melbourne, Australia -Resources. Elsevier. **55**, 1022–1029 (2011)
9. Yamashita, S., Watanabeb, R., Shimatanic, Y.: Smart adaptation to flooding in urban areas. Elsevier. **118**, 1096–1103 (2015)
10. Matos, C., Santos, C., Pereira, S., Bentesa, I., Imteazc, M.: Rainwater storage tank sizing: case study of a commercial building. Elsevier. **2**, 91–100 (2013)

Artificial Intelligence

Skin Disease Detection Using a Hybrid Deep Learning Model

Sharada Mani Dalu(✉), M. Anil, R. Gayathri, Ch. Akash, V. Mounika, and Ch. Silpa

QIS College of Engineering and Technology, Ongole, AP, India
sharadadalu1234@gmail.com

Abstract. Skin diseases represent a worldwide concern in terms of public health, and early detection is critical for effective therapy. Automated dermatological analysis is a potential answer with deep learning, with increased accuracy and efficiency. In this work, a new model is proposed with a feature extraction using EfficientNetB0 and a custom Convolutional Neural Network (CNN) for high-classification performance. Squeeze-and-Excitation (SE) blocks and techniques such as random flipping, rotation, zoom, contrast, and translation for enhancing generalization capabilities are utilized in the model. EfficientNetB0 is partially fine-tuned with all but its last 50 layers frozen for a proper feature transfer and task-specific training balancing. It utilizes scheduled learning rate decay, early stopping, and model checkpointing with an Adam optimizer for efficient training and minimizing overfitting. For testing, a disease skin dataset is utilized, and performance evaluation is conducted in terms of accuracy, precision, recall, and F1-score, and analysis through a confusion matrix is performed for evaluation of proposed model performance. Robust classification accuracy in experimental analysis proves efficiency in proposed hybrid model for disease skin detection, and its real-life clinical application is a positive sign with its potential to aid in quick and accurate diagnosis for dermatologists.

Keywords: Automated Dermatology System · Skin Disease Detection · Hybrid Model

1 Introduction

Skin disease classification is important research in the field of healthcare because of the epidemic nature and diversity of skin diseases, some kinds can threaten human health if it is not detected accurately and quickly [1]. Current diagnostic methods are heavily dependent on dermatologists, leading to inconsistencies where medical resources are scarce. Consequently, automated skin disease detection systems became a useful and implemental resource to help clinicians and increase the diagnostic accessibility. In the deep learning era, convolutional neural networks (CNNs) have proven to be extremely effective at image classification tasks. On the other hand, standard CNN architectures might not be able to take into account all subtle and more complex features of skin conditions required to distinguish similar ones. Pre-trained architectures such as EfficientNetB0 are effective strategies to further improve model performance when used

H. P. Dasari et al. (Eds.): NITK-CREST 2025, PTT, pp. 291–301, 2026.
https://doi.org/10.1007/978-981-95-3218-6_20

in conjunction with appropriate custom CNN layers, partially because the individual pre-trained exploration of feature representations enables specialization for skin disease classification. This research aims to solve the problem of automated accurate scalable skin disease classification. The final model utilizes EfficientNetB0 in conjunction with the custom CNN module to achieve high accuracy, and utilize powerful feature extraction to help close the gap in dermatological diagnosis. Essential training techniques, such as data caching, prefetching, and Overfitting prevention is also an integral part of this work to ensure that the embedded model performs effectively and reliably across various skin disease datasets and use cases.

The growing global demand among practicing physicians for accessible, accurate, and reliable tools to diagnose skin diseases serves as the primary motivation behind this research work. Millions of people from all over the globe are affected by skin diseases, and unnecessary delays or wrong diagnosis can impose fatal health conditions. Access to trained dermatologists is challenging in many parts of the world, especially where health resources are limited, and there is a gap between soft diagnosis and treatment for a variety of skin disorders. As deep learning-based artificial intelligence has shown great promise in recent years, it may provide a feasible tool to help clinicians and improve diagnostic scissors in the medical practice. This research harness the power of deep learning by developing an automated skin disease detection model to aid clinicians, reduce diagnostic variability and increase early detections. By the feature extraction of EfficientNetB0 pretrained model with and custom CNN, it gives us a new designed hybrid model to catch complex characteristics for skin disease images which can provide us an informative classification. This research aims to inform clinical support and build a scalable, flexible system that enables access to quality diagnostic tools for decentralised settings thereby ultimately enhancing patient outcomes.

2 Literature Survey

In [1] Due to varying causes, skin disorders are generally common and thus image processing offers an inexpensive way to diagnose them. The method extracts features from a digital image via a pretrained CNN and uses a Multiclass SVM to classify them. The process identifies the kinds, distributions, and severities of skin disease. Three skin diseases are diagnosed with 100% accuracy, and the method is easy, fast, and cheap.

In [2] On the other hand, state-of-the-art deep learning algorithms promise accurate, touch-free dermatological disease diagnosis but only for non-mobile platforms. Such algorithms include a modified-MobileNet architecture with hybrid loss function, achieving an accuracy of 94.76%. The method was therefore incorporated into an easy-to-use mobile application.

In [3] If undetected, skin cancer advances aggressively; therefore, timely detection is of very high priority. Image pre-processing and Convolutional Neural Networks are generally used here for the detection enhancement of BCC, SCC, and melanoma. The R-CNN algorithm claims skin cancer classification with 84.32% accuracy and very powerfully applied to large datasets.

In [4] Skin cancer is treated as fatal but can be saved with early detection. Machine learning techniques such as k-nearest neighbors, support vector machines, and CNNs provide alternatives to expensive biopsies in aiding diagnosis and reducing the workload of specialists. This paper reviews the methods, their performance, and issues with skin cancer detection.

In [5] Skin cancer is caused by aberrant proliferation of cells due to mutations in DNA. Early detection improves chances of survival, but manual image analysis is costly and tedious. In this study, we introduce the JAEO-LeNet method with bilateral filtering for noise removal, TransUNet for lesion segmentation, and LeNet for cancer detection. The proposed JAEO algorithm optimizes the weights of LeNet to achieve 91.99% accuracy, 90.95% sensitivity, and 92.13% specificity using the SIIM-ISIC Melanoma dataset.

In [6] This research studied human papillomavirus (HPV) types in skin cancers occurring in patients with renal transplants. In 65% of the biopsies, HPV DNA was found and a new type, RTRX7, was found in benign, premalignant, and malignant lesions. This new type RTRX7 exhibits 81% DNA and 84% amino acid homology to HPV12.

In [7] Skin ailments affect health care systems, for accurate diagnosis is vital to patient care. It could utilize, among others, machine learning in which algorithms such as k-nearest neighbors, SVMs, and CNN could significantly transform automated skin disease detection. Further, RNNs, GANs, and attention mechanisms will be used to improve diagnostic accuracy. These have been summarized in this review as improvements to the practice of dermatology and patient care.

In [8] This review evaluates 64 deep learning algorithms for skin disease detection, showing their satisfactory diagnostic accuracy for acne, psoriasis, and eczema. It discusses the problems faced by the models and the possible future scope for AI in dermatology.

In [9] Skin cancer classification optimization using deep learning models incorporates the Manta Rays Foraging Optimizer (MRFO). The MRFO-based model outperforms classical algorithms by achieving very high accuracies, 99.43% (ISIC) and 99.96% (PH2), with very low losses. Thus, its application in skin cancer early detection is very pertinent and, in turn, will promote better patient outcomes.

In [10] This study developed a CNN-based DeepLabv3+ model for skin image segmentation which attained 90% sensitivity and specificity in differentiating skin diseases. The accuracy-enhancement of CAD performance through segmentation of the dataset over original images versus lessening of physician annotation workload.

3 System Architecture and Methodology

The architecture of the proposed automated skin disease classification work is designed to efficiently process dermatological images using deep learning techniques. The system begins with data ingestion and preprocessing, it loads the skin disease dataset from labeled directories, splits them into training, validation, and test sets, applies data augmentation techniques such as rotation, flipping, and zooming to increase variability in the dataset, thus helping the model generalize better, and resizes the images to 224x224 pixels to match the model, and caching and prefetching are activated for fast data loading in training. The heart of architecture is the model design, which applies a hybrid approach. The pre-trained EfficientNetB0 model, strong in feature extraction, is used without its top layers. All the layers of EfficientNetB0 are frozen to retain the learned features while preventing overfitting. Adding the CNN module that will have the form of Conv2D, Max Pooling, followed by Batch Normalization and Dropout in order to sharpen the feature of the data; Global Average Pooling followed by two dense layers to give out multiclass classification results with a softmax output layer for classification; Compiling the model using Adam Optimizer, along with sparse categorical cross-entropy loss function while incorporating early stopping in validation to prevent overfitting. The model is then tested on the test set after training, and the outputs are used for class predictions with softmax. It has an optional user interface and deployment where images are sent over a web interface for uploading. Then, a Flask backend analyzes the images and returns predictions. Visualization tools that can be used, such as Grad-CAM, can allow for interpretability using which areas of the image are affecting the model predictions. This system offers an efficient solution for skin disease detection in clinical environments (Fig. 1).

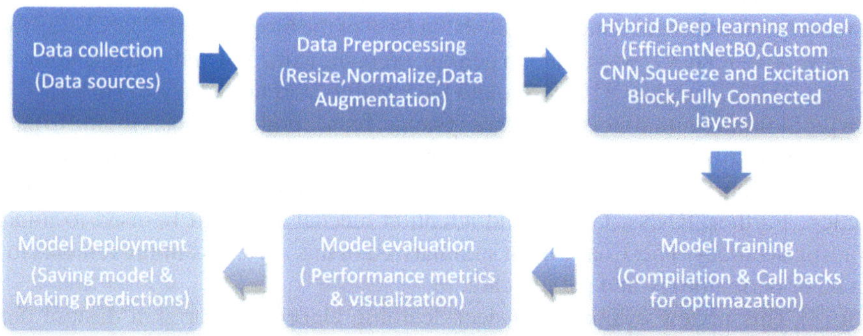

Fig. 1. System Architecture

3.1 Methodology

Advanced Data Augmentation

This work leverages effective data augmentation techniques to enhance the model's performance and generalization in skin disease detection. Medical datasets are very few in number of images collected and they are often highly imbalanced, therefore the variables that help to prevent overfitting are added through augmentation. This can be achieved through the application of different transformations such as Random Flip, Random Rotation, Random Zoom, Random Contrast, and Random Translation using tf.conversion.cake import kars in order to cater for occurrence in orientation, angle, distance, on the contrary, light and localization during training. With the flipping of the images, it would gain the left-to-right recognition, whereas different angles and distances can be accounted for by rotating and zooming in. Brightness changes and transposing positions will also contribute very much to the picture's robustness regarding brightness and location changes. Adding all these augmentations in real-time, rather than just memorizing the patterns, the model learns generalized features. This is indeed a promising step in further improving classification accuracy under which the model can be applied to real images. Therefore, even slight changes in the input will lead to true prediction by the model. In simple words, advanced-data augmentations build the model reliability towards precise diagnosis of skin diseases.

EfficientNetB0 for Feature Extraction

EfficientNetB0 is the advanced CNN for extracting multi-level, hierarchical features from images. It can be fine-tuned towards different special tasks or applications like skin diseases classification since it has been pre-trained on ImageNet. It reaches a very high accuracy, whilst being very computationally efficient. This model reduces the need for manual features engineering and therefore reduces the training time by transfer learning. In this work, EfficientNetB0 is loaded without its top layers, and only the last 50 layers are available for training, while all previous layers are kept frozen. It helps in retaining the general knowledge while making it more specific towards skin disease datasets. A set of input images of a size 224 x 224 pixels are used to perform the multi-scale feature extraction for classification. The process of fine-tuning the model would help in improving the capability of differentiating various skin diseases among mankind. It improves both the accuracy and robustness of the diagnosis as it is being applied to the real-world scenario. The establishment of the proposed model is then further fortified by EfficientNetB0 so that a high degree of efficiency can be achieved (Fig. 2).

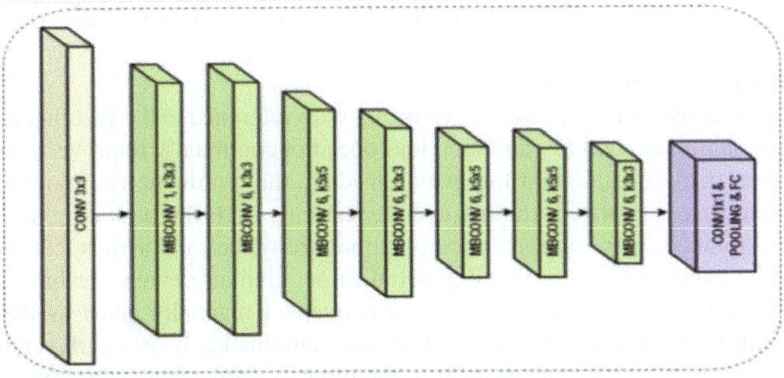

Fig. 2. EfficientNetB0 Architecture

Custom CNN Layers

In order to augment the characteristic extraction, more CNN layers are added to EfficientNetB0, also it includes Squeeze-and-Excitation for dynamically amplifying and reducing the distracting non-essentials contributing toward enhancing feature representations at classifying dermatological diseases. In this work, the output features extracted from EfficientNetB0 are passed through a Conv2D layer, followed by Batch Normalization and Dropout, as part of the fine-tuning process for the feature vector. The recalibration of the SE Block feature maps is done using Global Average Pooling, followed by two fully connected layers and a ReLU activation function ensuring the model attends to the most significant features. To direct the model towards the specific important regions in skin disease images improves its interpretability and accuracy in classification. Thus, adding these layers to the model makes it able to withstand and be more efficient in detecting complex shapes, leading to better generalization and overall performance in real-time applications (Fig. 3).

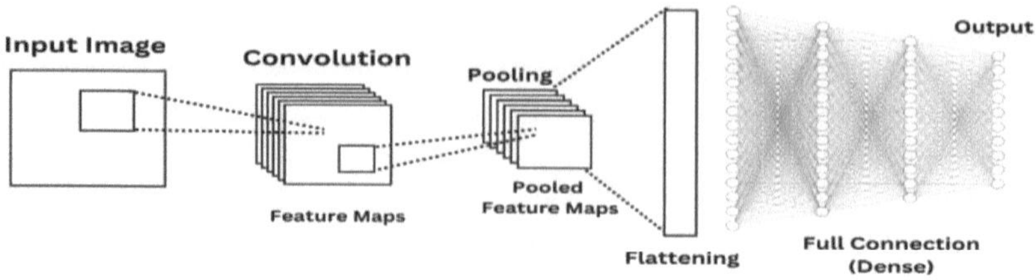

Fig. 3. CNN Architecture

Regularization Techniques (L2 Regularization & Drop Block)
In this work, L2 regularization and DropBlock are employed to prevent overfitting and enhance model generalization.L2 Regularization keeps weights small in dense layers, ensuring smooth generalization. DropBlock randomly drops regions of feature maps, forcing the model to learn more distributed features. These techniques enhance robustness, preventing reliance on specific neurons. By applying them to CNN and dense layers, the model becomes more confident in skin disease classification. Ultimately, they improve accuracy, precision, and adaptability across various skin disease images.

Early Stopping (Training Optimization)
Stopping the early training prevents overfitting with reference to the fact that it eventually stops training as soon as validation loss does not continually improve. This means that it keeps track of validation loss which leads to the termination of training after a certain number of stagnated training epochs restoring with the best weights for optimum performance. This way saves computational resources and avoids learning noise effect while balancing accuracy and generalization. Unlike previous designs, wherein the condition of stopping the training process is either manually given by the user or only specific types of research diseases, this study automatically uses early stopping for training when there is no improvement in the model's performance, in other words, it ensures that the model does well in skin disease data that has not been uncovered.

Softmax Activation (Class Probability Prediction)
The probability scores are obtained through the softmax layer over the logits, thus interpreting the model's prediction better. The outputs are normalized into the form of confidence scores per class. The confidence score indicates the reliability of predictions made by the model; this will matter for medical decisions since the confidence in a decision can affect diagnosis or treatment.The softmax function thus ensures unambiguous classification based on probability in this work, thereby enhancing the trustworthiness of skin disease detection.

Adam Optimizer (Efficient Learning)
Learned rates are adjusted smoothly in Adam Optimizer to enable training in a stable and efficient manner. Adam is a combination of RMSProp and SGD with momentum. Thus, noisy gradients are handled with ease while acceleration of convergence is achieved. Such a property renders the training insensitive to choice of hyperparameters, which is critical in the application of medical image classification. Hence, the Adam optimizer

reliably controls gradient updates to achieve optimal accuracy in skin disease detection within this work.

Deployment Strategy (Flask API)
The real-time skin disease detection model is hosted on the web using the Flask framework, to serve the Internet clients. Users can upload an image for processing, and classification results are returned promptly. This practical inference of the model would benefit dermatologists, researchers, and patients alike. The Flask API manages image uploading for model prediction and results serving through RESTful endpoints allowing Grad-CAM and SHAP visualizations to the user for explainability.

4 Experimental Evaluation and Result Analysis

The experiments were performed under an arrangement that included proper training, validation, and testing, coupled with standard accuracy, loss, precision, sensitivity, and F1-score assessments. The dataset ratio of training to validation was 80/20, thus helping the model learn and generalize efficiently. Caching and prefetching significantly eased the processing of the data on the pipeline. A hybrid mode combining EfficientNetB0 and custom CNN layers was fine-tuned to 20 epochs with early stopping based on validation loss to avoid overfitting. The Adam optimizer using sparse categorical cross-entropy loss was responsible for steering the model toward the best learning rate. Training and validation accuracy and loss were displayed per epoch in order to visualize model performance, and more importantly, early stopping helped reduce the risk of overfitting when the validation loss didn't improve for three epochs in a row, thus stopping training.

Dataset Preparation
The dataset is divided into separate folders for each category of skin disease and kept inside the training and testing directories. The images must be resized to a regular 224x224 pixels to meet the input requirements of the model. The dataset is further divided into a training set and validation set, where 80% of the actual images are used for training and 20% for validation to ensure that the model generalizes sufficiently. Caching and prefetching strategies are implemented to optimize data loading and minimize latency within the data processing pipeline.

Model Training and Validation
The model adopts a hybrid architecture that employs EfficientNetB0 as a feature extractor and specific custom CNN layers on the model for further pattern recognition and classification. EfficientNetB0 pretrained weights are then initialized with ImageNet-sourced weights for transfer learning. Also custom layers with Conv2D, Max Pooling, and Batch Normalization improve the feature capturing of the model for skin disease classification. The dropout regularization is introduced on some layers during training to reduce overfitting. The Sparse Categorical Cross-entropy loss function is used with respect to the Adam optimizer which leads to an effective gradient-based learning process. The training concedes whenever the validation loss fails to calibrate a minimum value three epochs in a row, keeping the parameter values of the ideal model. The training

and validation accuracies and losses are logged per epoch to monitor learning progress and control overfitting..

Performance Evaluation Metrics

MS uses accuracy, precision, recall, and F1-score in the evaluation of performance regarding both generalization and class-specific performance. Plotting accuracy and loss values for both training and validation datasets enables the visualization of performance trends over epochs. Post-training, the model is tested on a set of unseen images, achieving test accuracy of about 98%, with mean precision, recall, and F1 scores of about 0.9839, 0.983, and 0.983, respectively. The use of Grad-CAM further aids in interpretation by revealing critical regions the model attends to during prediction and, thus, validating what features of skin diseases are important for the model.

Confusion Matrix

A Confusion Matrix comprises a table of entries that summarizes the performance of a model in terms of the number of True Positives (TP), False Positives (FP), True Negatives (TN), and False Negatives (FN).

Confusion Matrix

	Actually Positive (1)	Actually Negative (0)
Predicted Positive (1)	True Positives (TPs)	False Positives (FPs)
Predicted Negative (0)	False Negatives (FNs)	True Negatives (TNs)

Precision

Precision represents the accuracy of the positive predictions made by the model. It is mainly about reducing the number of FPs, i.e. False Positives.

$$\text{Precision} = \text{TP}/(\text{TP} + \text{FP}) \tag{1}$$

Recall

Recall would check how much of the actual positive cases (disease) had been detected at a correct level. It is concerned with the reduction of FNs.

$$\text{Recall} = \text{TP}/(\text{TP} + \text{FN}) \tag{2}$$

F1 Score

F1 score is a single score combining Precision and Recall. Both false positives and false negatives are relevant when the cost of a mistake become very high.

$$\text{F1 Score} = 2 \times (\text{Precision} \times \text{Recall})/(\text{Precision} + \text{Recall}) \tag{3}$$

5 Result Analysis

On the base of the facts gathered, it was inferred that training and validation accuracy has been constantly improving: training accuracy on the 25th epoch was about 99.69% and validation accuracy, about 98.37%. The validation loss kept lowering over time, indicating efficient learning and confirming the model's robustness for unseen data.Upon testing on independent images, the model showed around 98% accuracy with precision, recall, and F1-scores averaging 0.98, 0.98, and 0.98 respectively which are shown in Fig. 4, highlighting balanced predictive capability across skin disease categories. Post-training,Shap was used to produce interpretability heatmaps which verified the model burned crucial areas of images corresponding to disease characteristics. Still rather strong overall performance! Misclassifications are then rather common in classes where examples are not present in sufficient quantities. Identifying those misclassifications would then provide future avenues of improvements such as targeted data augmentation or imbalanced class handling techniques that could improve classification accuracies and remove the remaining weak spots. All in all, this hybrid model propounds strong potential into the real-world applications of automated skin disease diagnosis, as long as these are validated in the future against wider datasets (Figs. 5 and 6).

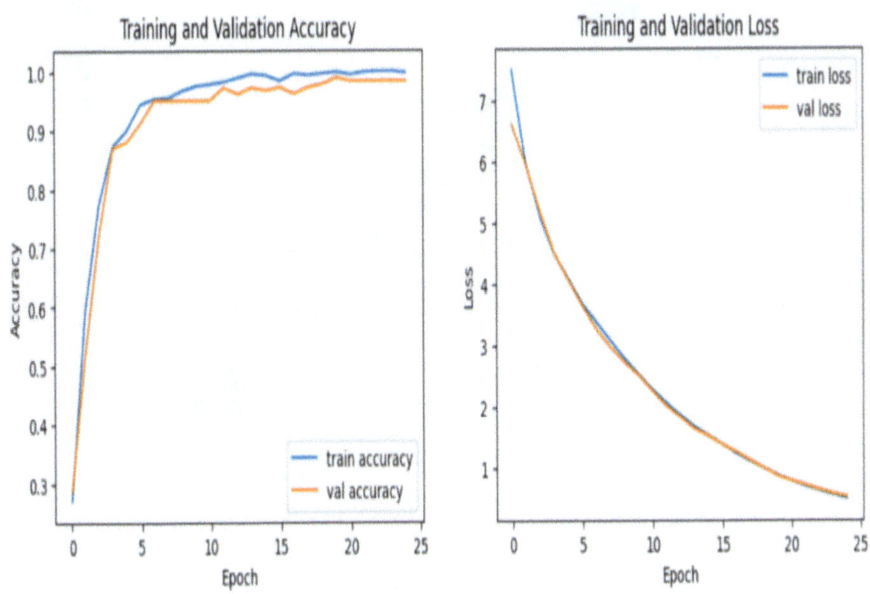

Fig. 4. Training and Testing Accuracy & Validation

```
1/1 ━━━━━━━━━━━━━━━━━━━━ 7s 7s/step
1/1 ━━━━━━━━━━━━━━━━━━━━ 2s 2s/step
1/1 ━━━━━━━━━━━━━━━━━━━━ 2s 2s/step
1/1 ━━━━━━━━━━━━━━━━━━━━ 2s 2s/step
1/1 ━━━━━━━━━━━━━━━━━━━━ 3s 3s/step
1/1 ━━━━━━━━━━━━━━━━━━━━ 5s 5s/step
Precision: 0.9839
Recall: 0.9837
F1-score: 0.9837
Accuracy: 0.9837
1/1 ━━━━━━━━━━━━━━━━━━━━ 2s 2s/step
1/1 ━━━━━━━━━━━━━━━━━━━━ 2s 2s/step
1/1 ━━━━━━━━━━━━━━━━━━━━ 3s 3s/step
1/1 ━━━━━━━━━━━━━━━━━━━━ 2s 2s/step
1/1 ━━━━━━━━━━━━━━━━━━━━ 2s 2s/step
1/1 ━━━━━━━━━━━━━━━━━━━━ 2s 2s/step
```

Fig. 5. Performance Evaluation Metrics

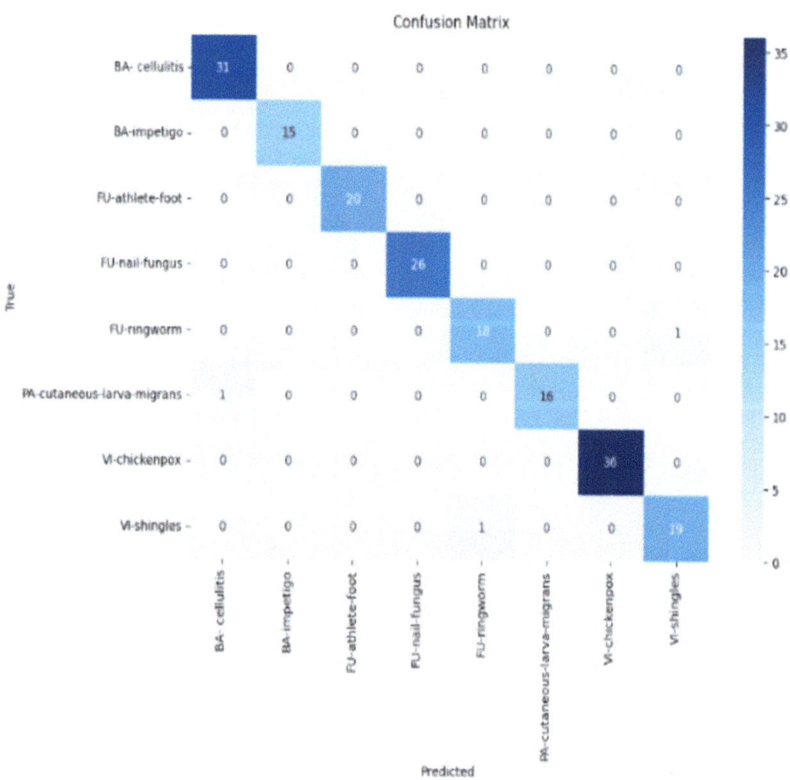

Fig. 6. Confusion Matrix

6 Conclusion

The developed hybrid deep learning model has shown great performance with 98% accuracy in skin disease classification, making it an excellent candidate for an effective and reliable tool in automation for dermatological diagnosis. It combines the power of the EfficientNetB0 pre-trained model with custom CNN layers to enable the learning of complex patterns from images of skin diseases, thus providing for robust feature extraction and classification. The regularization techniques of dropout and early stopping have prevented overfitting, allowing the model to generalize well on unseen data. This work demonstrates how deep learning can advance medical science by improving the efficiency of skin disease classification.

References

1. Alenezi, N.S.A.: A method of skin disease detection using image processing and machine learning. Proc. Comp. Sci. **163**, 85–92 (2021). https://doi.org/10.1016/j.procs.2019.12.090
2. Goceri, E.: Diagnosis of skin diseases in the era of deep learning and mobile technology. Comput. Biol. Med. **134**, 104458 (2021). https://doi.org/10.1016/j.compbiomed.2021.104458
3. Kavitha, C., Priyanka, S., Praveen Kumar, M., Kusuma, V.: Skin cancer detection and classification using deep learning techniques. Proc. Comput. Sci. **235**, 2793–2802 (2024). https://doi.org/10.1016/j.procs.2024.04.264
4. Bhatt, H., Shah, V., Shah, K., Shah, R., Shah, M.: State-of-the-art machine learning techniques for melanoma skin cancer detection and classification: a comprehensive review. Intell. Med., 180–190 (August 2023). https://doi.org/10.1016/j.imed.2022.08.004
5. Babu, R.R., Philip, F.M.: Optimized deep learning for skin lesion segmentation and skin cancer detection. Biomed. Sign. Process. Contr. (2024). https://doi.org/10.1016/j.bspc.2024.106292
6. Höpfl, R., Bens, G., Wieland, U., Petter, A., Zelger, B., Fritsch, P., Pfister, H.: Human papillomavirus DNA in non-melanoma skin cancers of a renal transplant recipient: detection of a new sequence related to epidermodysplasia verruciformis associated types. J. Invest. Dermatol. **108**, 53–56 (1997) https://www.sciencedirect.com/science/article/pii/S0022202X15427770
7. Vayadande, K., Amol, A.B., Pawar, R.G., Joshi, D.J., Bailke, P.A., Lohade, O.: Innovative approaches for skin disease identification in machine learning: A comprehensive study. Oral Oncol. Rep. (2024) https://www.sciencedirect.com/science/article/pii/S2772906024002115
8. Choy, K., et al.: Deep learning models across the range of skin disease: a systematic review. Nat. Digit. Med. (2024) https://www.nature.com/articles/s41746-024-01033-8
9. Adamu, S., HithamAlhussian, N.A., Abdulkadir, S.J., Alwadin, A., Abdullahi, M., Garba, A.: Unleashing the power of Manta Rays Foraging Optimizer: A novel approach for hyper-parameter optimization in skin cancer classification. Biomed. Sign. Process. Contr. (2024) https://www.sciencedirect.com/science/article/pii/S1746809424009133
10. Yanagisawa, Y., Shido, K., Kojima, K., Yamasaki, K.: Convolutional neural network-based skin image segmentation model to improve classification of skin diseases in conventional and non-standardized picture images. J. Dermatol. Sci. **109**, 30–36 (2023) https://www.sciencedirect.com/science/article/abs/pii/S0923181123000051

6 Conclusion

Author Index

H. P. Dasari et al. (Eds.): NITK-CREST 2025, PTT, p. 303, 2026.
https://doi.org/10.1007/978-981-95-3218-6

MIX
Papier aus verantwortungsvollen Quellen
Paper from responsible sources
FSC® C105338

If you have any concerns about our products,
you can contact us on
ProductSafety@springernature.com

In case Publisher is established outside the EU,
the EU authorized representative is:
Springer Nature Customer Service Center GmbH
Europaplatz 3, 69115 Heidelberg, Germany

Printed by Libri Plureos GmbH
in Hamburg, Germany